Studies in Surface Science and Catalysis 18

STRUCTURE AND REACTIVITY OF MODIFIED ZEOLITES

Studies in Surface Science and Catalysis 18

STRUCTURE AND REACTIVITY OF MODIFIED ZEOLITES

Proceedings of an International Conference, Prague, July 9—13, 1984

Editors

P.A. Jacobs
Laboratorium voor Oppervlaktescheikunde, Katholieke Universiteit Leuven, Leuven, Belgium

N.I. Jaeger
Forschungsgruppe Angewandte Katalyse, Universität Bremen, Bremen, F.R.G.

P. Jírů
J. Heyrovský Institute of Physical Chemistry and Electrochemistry, Czechoslovak Academy of Sciences, Prague, Czechoslovakia

V.B. Kazansky
N.D. Zelinsky Institute of Organic Chemistry, Academy of Sciences of the U.S.S.R., Moscow, U.S.S.R.

and

G. Schulz-Ekloff
Forschungsgruppe Angewandte Katalyse, Universität Bremen, Bremen, F.R.G.

ELSEVIER   Amsterdam — Oxford — New York — Tokyo   1984

ELSEVIER SCIENCE PUBLISHERS B.V.
Molenwerf 1
P.O. Box 211, 1000 AE Amsterdam, The Netherlands

Distributors for the United States and Canada:

ELSEVIER SCIENCE PUBLISHING COMPANY INC.
52, Vanderbilt Avenue
New York, NY 10017

Library of Congress Cataloging in Publication Data
Main entry under title:

Structure and reactivity of modified zeolites.

   (Studies in surface science and catalysis ; 18)
   Includes index.
   1. Zeolites--Congresses. 2. Catalysts--Congresses.
I. Jacobs, Peter A. II. Series.
TP159.C3S77 1984   660.2'995   84-7963
ISBN 0-444-42351-6

ISBN 0-444-42351-6 (Vol. 18)
ISBN 0-444-41801-6 (Series)

Printed in The Netherlands

V

CONTENTS

Studies in Surface Science and Catalysis IX

Preface XI

Modern methods of zeolite research (L.V.C. Rees) 1

Adsorption of xenon: A new method for studying the crystallinity
of zeolites (M.A. Springuel-Huet, T. Ito and J.P. Fraissard) 13

Solid state ^{29}Si and ^{27}Al NMR studies of decationized and
dealuminated zeolites (G. Engelhardt, U. Lohse, M. Mägi and
E. Lippmaa). 23

XPS study of transition metal complexes in zeolites (E.S. Shpiro,
G.V. Antoshin, O.P. Tkachenko, S.V. Gudkov, B.V. Romanovsky and
Kh.M. Minachev). 31

Spectroscopic study of the surface coordination of divalent
copper on dehydrated A zeolites (D. Packet and R.A. Schoonheydt) 41

Studies on the modified Claus reaction over alkaline faujasites
by simultaneous infrared, kinetics and ESR measurements (H.G. Karge,
Y. Zhang, S. Trevizan de Suarez and M. Ziołek) 49

The study of Broensted and Lewis acidity of decationized and dealuminated zeolites by IR-diffuse reflectance spectroscopy and
quantum chemistry (V.B. Kazansky). 61

Mechanism of radical formation in olefin adsorption on high-
silica zeolites and the problem of the synthesis of aromatic
structures (A.V. Kutcherov and A.A. Slinkin) 77

Quantum-chemical study of zeolite structure stability.
Comparative discussion of zeolites and borolites (A.G. Pelmentschikov,
G.M. Zhidomirov, D.V. Khuroshvili and G.V. Tsitsishvili) 85

Carbocation formation in zeolites. UV-VIS-NIR spectroscopic
investigations on mordenites (P. Fejes, H. Förster, I. Kiricsi
and J. Seebode). 91

Quantum chemical study   of the characteristics of molecules
interacting with zeolites (S. Beran) 99

New routes in zeolite synthesis (H. Lechert) 107

Study of mordenite crystallisation III: Factors governing mordenite
synthesis (P. Bodart, J.B. Nagy, E.G. Derouane and Z. Gabelica). 125

Structural peculiarities and stabilization phenomena of aluminium
deficient mordenites (H.K. Beyer, I.M. Belenykaja, I.W. Mishin and
G. Borbely). 133

On catalytic activity of synthetic offretite in conversion reactions of methanol and o-, m-, p-xylenes (G.V. Tsitsishvili,
Ts. M. Ramishvili and M.K. Charkviani) 141

Synthesis and study of properties of ZSM-II type silicalites of
I-VIII group elements (K.G. Ione, L.A. Vostrikova, A.V. Petrova
and V.M. Mastikhin). 151

A new route to ZSM-5 zeolite: Synthesis and characterization
(E. Moretti, G. Leofanti, M. Padovan, M. Solari, G. de Alberti
and F. Gatti). 159

Synthesis of gallosilicate and alumogermanate zeolites and investigation of their activity in the reaction of alcohol dehydration (Z.G. Zulfugarov, A.S. Suleimanov and Ch.R. Samedov) 167

Stereochemical effects in shape-selective conversion of hydrocarbons on zeolites (P.A. Jacobs, M. Tielen and R.C. Sosa) 175

Shape-selective effects on the metal catalyzed reactions of
prehnitene over zeolites (J.A. Martens and P.A. Jacobs). 189

Relation between acidic properties of ZSM-5 and catalyst performance of methanol conversion to gasoline (T. Inui, T. Suzuki,
M. Inoue, Y. Murakami and Y. Takegami). 201

Influence of Ni, Mg and P on selectivity of ZSM-5 class zeolite
catalysts in toluene-methanol alkylation and methanol conversion
(M. Derewiński, J. Haber, J. Ptaszyński, V.P. Shiralkar and
S. Dżwigaj). 209

Reaction of small amounts of methanol on HZSM-5, HY and modified
Y zeolites (L. Kubelková, J. Nováková and P. Jírů) 217

On the formation of hydrocarbon chains in the aromatization of
aliphatic olefins and dienes over high-silica zeolites
(G.V. Isaguliants, K.M. Gitis, D.A. Kondratjev and
Kh.M. Minachev) 225

Conversion of linear butenes to propylene on H-ZSM-5 zeolites:
Effect of reaction parameters and zeolite morphology on catalytic
activity (F. Colombo and G. de Alberti) 233

Study of ethylene oligomerization on Brönsted and Lewis acidic
sites of zeolites using diffuse reflectance IR spectroscopy
(L.M. Kustov, V.Yu. Borovkov and V.B. Kazansky). 241

Various states of Cr ions in Y and ZSM-5 zeolites and their
catalytic activity in ethylene polymerization (B. Wichterlová,
Z. Tvarůžková, L. Krajcíková and J. Nováková) 249

Metal carbonyl compounds entrapped within zeolite cavities.
Preparation, characterization and catalytic properties
(F. Lefebvre, P. Gelin, B. Elleuch, Y. Diab and Y. Ben Taarit). 257

Aromatization of ethane on metal-zeolite catalysts
(O.V. Bragin, T.V. Vasina, Ya.I. Isakov, N.V. Palishkina,
A.V. Preobrazhensky, B.K. Nefedov and Kh. M. Minachev). 273

Shape selective isomerization and hydrocracking of naphthenes
over Pt/HZMS-5 zeolite (J. Weitkamp, P.A. Jacobs and
S. Ernst) 279

Liquid phase synthesis of aromates and isomers on polyfunctional
zeolitic catalyst mixtures (H. Nguyen-Ngoc, K. Müller and M. Ralek). . . . 291

Concepts of reduction and dispersion of metals in zeolites
(N.I. Jaeger, P. Ryder and G. Schulz-Ekloff) 299

Local structure and bonding in zeolites by means of quantum
chemical ab initio calculations: Metal cations, metal atoms
and framework modification(J. Sauer, H. Haberlandt and W. Schirmer). . . . 313

On the bifunctional action of modified zeolites Y containing
nickel (H. Bremer, W.P. Reschetilowski, F. Vogt and K.-P. Wendlandt) . . . 321

Correlation between the structure of reduced transition metal
zeolite systems and the catalytic reactivity in benzene hydrogenation reactions (V. Kanazirev, V. Penchev, Chr. Minchev,
U. Ohlerich and F. Schmidt). 329

Characterization of X-type zeolites containing metallic nickel
particles by dielectric relaxation (J.C. Carru, D. Delafosse
and M. Briend) 337

Features of an electron state and catalytic properties of the
L-type platinum-containing zeolite (S.V. Gagarin, V.S. Komarov,
I.I. Urbanovich, T.I. Gintovt and Yu.A. Teterin) 345

Study of the dispersion of zeolite supported nickel in dependence
on the zeolite type and the reaction medium (N.P. Davidova,
M.L. Valcheva and D.M. Shopov) 353

Magnetic investigation of metallic iron clusters in zeolites
(J.M. Jabloński) 361

Hydrogenation of CO and CO_2 over stabilized NiY catalysts
(V. Patzelová, A. Zukal, Z. Tvarůžková and O. Maliček) 367

Author Index 375

Studies in Surface Science and Catalysis

Volume 1   **Preparation of Catalysts I.** Scientific Bases for the Preparation of Heterogeneous
Catalysts. Proceedings of the First International Symposium held at the Solvay
Research Centre, Brussels, October 14—17, 1975
edited by **B. Delmon, P.A. Jacobs and G. Poncelet**

Volume 2   **The Control of the Reactivity of Solids.** A Critical Survey of the Factors that Influence the Reactivity of Solids, with Special Emphasis on the Control of the Chemical Processes in Relation to Practical Applications
by **V.V. Boldyrev, M. Bulens and B. Delmon**

Volume 3   **Preparation of Catalysts II.** Scientific Bases for the Preparation of Heterogeneous
Catalysts. Proceedings of the Second International Symposium, Louvain-la-Neuve,
September 4—7, 1978
edited by **B. Delmon, P. Grange, P. Jacobs and G. Poncelet**

Volume 4   **Growth and Properties of Metal Clusters.** Applications to Catalysis and the Photographic Process. Proceedings of the 32nd International Meeting of the Société de
Chimie physique, Villeurbanne, September 24—28, 1979
edited by **J. Bourdon**

Volume 5   **Catalysis by Zeolites.** Proceedings of an International Symposium organized by the
Institut de Recherches sur la Catalyse — CNRS — Villeurbanne and sponsored by the
Centre National de la Recherche Scientifique, Ecully (Lyon), September 9—11,
1980
edited by **B. Imelik, C. Naccache, Y. Ben Taarit, J.C. Vedrine, G. Coudurier and
H. Praliaud**

Volume 6   **Catalyst Deactivation.** Proceedings of the International Symposium, Antwerp,
October 13—15, 1980
edited by **B. Delmon and G.F. Froment**

Volume 7   **New Horizons in Catalysis.** Proceedings of the 7th International Congress on
Catalysis, Tokyo, 30 June—4 July 1980
edited by **T. Seiyama and K. Tanabe**

Volume 8   **Catalysis by Supported Complexes**
by **Yu.I. Yermakov, B.N. Kuznetsov and V.A. Zakharov**

Volume 9   **Physics of Solid Surfaces.** Proceedings of the Symposium held in Bechyňe, Czechoslovakia, September 29—October 3, 1980
edited by **M. Láznička**

Volume 10   **Adsorption at the Gas—Solid and Liquid—Solid Interface.** Proceedings of an International Symposium held in Aix-en-Provence, September 21—23, 1981
edited by **J. Rouquerol and K.S.W. Sing**

Volume 11   **Metal-Support and Metal-Additive Effects in Catalysis.** Proceedings of an Interna-
tional Symposium organized by the Institut de Recherches sur la Catalyse — CNRS —
Villeurbanne and sponsored by the Centre National de la Recherche Scientifique,
Ecully (Lyon), September 14—16, 1982
edited by **B. Imelik, C. Naccache, G. Coudurier, H. Praliaud, P. Meriaudeau,
P. Gallezot, G.A. Martin and J.C. Vedrine**

Volume 12   **Metal Microstructures in Zeolites.** Preparation — Properties — Applications.
Proceedings of a Workshop, Bremen, September 22—24, 1982
edited by **P.A. Jacobs, N.I. Jaeger, P. Jiru and G. Schulz-Ekloff**

Volume 13   **Adsorption on Metal Surfaces.** An Integrated Approach
edited by **J. Bénard**

Volume 14   **Vibrations at Surfaces.** Proceedings of the Third International Conference, Asilomar,
California, U.S.A., 1—4 September 1982
edited by **C.R. Brundle and H. Morawitz**

Volume 15   **Heterogeneous Catalytic Reactions Involving Molecular Oxygen**
by **G.I. Golodets**

Volume 16   **Preparation of Catalysts III.** Scientific Bases for the Preparation of Heterogeneous
Catalysts. Proceedings of the Third International Symposium, Louvain-la-Neuve,
September 6—9, 1982
edited by **G. Poncelet, P. Grange and P.A. Jacobs**

x

Volume 17   **Spillover of Adsorbed Species.** Proceedings of the International Symposium,
             Lyon-Villeurbanne, September 12–16, 1983
             edited by **G.M. Pajonk, S.J. Teichner and J.E. Germain**

Volume 18   **Structure and Reactivity of Modified Zeolites.** Proceedings of an International
             Conference, Prague, July 9–13, 1984
             edited by **P.A. Jacobs, N.I. Jaeger, P. Jírů, V.B. Kazansky and G. Schulz-Ekloff**

PREFACE

Progress in the practical application of zeolites in catalysis will largely depend on a better understanding of their structural stability, the nature and strength of acid sites, and their bifunctional action. The papers contributed to the Conference on Structure and Reactivity of Modified Zeolites in Prague, which are collected in this volume, reflect the current ideas on these problems and present many new results.

The descriptions of the methodology, used in the investigation of structures, interactions and intermediates, include frequency response diffusion, Mössbauer spectroscopy, Szilard-Chalmers type recoils, ^{129}Xe NMR, ^{29}Si NMR, ^{27}Al NMR, XPS ESR, UV-VIS, IR and quantum chemical calculations. The reports on synthesis and modification procedures focus on Y, mordenite, offretite, ZSM-5, (Be, B, Al, Ga)-ZSM type and on GaSi and AlGe analogues of X. Metal complexes, cations and particles of Pt, Pd, Rh, Ru, Cu, Ni, Co, Fe and Cr are studied with respect to their structure and function, e.g. in hydrogenation reactions. The aromatization of ethane on metal loaded high-silica zeolites that is described is an important example of bifunctional action. The shape selectivity in acid catalyzed reactions is mainly investigated for the conversions of methanol, olefins and methylbenzenes.

The Organizing Committee felt obliged to carry out editorial changes in the papers where obvious typing errors or vagueness affected the clarity and understanding, and we apologize for not being able in each case to obtain an authorization in time. Our alternative was not to publish these papers, which would not have been justified in view of the quality of the published data.

The Organizing Committee are very grateful to all authors and hope that lively discussions will ensue from the publication of these papers.

MODERN METHODS OF ZEOLITE RESEARCH

L.V.C. Rees

Physical Chemistry Laboratories, Imperial College of Science and Technology,

London, SW7 2AY.

ABSTRACT

In this paper two frequency response methods will be described for determining the diffusion coefficients of sorbates which are in sorption equilibrium with the zeolite phase. Mossbauer spectroscopy, which involves the emission of recoilless, soft gamma photons, will be shown to provide useful information on cation/framework oxygen coordinations and interactions. The interaction of sorbate molecules with these Mossbauer active cations can also be determined. Finally the Szilard-Chalmers type recoils which cations are subjected to on emission of hard gamma photons following capture of a neutron by the cation nucleus is used to assess the site preferences of cations in zeolite framework as a function of calcination temperature.

FREQUENCY RESPONSE DIFFUSION COEFFICIENTS.

A new method of determining the diffusion coefficients of sorbates in the micropores of zeolites has recently been developed [1]. In this method the volume of a chamber, containing the permeating gas in equilibrium with the sorbent, is varied sinusoidally with an amplitude of only a few percent. The phase angle and amplitude of the pressure change of the gas phase induced by the volume variation are measured as a function of the frequency of the driving oscillations. Characteristic functions, representing in-phase and out-of-phase components of the pressure variations, can be calculated on the basis of Fick's law of diffusion in (i) a plane sheet and (ii) an isotropic sphere as models for one- and three- dimensional channels in zeolites respectively.

The sinusoidal volume change may be described by

$$V = V_e (1-ve^{i\omega t})$$ (1)

where V_e denotes the volume at equilibrium, v is the relative amplitude of the volume variation and ω is the angular velocity of the sinusoidal generator. The perturbed gas pressure, P, and the concentration, C, of the diffusing sorbate in the zeolite are expressed in general, in a periodic steady state, by

$$P = P_e \left\{ 1 + pe^{i(\omega t + \phi)} \right\}$$ (2)

2

and

$$C = C_e \left\{ 1 + \gamma e^{i(\omega t + \phi - \chi)} \right\} \qquad (3)$$

Both the relative amplitudes p and γ and the phase lags ϕ and χ depend on ω.

Assuming that the diffusion coefficient, D, is constant, since the perturbation to the system is small, Fick's second law can be solved for the above boundary conditions. For diffusion down channels of length L the solution for the one-dimensional case leads to the following two equations

$$(v/p) \cos \phi - 1 = K\delta_c \qquad (4)$$

$$(v/p) \sin \phi = K\delta_s \qquad (5)$$

where δ_c and δ_s are defined by

$$\delta_c = \frac{1}{\eta} \left\{ \frac{\sinh \eta + \sin \eta}{\cosh \eta + \cos \eta} \right\} \qquad (6)$$

and

$$\delta_s = \frac{1}{\eta} \left\{ \frac{\sinh \eta - \sin \eta}{\cosh \eta + \cos \eta} \right\} \qquad (7)$$

and $\eta = (\omega L^2/2D)^{\frac{1}{2}}$. The coefficient K is related to the gradient of the sorption isotherm at $P = P_e$.

For diffusion in isotropic spheres of radius, a, we once again obtain Eqn. 4 and 5 but δ_c and δ_s are now defined by

$$\delta_c = \frac{3}{\eta} \left\{ \frac{\sinh \eta - \sin \eta}{\cosh \eta \quad \cos \eta} \right\} \qquad (8)$$

and

$$\delta_s = \frac{6}{\eta} \left[\frac{1}{2} \left\{ \frac{\sinh \eta + \sin \eta}{\cosh \eta - \cos \eta} \right\} - \frac{1}{\eta} \right] \qquad (9)$$

and $\eta = (2\omega a^2/D)^{\frac{1}{2}}$

By determining the phase lag, ϕ as a function of ω the in-phase and out-of-phase components $(v/p) \cos \phi - 1$ and $(v/p) \sin \phi$ can be plotted against ω. From Eqn. 4 and 5 the best fit to the experimental data is obtained. From Eqn. 6 and 7 or 8 and 9 η is derived and hence D . By carrying out similar measurements at different equilibrium pressures, P_e, the sorption isotherm may also be obtained.

Yasuda [1] used the frequency response technique to measure the sorption and diffusion of krypton in sodium mordenite at -20°C over angular velocities ranging from 1 to 100 rad/min. Using Eqn. 6 and 7 for one-dimensional diffusion Yasuda found that $(D/L^2) = 5.3 \times 10^{-3}$ s^{-1} at equilibrium pressures around 300 Pa. but he did not attempt to estimate the length of the diffusion channels, L , and so did not obtain D. However, the order of magnitude of D/L^2 was comparable to that determined by more standard methods [2].

Yasuda [1] found that the in-phase and out-of-phase components were not asymptotic at high frequencies as predicted by theory. He interpreted this difference as resulting from a very fast intercrystalline, three-dimensional diffusion. However, the apparatus he used could not operate at high enough angular velocities to determine the magnitude of this fast, intercrystalline diffusion coefficient.

Betemps [3] has also recently determined diffusion coefficients of sorbates in zeolites using a method similar in nature to the frequency response technique described above. Betemps obtained an impulse response function of the equilibrium gas/sieve pair through the cross-correlation between a pseudo random binary variation of the gas phase volume and the resulting pressure variations. Using this technique the diffusion coefficient of argon in zeolite 5A at 298 K was found to be 1.66×10^{-14} m^{2}s^{-1} which agrees reasonably well with the corresponding coefficient obtained by Ruthven and Derrah [4] from sorption uptake rates.

MOSSBAUER SPECTROSCOPY

Mossbauer spectroscopy is concerned with the recoil-free emission and re-absorption of soft gamma photons by nuclei which belong to atoms or ions which are firmly bound in crystalline lattices [5]. In solids the lattice binding energies ($\sim$10eV) are much higher than the recoil energies of $\sim 10^{-2}$ eV associated with the emission or absorption of soft gamma photons. Momentum is conserved during a Mossbauer gamma transition. However, in a fraction of these events, (f), the whole crystal recoils and in a fraction of events,

(1-f), the energy is taken up by the phonon vibrational levels of the lattice. When the whole crystal recoils the recoil energy is small compared with the natural line width and hence transitions will be at the full gamma energy with a line width given by the Heisenberg principle.

Many elements have Mossbauer active isotopes e.g. ^{57}Fe, ^{119}Sn, ^{40}K, ^{73}Ge, ^{83}Kr, ^{121}Sb, ^{125}Te, ^{127}I, ^{129}Xe, ^{133}Cs, ^{133}Ba, several transition-metal, rare-earth and actinide elements. Apart from ^{57}Fe and ^{119}Sn it is usually necessary to cool both the source of the recoilless gamma photons and the absorber to liquid N_2 temperatures and, in many cases, to liquid He temperature before f becomes large enough to give useful spectra. Of all the Mossbauer spectra published in the literature the great majority are concerned with ^{57}Fe. The 14.414 keV gamma photon emitted by ^{57}Fe is ideal for Mossbauer spectroscopy. It has low energy and hence significant recoil free fractions, f, even above room temperature. The decay time of the first excited level of the ^{57}Fe nucleus of $t_{\frac{1}{2}}$ = 99.3 ns leads to line widths of 4.594×10^{-9} eV. The 14.414 keV level is efficiently populated via the electron capture decay of ^{57}Co. The source of the recoilless gamma radiation is usually ^{57}Co diffused into a cubic lattice e.g. metallic Pt. Since the half-life of ^{57}Co is 270 days sources have useful lives of at least three years.

The great energy resolving power available using the Mossbauer effect (eg. ^{57}Fe it is 1 part in 3×10^{12}) allows an investigation of the interaction of the surrounding electrons with the nucleus and hence of the chemical environment to be made. The parameters obtained are the chemical isomer shift (IS), the quadrupole splitting (QS) and the magnetic hyperfine interaction. The isomer shift is given by

$$IS = constant \left\{ |\psi_s(0)_A|^2 - |\psi_s(0)_S|^2 \right\} \tag{10}$$

where $|\psi_s(0)|^2$ is the s-electron density at the centre of the nucleus of the absorber, A , and source, S , respectively. The constant is negative in the case of ^{57}Fe and, hence, the isomer shift decreases in an absorber as the s-electron density at the absorber nucleus increases. The isomer shift is useful in the case of iron in assessing the valence and spin state of the absorber atom, as shown in Table 1.

TABLE 1

Typical isomer shift values for ^{57}Fe. (IS quoted in mm s^{-1} w.r.t. natural iron foil)

Valence State	Electronic Configuration	High Spin	Low spin
Fe(I)	[Ar] $3s^2 3d^7$	1.9 - 2.1	0.2 - 0.4
Fe(II)	[Ar] $3s^2 3d^6$	0.8 - 1.5	-0.2 - 0.4
Fe(III)	[Ar] $3s^2 3d^5$	0.2 - 0.6	-0.2 - 0.3
Fe(IV)	[Ar] $3s^2 3d^4$	-0.1 - 0.1	0.1 - 0.2
Fe(O)	[Ar] $3s^2 3d^6 4s^2$	0	

Generally iron in tetrahedral sites have lower isomer shifts than in octahedral sites due to increased covalency effects.

Nuclei with spin states greater than I = 1/2 are non-spherical and interact with non-cubic electric fields to remove some of the degeneracy of the nuclear spin states. This gives rise to multiple peak spectra and results in experimentally accessible quadrupole splittings. For ^{57}Fe the ground state has spin I = 1/2 and is unsplit in an electric field gradient (efg) but the excited state has spin I = 3/2 and is split in non-cubic electric fields into two energy levels. These energy levels are given by

$$E_{\pm} = \pm \frac{e^2 qQ}{4} [1 + \eta^2/3]^{\frac{1}{2}} \tag{11}$$

where the efg is expressed in terms of two parameters V_{zz} = eq and $\eta = [V_{xx} - V_{yy}]/V_{zz}$ is an asymmetry parameter. Q is the quadrupole moment of the nucleus in the excited state. If the Mossbauer atom is in a site of axial symmetry (ie $V_{xx} = V_{yy}$) then QS = $e^2 q Q/2$. For ^{57}Fe, the interaction of the quadrupole moment with the efg leads in most cases to symmetric doublets. Only the magnitude of the QS can normally be measured but it is possible by applying external magnetic fields to measure the sign of V_{zz}.

If there is a net magnetic field at the nucleus during the Mossbauer event then the nuclear spin degeneracy of both the ground and excited states are completely removed and multi-peak spectra result. The eigenvalues are given by

$$E_m = -g_N \beta_N H m_I \tag{12}$$

where m_I is the z-component of the nuclear spin, g_N is the nuclear g-factor, β_N is the nuclear magneton and H is the effective magnetic field at the nucleus. In the case of ^{57}Fe the resulting spectra contain 6 equally spaced lines (selection rate only allows $\Delta m_I = 0, 1$) with relative intensities for powdered samples of 3:2:1:1:2:3. The magnetic field at the nucleus may arise from a) an externally applied magnetic field, b) within the atom itself or c) from exchange interactions within the crystal. The internal magnetic field results from (i) the Fermi contact term and (ii) terms due to the interaction of the nucleus with the spin moment and orbital magnetic moment of the atom.

We have studied the Mossbauer spectroscopy of ^{57}Fe zeolites in depth and obtained much information on the location, chemical state and effect of sorbate molecule interaction on the iron cations present in the zeolites. When ferrous cations are ion exchanged into zeolite Y at room temperature Mossbauer spectroscopy at room temperature shows that the cations are very loosly associated with the framework oxygens when the zeolite is fully hydrated [6]. The very low f factor obtained indicated a strong preference of the ferrous ions for the zeolitic water molecules and a rapid diffusion of these hydrated ions in the channels of the zeolite. On dehydration at $360^{\circ}C$ for 24 h under vacuum a strong Mossbauer signal was obtained at room temperature which comprised of two quadrupole split doublets. One doublet was ascribed to Fe^{2+} ions sited in the hexagonal prism sites (site A, [7]) of zeolite Y. The other doublet had parameters which are very unusual for Fe^{2+} species. We proposed that these ferrous ions were sited in the puckered six-ring separating the sodalite cage from the supercage i.e. site G, [7]. We tested this hypothesis by next studying the Mossbauer spectra of ferrous ions exchanged into zeolite A where the puckered six-ring was expected to be the only site occupied by the ferrous ions. The first interesting feature of these spectra [8] was the very large f factor obtained at room temperature from the fully hydrated zeolite indicating that ferrous ions were coordinated to framework oxygens rather than zeolitic water molecules. The much higher negative charge density on the framework oxygens of zeolite A compared with zeolite Y must be the explanation of these findings. On dehydration of zeolite A the Mossbauer signal actually was degraded by 20% indicating the presence of a new loose mode of vibration for the ferrous ions perpendicular to the plane of the six-ring and the parameters obtained (IS = 0.835 mms^{-1}, QS = 0.469 mms^{-1}) were very similar to those found for ferrous ions in site G of dehydrated zeolite Y. The very small QS observed results from a near cancellation of the large, positive efg from the valence component (i.e. the 6th 3d electron) with the large, negative efg from the negatively charged oxygen framework ligands. Because of the puckered nature of the six-ring only the three nearest-neighbour, oxygens in a planar, trigonal configuration dominate the lattice component. Normally ferrous ions are in tetrahedral or octahedral coordination with ligands and the lattice term is then much smaller than the valence term.

When adsorbate molecules are now sorbed onto these ferrous ions located in site A [7] of dehydrated zeolite A the coordination becomes more symmetrical, i.e. tetrahedral-like at low loadings or octahedral-like at high sorbate loadings, and the QS increases to the more normal values for these coordinations. Mossbauer spectroscopy of ferrous zeolite A loaded with various sorbates is a very sensitive method of studying sorbate-cation interactions. We have shown, for example, that in partially hydrated (Na^+, Fe^{2+})-A zeolite the following equilibrium exists

$$(Z - 0)_3 Fe^{2+} - H_2 0 \rightleftharpoons (Z - 0)_3 Fe^{2+} - OH^- + H^+$$

with proton hopping between framework oxygens and the hydroxyl group on the ferrous cation occuring with a half-life greater than 10^{-7} s (the Mossbauer event time) at room temperature. On cooling to liquid N_2 temperatures the equilibrium is shifted strongly to the left. In the fully hydrated zeolite the proton hopping is too fast to observe in the 10^{-7} s time scale but on cooling, the spectra indicate the two different ferrous environments represented in the above equilibrium. Exchange of the Na^+ ions by Ca^{2+} and NH_4^+ to produce (Ca^{2+}, Fe^{2+})-A and (NH_4^+, Fe^{2+})-A was found to slow down the proton hopping and the above two different Fe^{2+} species were observed in the spectra of even the fully hydrated materials at room temperature [9, 10].

Adsorption of ethylene in (Na^+, Fe^{2+})-A at ethylene: Fe^{2+} loadings well below one gave spectra which showed only one doublet at room temperature but on cooling below 200 K two doublets began to be resolved, one doublet with the parameters of the bare ferrous in site A and the other with parameters consistent with the distorted tetrahedron resulting from sorption of an ethylene molecule on the ferrous cation [11]. Once again these results demonstated that at room temperature the ethylene molecules were hopping from one ferrous ion to the next with residence times much shorter than 10^{-7} s. On cooling to ∿200K the residence time was 10^{-7} s. Analysis of the spectra as a function of temperature gave approximate energies of activation for the ethylene jump of 23 k J mol^{-1}.

The Mossbauer spectra of ferrous exchanged Na- and Ca-A zeolites showed that the ferrous ions could be oxidised to ferric by dry oxygen at 360°C or wet air at ambient temperatures [12]. On dehydration under vacuum of the fully hydrated ferric forms some 20% of the ferric ions were reduced to ferrous. The ferric ions could also be reduced to ferrous by treatment with hydrogen at 360°C for 4 h.   No metallic iron was observed after these treatments with hydrogen.   The co-cation present in the zeolite was found to have a profound influence on these reactions.

The Mossbauer spectrscopy of ferrous exchanged zeolite L has also been studied [13].   The ferrous ions were found to be coordinated only to the zeolitic water molecules after exchange of the ferrous ions into the zeolite at room temperature.   On dehydration at temperatures of 360°C and above the ferrous ions were found to be located in both the hexagonal prism sites (site A) and in site E.   In this latter site the ferrous ions are coordinated to four framework oxygens in a square-planar complex.   As found in dehydrated zeolite A the lattice contribution to the efg is large and a partial cancellation of the efg produced by the valence electron results in low IS and QS values (IS = 0.95 mms^{-1}, QS = 0.66 mms^{-1}).

The spectra of dehydrated, ferrous zeolite L were taken after additions of various sorbates (e.g. water, ethanol, propan-2-ol, 2-methyl propan-2-ol, methyl cyanide and triethylamine) had been made [14]. The low QS doublet due to ferrous ions in site E was found to be progressively reduced in intensity when most of the above sorbates were titrated into the zeolite with the formation of a wide QS doublet due to ferrous ions now sited in site D coordinated to framework oxygens and one sorbate molecule.   The loading of sorbate corresponding to complete removal of ferrous ions from site E could, therefore, be determined.

The Mossbauer spectra of zeolites containing Fe(CO)$_5$, (C$_5$H$_5$)$_2$ Fe, ferrocene, ferrocyanides and nitroprussides sorbed in or on the zeolites have been recorded [15].   The effect of photochemical and thermal activation of these sorbed species was also followed.   The sorbates were, also, reduced in hydrogen between 300 and 500° C. Ferrocene, on reduction, gave a weak six-line spectrum due to α-iron metal along with a strong central doublet. Reduction of the iron carbonyl systems led to complete reduction to α-iron with quite large particles of 7.5 - 12.5 nm diameter. Reduction of the ferro-cyanides and nitroprussides also produced large particles of metallic iron.

In all of these studies it would seem that the iron aggregates formed on reduction are located on the external surfaces of the zeolites.

Gunser et al [16] reduced ferrous ions in zeolite A with sodium vapour and concluded from magnetic susceptibility and Mossbauer experiments that super paramagnetic iron particles were formed. It is possible that these small particles are residing in the zeolite cavities. Schmidt et al [17] decomposed $Fe(CO)_5$/NaY adducts and found that 97 wt % of the iron particles had diameters < 1.3nm. No Mossbauer spectra were reported, however, of these interesting superparamagnetic aggregates.

SZILARD-CHALMERS CATION RECOILS

When a nucleus captures a neutron ~ 8 MeV of binding energy is released as gamma radiation. The average recoil energy associated with this (η, γ) reaction is of the order of a few hundred electron volts. This recoil energy is more than sufficient to move an atom from its normal lattice location and is entirely different in character from the very small recoil energies associated with Mossbauer gamma emissions. Assuming a hard sphere model it is possible to calculate the mean free path, L_s, of the recoiling atom. L_s is found to be 1.72 nm for hydrated Na-X and 1.84 nm for hydrated Na-L. The mean free paths of the recoiling ^{24}Na cation (and all other cations) are similar to unit cell lengths of zeolites.

In some early studies the recoil of various uni-, di- and trivalet ions in zeolites X and Y from supercages to sodalite cage sites and the complementary recoils from sodalite cages and hexagonal prism sites to the supercages have been measured [18]. It was found in all cases when cations were sited in the supercages or in the sodalite cages before irradiation that $\sim 90\%$ of the cations on capture of a neutron, recoiled to supercage sites. Since the sodalite cages have some 10% of the free volume of the supercages per unit cell these recoils seemed to be strictly random in behaviour and independent of recoil energy or cation species. However, cations sited in the hexagonal prisms before irradiation recoiled with a 40-50% probability into the supercages on capture of a neutron.

These differences in recoiling probabilities made it possible to establish the preferences shown by the cations studied for the hexagonal prism sites (site A) as a function of calcination temperature, T_c, concentration and type of co-cation present in the structure. When $T_c > 400\,^{\circ}C$, Rb^+ and Cs^+ ions began to populate site A. When $T_c > 440\,^{\circ}C$ Ba^+ ions occupied site A in preference to sodalite cage sites. Co^{2+}, Zn^{2+} and Cu^{2+} ions all exhibited similar affinities for the "locked-in" sites (ie sodalite cages and hexagonal prisms) of zeolite X and Y. When any of these three ions had to compete with Na^+ ions for these "locked-in" sites at Tc > 500 $^{\circ}$C the divalent ions were always found in site A as long as the concentration of these divalent ions was kept below 8 ions per u.c.

In a more recent study the location of Li^+, Na^+, K^+, Rb^+, Cs^+, NH_4^+, Ag^+, Ca^{2+}, Sr^{2+}, Ba^{2+}, Co^{2+}, Cu^{2+}, Zn^{2+} and La^{3+} ions in zeolite L among the five cation sites in this zeolite has been ascertained. The effect of increasing the ion exchange temperature from 25° to 95°C and the effect of calcination at 650 $^{\circ}$C on these cation site distributions have been determined [19]. On exchanging these ions into K-L zeolite at 95°C some population of the hexagonal prism sites was found in all cases except Na^+ and NH_4^+. Only NH_4^+ ions were found however, to populate the cancrinite cage sites. On calcining these ion exchange forms (ion exchange carried out at 25°C) at 650°C a marked preference for hexagonal prism and cancrinite cage sites was now observed. When the total number of equivalents of cation per u.c. located in the hexagonal prism sites after calcination at 650°C is plotted against the cation diameter a maximum is observed at a cation diameter of $\sim$0.23 nm, suggesting that the hexagonal prisms attains its maximum stability when a cation of the size of La^{3+} is sited in its centre.

In a recent unpublished study the locations of La^{3+}, Gd^{3+}, Ce^{3+} and Yb^{3+} ions (which had been ion exchanged into K-L zeolite at room temperature) were determined after calcination of the zeolite at 650°C. After calcination all of the rare-earths showed some occupancy of the hexagonal prism sites, the order being La > Gd > Ce >> Yb. In all cases the total equivalence of cation in the hexagonal prism sites exceeded 2.0 per u.c. Supporting the conclusions above that cations of diameter $\sim$0.23 nm stabilise the framework charge more effectively when sited in hexagonal prisms. Only Gd^{3+} showed any preference for site E while La^{3+} was found in a small amount in the cancrinite cages.

REFERENCES

1. Y. Yasuda, J. Phys. Chem., 86 (1982) 1913-1917
2. P.E. Eberly, in J.A. Rabo (Ed), Zeolite Chemistry and Catalysis ACS Monogr. 171, 1976, pp 413-420
3. M. Betemps, in L.V.C. Rees (Ed), Proc. Fifth Int. Conf. Zeolites, Heyden, London, 1980, pp 526-534
4. D.M. Ruthven and R.I. Derrah, J. Chem. Soc. Faraday I, 71 (1975) 2031-2044
5. N.N. Greenwood and T.C. Gibb, Mossbauer Spectroscopy, Chapman and Hall Ltd., London 1971
6. B.L. Dickson, Ph.D. Thesis, University of London, 1973, pp 90-111

7. W.J. Mortier, Compilation of Extra Framework Sites in Zeolites,
    Butterworth Scientific Ltd., Guildford, 1982
8. B.L. Dickson and L.V.C. Rees, J. Chem. Soc. Faraday I, 70 (1974) 2038-
    2050
9. Zi Gao and L.V.C. Rees, Zeolites, 2 (1982) 79-86
10. Zi Gao and L.V.C. Rees, Zeolites, 2 (1982) 205-214
11. B.L. Dickson and L.V.C. Rees, J. Chem. Soc. Faraday I, 70 (1974) 2051-
    2059
12. Zi Gao and L.V.C. Rees, Zeolites, 2 (1982) 215-220
13. F.R. Fitch and L.V.C. Rees, Zeolites, 2 (1982) 33-41
14. F.R. Fitch and L.V.C. Rees, Zeolites, 2 (1982) 279-289
15. S.L. Suib, K.C. McMahon and D. Psaras in G.D. Stucky and F.G. Dwyer,
    (Eds.), Intrazeolite Chemistry, ACS Sym. Ser., 218, 1983, pp 301-317
16. W. Gunsser, J. Adlof and F. Schmidt, J. Magn. Magn. Mat., 15-18 (1980)
    pp 1115-1116
17. F. Schmidt, Th. Bein, U. Ohlerich and P.A. Jacobs, in D.H. Olson and
    A. Bisio, (Eds.), Proc. Sixth Int. Conf. Zeolite, Butterworth Scientific
    Ltd., Guildford, 1984
18. P.P. Lai and L.V.C. Rees, J. Chem. Soc. Faraday I, 72 (1976) 1809-1839
19. P.A. Newell and L.V.C. Rees, Zeolites, 3 (1983) 28-36

P.A. Jacobs et al. (Editors), *Structure and Reactivity of Modified Zeolites*
© 1984 Elsevier Science Publishers B.V., Amsterdam — Printed in The Netherlands

ADSORPTION OF XENON: A NEW METHOD FOR STUDYING THE CRYSTALLINITY OF ZEOLITES

M.A.SPRINGUEL-HUET[1], T.ITO[2] and J.P.FRAISSARD[1]

[1]Laboratoire de Chimie des Surfaces, associé au C.N.R.S. ERA 457.
Université Pierre et Marie Curie, Tour 55, 4 Place Jussieu 75230
Paris Cedex 5

[2]Research Institute for Catalysis, Hokkaïdo University, Sapporo 060 /Japan/.

ABSTRACT

We have shown that the chemical shift of xenon adsorbed on a Y or A zeolite depends on several factors, including the structure of the cages. We have used this fact to determine the number of well-formed cages in a zeolite of these types. This information on the crystallinity of these solids complements those obtained by x-ray diffraction or by the adsorption of certain molecules such as benzene.

INTRODUCTION

The central idea of this research was to find a non-reactive molecule, particularly sensitive to its environment and to interactions with other chemical species, which could serve as a probe for a new method for the determination of certain properties of solid catalysts, and, especially, zeolites. In addition, this probe should be detectable by NMR since this technique is particularly suitable for the study of electron perturbation in rapidly moving molecules.

We have shown that the NMR chemical shift of xenon adsorbed on a zeolite is very large. It depends on several factors: collisions between xenon atoms and the cage walls, the cations or other xenon atoms /ref. 1,2/, the number and nature of the compensating cations and the electric field they create /ref. 3/, the number of supported metal particles and the nature and concentration of supported metal particles as well as the nature and concentration of any gas prechemisorbed on the zeolite /ref.4/.

In the present article we wish to demonstrate the importance of this technique in the study of zeolite crystallinity. In particular, by counting the well-

-formed cages in a sample, the suggested method should make it possible to resolve the following question: is the amorphous part detected by x-rays really caused by the disordered state of a fraction of the solid or by crystallites which are too small to give rise to an x-ray diffraction spectrum?

EXPERIMENTAL

We have studied xenon adsorption on some well-crystallized zeolites: HY, NaY_x /where $\underline{x}$ represents the ratio of Si atoms to Al/, CaA, and on various samples consisting either of mixtures of these zeolites with each other or with amorphous solids. The samples KAT 80 and CMH 5 were supplied by the French Petroleum Institute /IFP/. The former is an $Na_{0.2}H_{0.8}Y$ zeolite, the second a mixture of KAT 80 and alumina.

All samples were pretreated as follows. A known amount of solid was placed in an NMR tube and degassed under vacuum $/10^{-3}Pa/$ at room temperature, then at a temperature rising slowly to $400^{\circ}C$ and maintained at this value for 8 h. The adsorption isotherms of xenon gas /99.99%/ were measured at $25^{\circ}C$ with a classical volumetric apparatus. NMR absorption of ^{129}Xe adsorbed on these samples was observed with a Bruker CXP 100 spectrometer at the frequency of 24.9 MHz. The radio frequency field pulse is repeated every one to three seconds which is sufficiently long when compared with the relaxation time. As the reference signal of ^{129}Xe that of Xe gas extrapolated to zero pressure according to Jameson's equation /ref.5/ is taken. All resonance signals of ^{129}Xe adsorbed on these solids were shifted to a higher frequency in relation to the reference. This is defined as the positive direction in this paper.

THEORY AND RESULTS

We have shown /ref.2/ that the NMR chemical shift of xenon adsorbed on a zeolite is given by the sum of three terms characteristic of each of the effects which a gas can undergo there:

$$\delta = \delta_o + \delta_s + \delta_E + \int_o^{\rho} /Xe\text{-}Xe/ \cdot d\rho \qquad \text{/1/}$$

- δ_o is the reference;
- δ_E is due to the electric field created by the cation;

$-\int_0^\rho \delta_{/Xe-Xe/} \cdot d\rho$, where ρ is the density of xenon adsorbed in the cavities, corresponds to the increase in shift caused by Xe-Xe collisions. $\delta_{/Xe-Xe/}$ corresponds to the effect of collisions between xenon atoms.

$- \delta_s$, due to collisions between xenon and cage walls, is expressed as

$$\delta_s = \delta_{/Xe-zeolite/} \cdot \rho_s$$

$\delta_{/Xe-zeolite/}$ is characteristic for collisions between the xenon and the walls. ρ_s, corresponding to density, depends only on the cage or channel structure.

At very low Xe concentrations, a cage cannot contain more than one adsorbed atom. The probability of a Xe-Xe collision equals practically zero. The motion of each atom is disturbed only by the cage walls. Consequently, the chemical shift $\delta_s + \delta_E$ obtained by extrapolation of the line $\delta = f/\rho/$ to $\rho = 0$ can be considered as characteristic of the zeolite with respect to xenon adsorption /see Fig.1/.

We have seen that δ_E is negligible in the case of NaY, and HY /ref.2/. In our opinion this is due to the rapid motion of Xe in the zeolite /at $25°C$/ which reduces to zero the effect of the fields F created by the different cations. In the case of CaA, $\delta /E/$ is also negligible except in the low Xe concentration range where the effect of Ca^{2+} causes a non-linear change of δ.

On the contrary, in the case of Y zeolites containing alkaline earth cations, the effect of the electric field becomes very important as soon as the degree of exchange with Na^+ exceeds 50%./The first 50% go into the sodalite cages which are inaccessible to xenon because of its size/ /ref.3/. However, whatever is the effect of E, there exists a single-valued relationship between the chemical shift δ and the concentration of Xe in the cages of the zeolite in question. Furthermore, at each value of δ the signal strength is proportional not only to the quantity adsorbed per cage but also, of course, to the number of cages.

Consequently, if the sample consists of a mixture of various solids /e.g. two zeolites of different structure, or a zeolite and an amorphous phase/ the adsorbed xenon spectrum must include as many components as there are different structures, the intensities of the components being a direct measure of the number of cages of each type /except the amorphous phases where the sig-

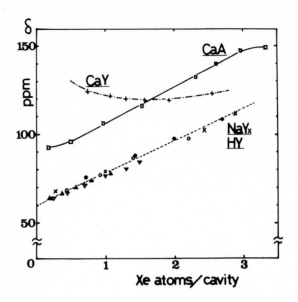

Fig.1. Chemical shift δ against the number of xenon atoms per cage. $\square$:CaA, $+$:CaY, $\blacktriangledown$:HY, NaY$_x$ ...x = /$\times$:1.28, $\bullet$:1.35, o :2,42, $\blacktriangle$:54,2/

nals are very broad and weak/. Let us look now at each case in more detail.

The case of NaY$_x$ and HY

We have said that the variation of δ with the number of Xe per cage is practically independent of the Si/Al ratio, x. The adsorption isotherm, however, depends greatly on the number of Na$^+$ cations. The curves of δ as function of the xenon pressure are therefore different to each x value. Now, this is the only type of variation which can be used. It is impossible to determine the xenon concentration per cage without knowing the number of cages, which is precisely what we are looking for.

1 - Let us suppose first of all that the reference sample /perfectly crystallized NaY$_x$ zeolite/, r, and the well-crystallized part of the sample, A, are identical /with the same value of x/. In this case the adsorption isotherms "per supercage" are identical and the curves of δ = f/ρ/ may be superposed on each other. For each given pressure /or each δ / the signal ratio, I$_r$/I$_A$,

is equal to the corresponding ratio of the number of supercages.

$$\frac{I_r}{I_A} = \frac{\text{number of cages, r}}{\text{number of cages, A}} \qquad /2/$$

2 - Let us suppose now that the values of $\underline{x}$ for $\underline{r}$ and $\underline{A}$ are different but that the value $\delta_s + \delta_E$ of δ at zero pressure is the same, which is very often the case for all NaY_x samples /see Fig.2; r = NaY Linde; A_1 = KAT 806/.

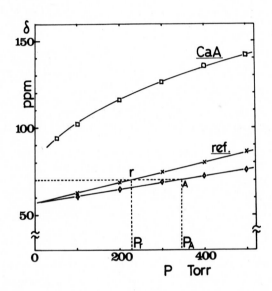

Fig. 2. Chemical shift δ against the pressure of xenon adsorption ✕ : reference NaY, ◆ : sample A_1 /KAT 806 from IFP/

Any line of the form δ = constant intersects the two δ = $f/\varphi/$ lines at the points $\underline{r}$ and $\underline{A}$ corresponding to two different pressures P_r and P_A, but at the same xenon concentration per supercage in the reference sample and in A. Consequently, for the same value of δ, equation 1 is still valid.

    Example 1: We prepared a mixture of 60% NaY + 40% CaA w/w at $25°C$. After desorption of water at $400°C$ and 10^{-5} Torr the mixture consisted of 58% NaY + 42% CaA. Whatever the xenon pressure, the spectrum of the adsorbed gas consists always of two well resolved components corresponding to CaA

and NaY, respectively /see Fig.3/. By comparing the intensity of the NaY
signal with that of the reference sample, one finds that the composition of the
mixture is exactly that given above to within $\pm$ 1%.

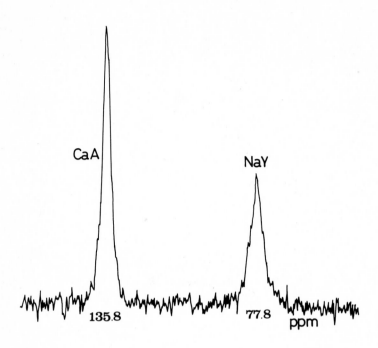

Fig. 3 Spectrum of xenon adsorbed on the NaY and CaA mixture.
Equilibrium pressure: 400 Torr.

3 - It often happens that the spectra of two zeolites NaY_x and HY_x display
slightly different $\delta_s + \delta_E$ values /see Fig. 4/. We have not yet completely
elucidated the reasons for these differences. We are, however, sure that
among other factors these reasons are the size of the crystallites, the more
or less rapid diffusion of the xenon from one crystallite to another, the presence of Al cations in the supercages after partial dealumination of HY_x, etc.
The variations of δ with the number of xenon atoms per cage are then parallel
/see Fig.4/ but it is obvious that in the $\delta = f/\varrho/$ diagram, any line representing δ = constant, will intersect the lines of the studied samples A and reference r at points which do not correspond to the same concentration of xenon

per cage.

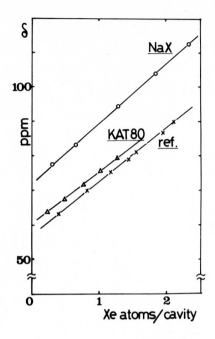

Fig. 4. Chemical shift δ against the number of xenon atoms per cage.

The simplest solution of the problem consists then in plotting the linear variation of the intensity I of the signal against δ /see Fig. 5/. If the sample A
is pure from the crystallographic point of view, the lines corresponding to
I = f/δ/ for A and r are parallel /r = NaY Linde and A = KAT 80/. In the case of a sample $\underline{A}'$ essentialy of the same kind but containing deformed or destroyed supercages, the slope of I = f/δ/ will be less than that one of A. We
have then

$$\frac{I_{A'}}{I_A} = \frac{\text{number of supercages in A}'}{\text{number of supercages in A}}$$

One sees therefore that related to any reference $\underline{r}$, the ratio of the slopes
will give the ratio of the numbers of well formed supercages.

Example 2: We have mixtures A_1' and A_2' corresponding after dehydration
to A_1' : 76% NaY + 24% Al_2O_3; A_2' : 50% KAT 80 + 50% Al_2O_3. Measuring the

intensity of the signal of xenon adsorbed on NaY with respect to the reference we obtained the proportions $77 \pm 2\%$ for $A_1^{'}$ and $53 \pm 3\%$ for $A_2^{'}$ /see Fig. 5/.

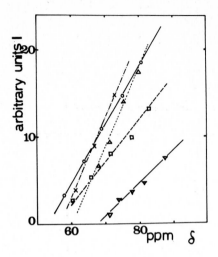

Fig. 5. Signal intensity I against the chemical shift δ . $\times$: reference $NaY_{2.42}$
$\triangle$: KAT 80, $\bigcirc$: NaY + Al_2O_3, $\square$: NaY + CaA, $\blacktriangledown$: KAT 80 + Al_2O_3

Other zeolites

The method can be applied further to zeolites other than NaY_x or HY_x when the variation of the chemical shift: $\delta = f/\rho/$, related to the crystalline part is the same as that of the reference sample. For example, by comparing the signal intensities of CaA pure and in the mixture of example 1 /see Fig.2/ one finds the exact composition of the latter.

CONCLUSION

The suggested method makes it possible to count the well-formed cages within zeolite samples, and therefore, to determine the degree of crystallinity of a zeolite independently of the crystallite size. It is more exact than x-ray diffraction which is more difficult to use quantitatively and which detects particles only if they are sufficiently large. It is also more precise than the methods involving adsorption of certain molecules such as benzene. These molecules are usually preferentially adsorbed by the well-formed part of the zeolite but the amounts adsorbed by the deformed or amorphous parts are not always negligible.

REFERENCES

1 T.Ito and J.P.Fraissard, Proc. Fifth Int. Conf. Zeolites, Naples, Heyden, 1981, 510-515.
2 T.Ito and J.P.Fraissard, J.Chem.Phys. 76 /1982/ 11, 5225-5229 .
3 T.Ito and J.P.Fraissard, in press.
4 L.C. de Menorval, J.P.Fraissard and T.Ito, J.Chem.Soc., Faraday Trans. I 78 /1982/ 403-410.
5 A.K.Jameson, C.J.Jameson and H.J.Gutowsky, J.Chem.Phys. 53 /1970/ 2310.

P.A. Jacobs et al. (Editors), *Structure and Reactivity of Modified Zeolites*
© 1984 Elsevier Science Publishers B.V., Amsterdam — Printed in The Netherlands

SOLID STATE ^{29}Si AND ^{27}Al NMR STUDIES OF DECATIONIZED AND DEALUMINATED ZEOLITES

G. ENGELHARDT[1], U. LOHSE[1], M. MÄGI[2] and E. LIPPMAA[2]

[1]Central Institute of Physical Chemistry, Academy of Sciences of the GDR, 1199 Berlin-Adlershof, German Democratic Republic

[2]Institute of Chemical and Biological Physics, Estonian Academy of Sciences, 200001 Tallinn, USSR

ABSTRACT

A short survey of the potentialities of ^{29}Si and ^{27}Al NMR in structure elucidation of dealuminated and decationized zeolites is given. Three topics of current interest are selected for demonstrating the application of the NMR technique: The preparation of zeolite HY by careful deammoniation of zeolites NH_4Y; the reexamination of the postulated 'reinsertion of aluminium' into the framework of dealuminated zeolites Y; and the dealumination of zeolite ZSM-5.

INTRODUCTION

Decationization and dealumination processes by thermal or chemical treatment of zeolites are of particular interest in preparing industrially important catalysts of high activity, selectivity and stability. The catalytic and other chemical properties of the modified zeolites are closely related to framework constituents and structure and, therefore, their detailed characterization is of fundamental interest.

High-resolution magic-angle spinning solid state ^{29}Si and ^{27}Al NMR have become outstanding tools in the structural study of zeolites (1,2) and can provide valuable information on various aspects of the constitution and structure of dealuminated zeolites (3-5). The aim of this paper is to give a short survey of the potentialities of the ^{29}Si and ^{27}Al NMR techniques in structure elucidation of dealuminated and decationized zeolites and to demonstrate the advantages of the NMR method by selected examples of current interest. It will be shown that i. the deammoniation of zeolites NH_4Y is generally accompanied by some degree of framework dealumination, ii. no reinsertion of Al into the framework of dealuminated zeolites Y by alkaline treatment occurs, and iii. zeolite H-ZSM-5 is partially dealuminated by calcination at 800 °C.

STRUCTURAL INFORMATION OBTAINABLE FROM ^{29}Si AND ^{27}Al NMR SPECTRA OF DEALUMINATED ZEOLITES

^{29}Si NMR

The power of ^{29}Si NMR in structural studies of zeolites relies mainly on the fact that separate peaks of distinct chemical shifts appear in the spectra for the five possible Si(nAl) building units in aluminosilicates (Si(nAl) denotes here a SiO_4 tetrahedron connected to n AlO_4 tetrahedra, and (4-n) other SiO_4 tetrahedra, where n = o-4). The quantitative relation between the peak intensities and the number of Si(nAl) units in the zeolite sample provides a simple possibility to determine the relative proportions of the various Si(nAl) units present in the sample, and the Si/Al ratio of the tetrahedrally bound aluminosilicate framework of the zeolite (6). The latter possibility is of particular interest in the case of dealuminated zeolites because it opens a simple way to determine the proportion of aluminium which remains in tetrahedral lattice positions after the dealumination process, and - if the overall aluminium content of the sample is known from e. g. chemical analysis - the amount of Al removed from the lattice.

Detailed information on the arrangement of Si and Al atoms on the tetrahedral sites of the zeolite framework can be derived by comparison of the relative populations of the various Si(nAl) units i. experimentally determined from the ^{29}Si NMR line intensities and ii. calculated from model structures with distinct Si, Al orderings (6-8). This procedure has been used for a detailed description of the Si, Al orderings in a series of progressively dealuminated Y zeolites and of the pathway of the stepwise elimination of Al atoms from the lattice (5b). From a more general point of view it seems of interest whether the Si, Al orderings in zeolites with varying Al contents are governed by certain rules, especially the rule of Loewenstein (9) which forbids AlOAl pairings and the rules of Dempsey (10) which state that the number of second nearest Al neighbours (AlOSiOAl linkages) in the zeolite framework is minimized. Recently, Vega (11) has presented results of Monte Carlo computer calculations of randomized Si, Al distributions of faujasite-type zeolites under the restriction of Loewenstein's and Dempsey's rules. We have compared the theoretical predictions with the experimental results obtained from the ^{29}Si NMR spectra of more than 50 directly synthesized or dealuminated zeolites X and Y covering Si/Al ratios of 1 up to about 70, now available from our own recent studies (5, 6 and unpublished results) and from the work of other authors (7, 8). From the comparison it follows that for Si/Al ratios in the range of 1.5 to 4 a pronounced tendency to avoid second nearest Al pairs exists, however, Dempsey's rule is not completely obeyed. At Si/Al ratios higher than 4 the Si,Al orderings approach a random distribution with the restriction of Loewenstein's rule. The latter restriction is generally valid over the whole range of Si/Al ratios.

Another important problem in structural studies of dealuminated zeolites is the presence of terminal SiOH groups due to lattice defects generated by the treatment of the zeolite. This problem may be successfully investigated using the ^{1}H-^{29}Si cross-polarization technique (CP) which gives strong and selective intensity enhancements of the ^{29}Si NMR peaks of Si atoms carrying OH groups (1, 3). In faujasite-type and other zeolites the chemical shifts of $(SiO)_3SiOH$ and $(SiO)_2Si(OH)_2$ groups coincide with the shifts of Si(3Al) and Si(1Al) units, respectively. Therefore, before the Si(nAl) line intensities of the direct FT spectrum are used for estimating the Si/Al ratio of modified zeolites, it should be tested by means of the CP spectrum whether SiOH units contribute to the Si(nAl) peaks. It should be noted that under certain conditions not only SiOH protons but also protons of other hydrogen containing constituents of the zeolite sample, e. g. H_2O or organic molecules, may give effective cross-polarization. In such cases usually all lines of the spectrum are equally intensity enhanced as it is observed for hydrated Y or other zeolites and tetrapropylammonium hydroxid containing zeolites ZSM-5 (12). On the other hand, acidic protons of structural SiOAl groupings are not discernible by CP technique because of the fast mutual exchange of the protons (5).

Microcrystalline samples with intact and well ordered aluminosilicate framework exhibit, in general, narrow and well resolved ^{29}Si NMR peaks for the various Si(nAl) units, whereas imperfections or partial destruction of the lattice cause line broadenings and overlapping peaks. In the limiting case of totally collapsed framework structures, i. e. amorphous products, a single broad peak without fine structure will be observed. The presence of amorphous silica, a typical by-product of hydrolytic damage of the zeolite lattice, may be identified in the ^{29}Si NMR spectra by a broadened peak or shoulder at about -110 ppm. In this way, lattice destructions of the thermochemically or otherwise treated zeolites may be recognized. To a smaller extent, line broadenings may also be observed if the lattice geometry is distorted during the thermo-chemical treatment or by occluded materials and by cation effects.

^{27}Al NMR

The particular advantage that ^{27}Al NMR affords in structural studies of dealuminated zeolites is that the technique can readily and quanti-tatively distinguish between tetrahedrally and octahedrally coordinated aluminium. For the two kinds of Al well separated lines with chemical shifts of about 60 ppm (tetrahedral Al) and 0 ppm (octahedral Al) appear in the spectra (4). Since the aluminium released from the tetrahedral sites of the zeolite lattice changes its coordination to octahedral, the content of non-framework Al can easily be determined from the intensity of the NMR peak at 0 ppm. Provided that no Al is removed from the sample, as it is the case in the thermochemical treatment of NH_4-exchanged zeolites (13), the degree of lattice dealumination can directly be followed by the intensity of the two peaks for tetrahedral and octahedral Al. If non-framework Al is extracted from the sample in the course of the dealumination reaction, e. g. in EDTA (14) or $SiCl_4$ treatments (15), the degree of dealumination can be determined from the tetrahedral Al peak intensity related to a standard sample of known Al content.

Difficulties may result from line broadenings caused by interactions of the ^{27}Al quadrupolar moment with electric field gradients induced by a non-symmetrical electron distribution around the ^{27}Al nucleus. In heavy cases the lines may be broadened beyond detection and errors may arise in the quantitative determination of the aluminium content from peak intensities. Therefore, it should be carefully tested whether the total Al content of a sample is reflected by the measured peak intensi-ties. Sometimes it is advantageous to increase the mobility of the Al containing species by hydration or by adding to the zeolite sample a complexing agent such as acetylacetone (16). With increasing mobility the NMR line will narrow and the intensity measurement becomes more reliable. Since the second order quadrupolar interaction is inversely proportional to the external magnetic field H_o, the application of a high-field NMR spectrometer with $H_o \gtrsim 7$ T is highly favourable.

Although ^{27}Al NMR is not, in general, as powerful as ^{29}Si NMR in in detecting subtle details of the zeolitic aluminosilicate framework structures, the technique is most valuable in detecting and estimating quantitatively octahedrally coordinated aluminium resulting from lattice dealumination or lattice destruction of chemically or thermally treated zeolites. Moreover, due to its high inherent sensitivity ^{27}Al NMR is capable of determining small contents of tetrahedral Al in ighly aluminium depleted zeolites which is hardly possible by means of ^{29}Si NMR because of its low sensitivity (17).

APPLICATIONS

Evidence for dealumination in the preparation of decationized zeolites HY

Some years ago, Kerr claimed that 'pure HY zeolite' can be prepared by shallow bed calcination of ammonium zeolite Y (18). The existence of a 'true' hydrogen form of zeolite Y containing only framework protons and devoid of any non-framework aluminium has been questioned by Flank et al. (19) and subsequently a lively pro- and contra-discussion has been conducted in J. Catal. and elsewhere (20, 21). ^{29}Si and ^{27}Al NMR should be particulary well suited to investigate this problem. Therefore, we have studied a series of 'HY' samples obtained by shallow bed (depth of the bed 1mm) calcination of NH_4-exchanged zeolites NaY (70% NH_4-exchange, Si/Al = 2.5) at constant temperatures between 150 and 300 °C for one week under vacuum. Even under these very mild

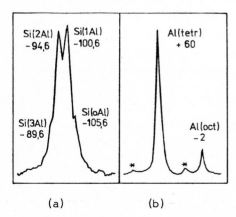

(a)   (b)

Fig. 1. ^{29}Si (a) and ^{27}Al (b) NMR spectra of NaHY zeolite obtained from NaNH$_4$Y by shallow bed calcination at 300 °C for one week. (Chemical shifts in ppm related to SiMe$_4$ (^{29}Si) and Al(H$_2$0)6^{3+} (^{27}Al). Asterisks denote spinning side bands.

conditions we observed in all samples a partial dealumination of the tetrahedral zeolite framework. The ^{29}Si and ^{27}Al NMR spectra of the

'HY' sample treated at 300 °C are shown in Fig. 1. The lattice dealumination is clearly demonstrated in the ^{29}Si NMR spectrum by the changes of the Si(nAl) peak intensities in comparison to the parent zeolite (5a) and the estimated Si/Al ratio of 2.7. The formation of octahedral non-framework Al is directly visible in the ^{27}Al NMR spectrum by the peak at -2 ppm, not present in the spectrum of the parent zeolite. We have, therefore, to conclude that the deammoniation of NH_4-exchanged zeolites Y is generally accompanied by a certain extent of dealumination of the tetrahedral lattice and consequently a 'pure HY zeolite' cannot be obtained in a simple way by thermal treatment of NH_4Y zeolites.

A reexamination of the reinsertion of aluminium in the framework of dealuminated zeolites Y as proposed by Breck and Skeels

As one result of their extended studies of stabilized zeolites Y Breck and Skeels (22) conclude that non-framework aluminium species, formed during calcination of zeolites NH_4Y, can be reinsered into the vacant sites of the dealuminated framework by treatment with NaOH solution. Since the reinsertion reaction seems to be doubtful for various reasons and has never been observed in our extensive investigations of dealuminated and stabilized zeolites Y, we prepared samples under almost the same conditions as described by Breck and Skeels and estimated the Si/Al ratio of their tetrahedral framework by ^{29}Si NMR. We started from a 85 % NH_4-exchanged zeolite NaY, the treatment under deep bed conditions at 400 °C yields sample A, slurrying of sample A in NaCl solution leads to the 'filtrate' sample B, and after titration of the slurried sample with NaOH to pH = 10 the 'slurry' sample C has been obtained. Whereas the Si/Al ratio increases from 2.5 in the parent zeolite to 3.6 in sample A - showing the framework dealumination by the thermal treatment - the Si/Al ratio is not changed at all when passing from sample A to samples B and C, and the ^{29}Si NMR spectra of the three samples are nearly identical. This shows clearly that no reinsertion of non-framework Al into the framework of sample C occurs during the alkaline treatment.

Dealumination of zeolite ZSM-5

The present interest in the catalytic and structural properties of
zeolite ZSM-5 prompted us to perform extensive NMR studies on the
structural modifications of various forms of ZSM-5 induced by thermo-
chemical treatment. For this purpose we prepared by standard methods
a Na, TPA-ZSM-5 with Si/Al of about 40 which was isotopically enriched
to about 90 % by ^{29}Si in order to increase the ^{29}Si NMR sensitivity.
The parent material was transformed into Na,H-ZSM-5 by calcination at
air for 2h at 600 °C, subsequently NH_4-exchanged and thermally treated
at 500 °C and 800 °C for different periods of time. A full description of
the results of our NMR studies will be presented elsewhere (23), what
follows is confined to the detection of framework dealumination during
the thermal treatment of the NH_4-ZSM-5. The ^{29}Si NMR spectrum of the
NH_4-ZSM-5 sample (see Fig. 2a) exhibits besides the strong Si(oAl)
peaks at -112.6 and -115.8 ppm three weak lines at -98.0, -104.4 and
-105.9 ppm. The first two lines correspond to silanol groups which
originate from lattice defects and are most probably present as
$(AlO)(SiO)_2SiOH$ and $(SiO)_3SiOH$ units, respectively. The line at -105.9 ppm
has to be assigned to Si(1Al) units. (24). After calcination of the
NH_4-ZSM-5 for 3h at 500 °C (Fig. 2b) the peaks at -98.0 and -104.4 ppm
disappear , showing the healing out of the lattice defects by condensation
of the silanol groupings. The Si(1Al) line remains essentially unchanged
after the treatment at 500 °C but decreases considerably by further
calcination of the sample for 2h at 800 °C. This shows unambiguously

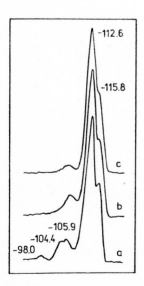

Fig. 2. ^{29}Si NMR spectra of zeolites ZSM-5
a - untreated NH_4-ZSM-5, b - calcined at
500 °C for 3h, c - calcined at 500 °C for
3h and at 800 °C for 2h. (Chemical shifts
in ppm related to $SiMe_4$)

the elimination of some Al from tetrahedral lattice positions.
From the peak intensities an increase of the Si/Al ratio to about 130
could be estimated. The formation of non-framework Al could be confirmed by the ^{27}Al NMR spectrum which shows besides the intense line at
55 ppm for tetrahedral framework Al, a weak line at 0 ppm for
octahedrally coordinated non-framework Al species.

REFERENCES

1 E. Lippmaa, M. Mägi, A. Samoson, M. Tarmak and G. Engelhardt,
   J. Amer. Chem. Soc., 103 (1981) 4992.
2 C.A. Fyfe, J.M. Thomas, J. Klinowski and G.C. Gobbi, Angew.
   Chem., 95 (1983) 257.
3 G. Engelhardt, U. Lohse, A. Samoson, M. Mägi, M. Tarmak and
   E. Lippmaa, Zeolites, 2 (1982) 59.
4 J. Klinowski, J.M. Thomas, C.A. Fyfe and G.C. Gobbi, Nature,
   296 (1982) 533.
5 G. Engelhardt, U. Lohse, V. Patzelova, M. Mägi and E. Lippmaa,
   Zeolites, 3 (1983) 233 (a) and 239 (b).
6 G. Engelhardt, U. Lohse, E. Lippmaa, M. Tarmak and M. Mägi,
   Z. anorg. allg. Chem. 482 (1981) 49.
7 J. Klinowski, S. Ramdas, J.M. Thomas, C.A. Fyfe and
   J.S. Hartman, J. Chem. Soc. Farad. Trans. 2, 78 (1982) 1025.
8 M.T. Melchior, D.E.W. Vaughan and A.J. Jacobsen, J. Amer. Chem.
   Soc. 104 (1982) 4859.
9 W. Loewenstein, Amer. Mineral., 39 (1954) 92.
10 E. Dempsey, G.H. Kühl and D.H. Olson, J. Phys. Chem., 73 (1969)
   387.
11 A.J. Vega, 'Intrazeolite Chemistry', ACS Symp. Ser., 218 (1983)
12 J.B. Nagy, Z. Gabelica and E.G. Derouane, Chemistry Lett. 1982,
   1105.
13 C.V. McDaniel and P.K. Maher, 'Molecular Sieves', Soc. Chem. Ind.
   London 1968, p. 186.
14 G.T. Kerr, J. Phys. Chem., 72 (1968) 2594.
15 K.K. Beyer and J. Belenykaja, 'Catalysis by Zeolites', Elsevier
   Amsterdam 1980, p. 203.
16 D. Freude, T. Fröhlich, H. Pfeifer and G. Scheler, Zeolites, 3
   (1983) 171.
17 J. Klinowski, J.M. Thomas, C.A. Fyfe, G.C. Gobbi and
   J.S. Hartman, Inorg. Chem., 22 (1983) 63.
18 G.T. Kerr, J. Catal., 15 (1969) 200.
19 W.H. Flank and G.W. Skeels, 'Proc. 5th Internat. Conf. Zeol.'
   Naples, Italy 1980, p. 344 and ref. cited therein.
20 G.T. Kerr, J. Catal. 77 (1982) 307.
21 G. W. Skeels, J. Catal., 79 (1983) 246.
22 D.W. Breck and G.W. Skeels, 'Proc. 5th Internat. Conf. Zeol.'
   Naples, Italy 1980, p. 335.
23 G. Engelhardt, B. Fahlke, M. Mägi and E. Lippmaa, in preparation.
24 J.B. Nagy, Z. Gabelica, E.G. Derouane and P.A. Jacobs,
   Chemistry Lett., 1982, 2003.

P.A. Jacobs et al. (Editors), *Structure and Reactivity of Modified Zeolites*
© 1984 Elsevier Science Publishers B.V., Amsterdam — Printed in The Netherlands

XPS STUDY OF TRANSITION METAL COMPLEXES IN ZEOLITES

E.S. SHPIRO, G.V. ANTOSHIN, O.P. TKACHENKO, S.V. GUDKOV,
B.V. ROMANOVSKY and Kh.M. MINACHEV
N.D. Zelinsky Institute of Organic Chemistry of the USSR Academy of Sciences, Moscow

ABSTRACT

Transition metal complexes (TMC) fixed in Y zeolites by "surface assembling" [Ni, Co, Cu, Ru phthalocyanines (Pc)] or by exchanging Na^+ for $[RuNO(NH_3)_4OH]^{2+}$ and $[Rh(NH_3)_5Cl]^{2+}$ cations have been investigated by XPS. MePc were established to be located in supercages as isolated complexes which interact with zeolitic OH groups via meso N atoms of the porphyrin ring. The stability of both MePc and complex cations under various conditions of zeolites pretreatment was compared. MePc were found to be quite stable in inert and reducing atmosphere while Ru and Rh complexes decomposed causing the stabilization of metals in intermediate and zero-valence states. XPS combined with a catalytic study showed that zeolites with TMC are of great interest as promising metal-complex catalysts or precursors of zeolite supported catalysts with high dispersion of metals.

INTRODUCTION

The synthesis and study of TMC in zeolites opens new possibilities for the use of zeolite catalysts. Because of the regular structure, high adsorption potential and some other properties of zeolites a large number of centres can be produced the characteristics of which are expected to be similar to those of individual homogeneous complexes. A lot of work, reviewed in (ref. 1,2), was devoted to solve this problem but in most cases the fixed complexes were unstable under catalytic conditions and a small amount of information was obtained on TMC distribution in the zeolite structure and on the influence of the latter upon the properties of TMC. This paper concerns an XPS investigation of Ni, Co, Cu and Ru phthalocyanines fixed in Y zeolites which were found (ref. 3,4) to exhibit some interesting catalytic properties. Both zeolites with MePc synthesized in the zeolite cavities in situ, and $[RuNO(NH_3)_4OH]^{2+}$ or $[Rh(NH_3)_5Cl]^{2+}$ -containing zeolites have been examined. The main problems involved

are: (a) the characterization of TMC; (b) the nature of the TMC-
-zeolitic framework interaction; (c) the effect of various pre-
treatments on the TMC stability and on the valence and coordination states of the metal. In addition, several TMC-containing
samples were tested in catalytic NO reduction. Some XPS data of
the samples studied have been reported earlier (ref. 5,6).

EXPERIMENTAL

The procedure of MePcY preparation consisted in (ref. 3,4):
(a) dehydration of exchanged forms in vacuo at $300-350^{\circ}$;
(b) exposure of the samples to phthalonitrile vapours at the same temperature; (c) elimination of the excess of the complexing
agent and removal of residual cations from zeolite by the exchan-
ge with NaCl solution. $\left[RuNO(NH_3)_4OH\right]^{2+}$ and $\left[Rh(NH_3)_5Cl\right]^{2+}$ cations were introduced into NaY via cationic exchange (ref. 5).
XP spectra were recorded with an ES 200B spectrometer (ref. 7).
The C 1s line (E_b = 285 eV) and the Si 2p line (E_b = 103 eV) were
used for energy calibration. The atomic ratios were determined
from integral intensities; the photoionization cross-sections
and corrections for escape depths and analyser transmission coefficients for photoelectrons with different E_k were taken into
account. The catalytic reaction NO+CO was performed in a static
reactor (ref. 5). NO labelled with $^{15}N^{18}O$ was used in some experiments.

RESULTS AND DISCUSSION
1. MePc synthesized in the zeolite matrix

Fig. 1 shows Ni $2p_{3/2}$ spectra of Ni in the samples studied,
as an example. The interaction of phthalonitrile with the cationic form of zeolites leads to an E_b shift of the main peak to
values which are characteristic of individual MePc (see Table 1).
These shifts indicate decreasing effective positive charges on
Ni and Co when compared to isolated cations (ref. 8). According
to correlations, the charge values are 0.35 for Ni and 0.37 for
Co, in agreement with those derived from EHM calculations of the
corresponding MePc (ref. 9). The changes of the shake up satellite structure and the line shape which are most pronounced for
Ni samples also indicate the MePc formation in zeolites. The
decreasing satellite intensity as well as line HWFM are caused by
decreasing Ni unpaired spin density due to the diamagnetic state
of NiPc (ref. 7). Since CoPc exhibits some paramagnetism, the

satellite structure of CoPcY spectrum did not change so dramatically.

The exposure of the Ru form to phthalonitrile also results in an appreciable shift of the Ru $3d_{5/2}$ spectrum from 281.1 eV to 284.2 eV, which can be attributed to Ru(III) complexes (ref. 5). This fact as well as the high N/Ru ratio presumably indicates the formation of RuPc. The Cu $2p_{3/2}$ E_b in CuPc and $Cu^{2+}Y$

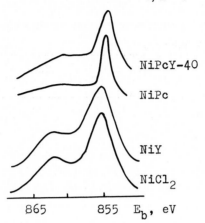

Fig. 1. XP Ni $2p_{3/2}$ spectra of the samples studied

TABLE 1

The parameters of XP Me $2p_{3/2}$ spectra for the samples studied

Sample	Ni				Co		
	E_b, eV	HWFM, eV	δ, eV[a]	β[b]	E_b, eV	HWFM, eV	β
MeY-13	855.6	2.7(2.4)[c]	5.7	1.63	781.7	2.9	1.89
MeY-30	855.7	2.9(2.3)	5.9	1.70	-	-	-
MeY-40	855.8	3.1(3.0)	6.0	1.57	781.7	3.0	2.12
MeY-60	856.2	2.9(2.3)	6.0	1.46	781.8	3.4	2.08
MePcY-13[d]	855.2	2.7(2.2)	6.2	1.40	780.8	2.2	2.17
MePcY-30	854.5	2.1(1.8)	7.3	1.88	-	-	-
MePcY-40	854.9	2.6(1.9)	6.8	2.26	780.6	2.7	2.12
MePcY-60	854.9	2.5(1.9)	7.5	3.24	780.8	2.1	2.08
MePc/NaY[f]	854.9	1.7(1.6)	6.5	-	-	-	-
MePc	854.8	1.5(1.4)	7.8	3.84	780.8	2.6	-

a) satellite-peak splitting; b) peak/satellite intensities ratio; c) HWFM without multiplet splitting contribution; d) the number denotes the exchange degree in the initial cationic form; f) prepared by sublimation of NiPc on NaY in vacuo at 500°.

are rather close, therefore the identification of CuPc in zeolite is not so unambiguous. However, the N 1s spectra and chemical analysis imply that a complex is formed, at least in the samples with high Cu content.

The parameters of the Me $2p_{3/2}$ spectra of Ni and Co cationic forms with various Me contents are rather similar but the differences became more pronounced after the reaction with phthalonitrile. The higher is the exchange degree, the closer becomes the satellite structure to that of the corresponding individual complex. This would suggest an increasing degree of Me complexing. The latter values can be estimated quantitatively from Me/N determinations (see Table 2). From these data it follows that (a) the number of cations bounded in Pc increased with the degree of exchange; (b) the number of residual cations, which reacted neither with phthalonitrile nor with NaCl during the reverse exchange was 2-3 per unit cell, independently of the Me content. Apparently, these cations occupy inaccessible S_I - sites (ref. 10).

TABLE 2

MePc contents in Y zeolites as determined by XPS

Sample	Me,[a) wt %	I_{Me}/I_N	MePc, %	The number of Me atoms per cell unit			
				total	extracted by NaCl	MePc	Me^{2+} residual
NiPcY-13	1.0	0.96	30	4	1	1	2
NiPcY-30	1.7	0.43	67	8	2	4	2
NiPcY-40	2.0	0.44	66	11	4	5	2
NiPcY-60	3.1	0.37	78	16	4	9	3
NiPc	-	0.29	100	-	-	-	-
CoPcY-13	1.0	1.93	20	4	1	1	2
CoPcY-40	1.8	0.86	44	11	5	3	3
CoPcY-60	3.0	0.62	61	16	5	7	4
CoPc	-	0.38	100	-	-	-	-

a) obtained by chemical analysis

The critical dimensions of Pc molecules are equal to 12-13Å, which is very close to the supercage diameter. This means that the number of MePc formed in a unit cell, if located inside the zeolite cavities appeared to be limited to 8. This is actually the case within the error limits (see Table 2) but some doubts still remain whether all complexes are inside the cavities or whether a part of them aggregates on the external surface. The detailed analysis of Me/Si and Me/Na intensity ratios gives evidence in favour of a preferential location of the MePc complexes inside the zeolite structure. The diagram in Fig. 2 demonstrates that the differences between surface and bulk composition did not exceed 25-30 % except of NiNaY-13, the external surface of which

is strongly enriched in NiPc. This may be due to Ni reduction
and subsequent Ni^o migration during vacuum treatment of the hydrated NiNaY-13 (ref. 7,8). The data for NiPc sublimated cn the
external surface of NaY also supported the conclusion about a
rather uniform MePc distribution over the zeolite structure. In
this case Me/Na and N/Si ratios exceed several times those of
MePcY with the same Ni content.

The nearly molecular dispersion of MePc, which can be achieved
in zeolite, facilitated an interaction between Pc and the zeolitic framework which may be displayed in the XP spectra. The coin-
cidence of Me 2p E_b for MePc and MePcY may suggest that the zeolitic framework did not affect the Me state in the complex.

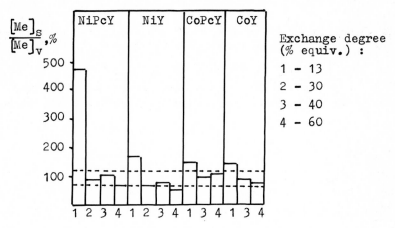

Exchange degree
(% equiv.) :

1 - 13
2 - 30
3 - 40
4 - 60

Fig. 2 Diagram illustrating the Me distribution in MeY and MePcY
samples (dashed lines limit the error range of $\pm$ 25 %).

On the contrary, N 1s spectra of MePcY greatly differ from those
of MePc. Instead of a narrow singlet (E_b=399.0 eV, HWFM=1.9 eV)
for MePc, a strongly broadened peak (up to 3.5 eV HWFM) or a
poorly resolved doublet appeared. Computer deconvolution of such
a spectrum for NiPcY-30 gave two components with the values of
399.0 and 400.5 eV. The appearence of non-equivalency of N atoms
in Pc may be due to the interaction between meso N atoms and
structural OH groups of the zeolitic framework. This assumption
is confirmed by similar E_b shifts of N 1s observed for the   NH
fragment in porphyrin (ref. 11) or during the proton adsorption
on a Pc molecule (ref. 12). The other reason of the changes in
N 1s spectra may be a distortion of the planar Pc molecule in
the supercage due to steric hindrances and to the strong electrostatic field of the lattice.

Low thermal stability of TMC is likely to be the major disadvantage which prevents the application of complex-based catalysts. An examination of TMC stability during zeolite pretreatment is therefore of great importance. After the heating of NiPcY-30 in He flow up to 600° no appreciable change of the XP spectra was observed. A small part of complexes desorbed at $400-450^\circ$ was probably located on the external surface. H_2 treatment at 350° also did not affect Ni $2p_{3/2}$ and N 1s spectra. When the temperature increased up to $400-450^\circ$, a Ni° line appeared in the Ni $2p_{3/2}$ spectrum and the N 1s intensity decreased drastically, which indicated a partial hydrogenolysis of the complex. Unlike Ni^{2+}-Y, no migration was observed in the case of reduced NiPcY. In this temperature range complexes of NiPc/NaY were desorbed from the surface without any decomposition. The treatment with O_2 affected MePcY at a lower temperature than with H_2 or He. The decrease of the intensity of the N 1s peak and the appearence of the spectrum of Ni $2p_{3/2}$ similar to that of Ni^{2+} cations at $200-350^\circ$ indicate a strong decomposition of NiPc. Both individual NiPc and NiPc/NaY samples are stable at temperatures as high as 300° and are decomposed under NiO formation at higher temperatures.

2. Transformations of complex cations in zeolites

A higher E_b of Me $3d_{5/2}$ levels for $\left[\text{RuNO(NH}_3)_4\text{OH}\right]^{2+}$ and $\left[\text{Rh(NH}_3)_5\text{Cl}\right]^{2+}$ introduced into zeolites in comparison with corresponding individual complexes may indicate ionic bonding between TMC and the zeolitic framework. These complexes, however, are gradually decomposed in the course of treatments of the corresponding zeolite samples in vacuo, H_2 or O_2, during which intermediate and zero-valence states of the metals are observed (see Fig. 3). According to XPS and IR data the coordination Ru-NO bond is partially maintained up to $200-300^\circ$. As a result of vacuum or H_2 treatments of MeY, metals are stabilized as Me^{2+} (Ru) or Me^{1+} (Ru, Rh). The RuY sample treated in H_2 at 450° exhibits spectra similar to those of Ru^{1+} species. However, these spectra correspond most likely to Ru clusters dispersed inside zeolite cavities and the +0.9 eV shift with respect to Me° is due to Ru interaction with electron acceptor sites of the framework (ref. 5,7). The absence of Ru migration to the external surface is indicated by a constant Ru/Si ratio. The conclusion about a high dispersion of metal produced by gradual reduction of the complex seems to be valid for Rh samples. Unlike treatment in vacuo and

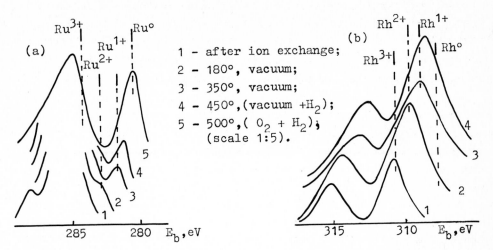

Fig. 3. XP Ru 3d+C 1s (a) and Rh 3d (b) spectra of RuY and RhY

in H_2, the decomposition of the Ru complex in O_2, followed by
the reduction in H_2, resulted in Ru^o migration to the external
surface and the formation of a metallic phase (see Fig. 3).

3. Catalytic properties of Ru containing zeolites in NO reduction with CO

XPS study of Ru zeolites prepared either via $\left[RuNO(NH_3)_4OH\right]$
Cl_2 exchange or by RuPc synthesis showed considerable differences in coordination and valence state of Me. In Pc, the metal
is surrounded by four N atoms and the axial coordination sites
are unsaturated. In case of RuY, the Me state changed from ionic to cluster or metallic one depending on the pretreatment
conditions. These differences greatly influenced the catalytic
properties of Ru zeolites in NO reduction. The higher activity
of RuPcY (see Table 3) in comparison with RuY, which contains
Ru^{2+} and Ru^{3+} ions under catalytic conditions, can be probably
explained by a higher reactivity of coordinatively unsaturated
RuPc. Reduction of RuY gives rise to predominant $Ru^{\delta+}$ clusters
formation which exhibited maximum activity. On the contrary, H_2
treatment of RuPcY had no effect on the initial reaction rate
and enhanced the final conversion only. According to XPS, Ru
atoms are formed in the course of RuPc decomposition under these
conditions but some active sites may be blocked by hydrogenoly-
tic products. The examination of the reaction mechanism by means
of $^{15}N^{18}O$ isotopes suggests that NO reduction over RuY proceeds
via NO dissociation and O_{ads}+CO interaction, the last step being

the rate-controlling one. The formation of strong complexes between Ru active sites and molecularly adsorbed NO appear to be responsible for the deactivation of the Ru form at low temperatures. The lower final conversion observed for RuPcY may be explained by the additional poisoning effect of CO (ref. 3).

TABLE 3

Catalytic properties of Ru containing zeolites in NO reduction with CO ($300°$, 15 torr)

Sample	Pretreatment conditions	Ru state before catalysis	$W_0 \times 10^{-18}$, $\frac{molecules}{g\ sec}$	$\mathcal{L}_{max}$ [a], %	Ru state under catalysis
RuY	$350°$,vac.	$Ru^{3+}, Ru^{\delta+}$	2.1	50	Ru^{3+}, Ru^{2+}
RuPcY	$350°$,vac.	RuPc	6.6	30	
RuY	$450°$,vac.+H_2	$Ru^{\delta+}$	14.2	90	$Ru^{\delta+}(60\%)$
RuPcY	$450°$,vac.+H_2	RuPc,$Ru°$	7.0	70	–

a) final conversion.

CONCLUSION

XPS provided new quantitative data on the surface layers of zeolites modified with TMC. This information can be related not only to the external surface but to the zeolite structure as well. The good agreement between XPS and chemical analysis data concerning Me/Na and Si/Al values confirmed this conclusion. XPS revealed some fine features of TMC properties in zeolites. Phthalocyanine complexes being isolated in supercages maintain to a great extent their individual characteristics. The zeolite acts as a solid solvent affecting the meso N atoms of the chelate unit. Although MePc loses some of its stability when introduced into zeolites, it remaines sufficiently stable, so that these samples can be used as catalysts at elevated temperatures. In addition to the reaction NO+CO, MePcY exhibits activity in dehydrogenation of cyclohexane to benzene and of benzene to styrene (ref. 3,4,13), their behaviour being different from that of conventionally prepared metal zeolite catalysts. The decomposition of Ru and Rh complex cations in vacuo and in H_2 leads to Me stabilization in low valence states or highly dispersed clusters. According to XPS, zeolite cavities contained Ru clusters even at $550-600°$. The high efficiency of Ru containing zeolite catalysts in NO+CO and CO+H_2 (ref. 14) reactions is probably explained by the presence of a large number of active sites with special electronic properties.

REFERENCES

1 J.H. Lunsford, Catal. Rev., 12 (1975) 137-162.
2 Y. Ben Taarit, M. Che, B. Imelik (Ed.), Catalysis by Zeolites, Elsevier, Amsterdam, 1980, pp. 167-193.
3 V.Yu. Zaharov, B.V. Romanovsky, Vestnik MGU, Ser. Khim., 18 (1977) 143-145.
4 M.V. Gusenkov, V.Yu. Zaharov, B.V. Romanovsky, Nephtekhimia, 18 (1978) 105-108.
5 O.P. Tkachenko, E.S. Shpiro, G.V. Antoshin, Kh.M.Minachev, Izv. An SSSR, Ser. Khim., 6 (1980) 1249-1256.
6 S.V. Gudkov, E.S. Shpiro, B.V. Romanovsky, G.V. Antoshin, Kh. M. Minachev, Izv. An SSSR, Ser. Khim., 11 (1980) 2448-2451.
7 Kh.M. Minachev, G.V. Antoshin, E.S. Shpiro. Photoelectron Spectroscopy and its Application in Catalysis, Moscow, "Nauka", 1981, 213 pp.
8 Kh.M. Minachev, G.V. Antoshin, E.S. Shpiro, Yu.A. Yusifov, Proc. VIth Intern. Congress on Catal., London, 1977, Chem. Soc., v.2, 621-632.
9 M.V. Zeller, R.G. Hayes, J. Amer. Soc., 95 (1973) 3855-3860.
10 P. Gallezot, Y. Ben Taarit, J. Phys. Chem., 71 (1975) 652-656.
11 J.P. Macquet, M.M. Millard, T. Theophamidis, J. Amer. Soc., 100 (1977) 4741-4746.
12 T. Kawai, M. Soma, Y. Matsumoto et al., Chem. Phys. Letters, 37 (1976) 378-382.
13 S.V. Gudkov, B.V. Romanovsky, E.S. Shpiro, G.V. Antoshin, Kh.M. Minachev, Vestnik MGU, Ser. Khim., Suppl. 1980, 12pp.
14 H.H. Nijs, P.A. Jacobs, J.B. Uytterhoeven, Chem. Commun., 4 (1979) 180-181.

P.A. Jacobs et al. (Editors), *Structure and Reactivity of Modified Zeolites*
© 1984 Elsevier Science Publishers B.V., Amsterdam — Printed in The Netherlands

SPECTROSCOPIC STUDY OF THE SURFACE COORDINATION OF DIVALENT COPPER ON DEHYDRATED A ZEOLITES

D. Packet and R.A. Schoonheydt

Laboratorium voor Oppervlaktescheikunde, Katholieke Universiteit Leuven, Kardinaal Mercierlaan 92, B-3030 Leuven, Heverlee (Belgium)

ABSTRACT

Dehydrated CuNaA, CuKA and CuCsNaA zeolites have been examined by EPR and diffuse reflectance spectroscopy (DRS). The Cu(II) ion is coordinated to the oxygen atoms of the hexagonal window connecting the α- and β-cages. The position of the Cu(II) ion with respect to the plane of the oxygen six-ring depends on the α-cage crowding by the major cation. In the case of CuNaA the Cu(II) ion is bound nearly in-plane. In the CuKA and CuCsNaA samples the cupric ions are displaced towards the β-cages.

INTRODUCTION

Divalent copper exhibits electronic and electron paramagnetic resonance (EPR) spectra that are very characteristic of its molecular and crystal environment. It can be used as a probe in the study of cation sites in zeolites (ref.1, ref.2). Until now attention has been given almost exclusively to X- and Y-type zeolites. Klier et al. reported identical electronic spectra for dehydrated CuNaA and CuNaY (ref.3). The observed d-d absorption bands in the region 10000 cm^{-1} - 17000 cm^{-1} are ascribed to the $E' \rightarrow E''$ and $E' \rightarrow A_1$ transitions of Cu^{2+} on a trigonal site (D_{3h} symmetry) (ref.4). Single-crystal X-ray diffraction techniques on vacuum-desolvated Cu_8A reveal that most of the cupric ions are located in the plane of the oxygen six-ring (D_{3h} symmetry) (ref.5). EPR and electron spin echo modulation (ESEM) experiments have been performed on samples with very small Cu^{2+} loading (ref.6,ref.7). Thus dehydrated CuNaA exhibits a signal with the following spin-Hamiltonian parameters : $g_{\parallel}$ = 2.3715, $A_{\parallel}$ = 1.23 x 10^{-2} T, $g_{\perp}$ = 2.068 and $A_{\perp}$ = 1.99x10^{-3} T (ref. 6). It is ascribed to Cu^{2+} coordinated to the hexagonal ring. This idea is refined by ESEM : Cu^{2+} is located in the β-cage in the presence of Na^+ but in the α-cage when the major cation is K^+ or Cs^+ (ref.7).

There remain 2 problems with these data : (1) it is not known in how far results obtained at high Cu^{2+} loadings (X-ray diffraction, electronic spectroscopy) are comparable to those from samples with small loadings (EPR,

ESEM); (2) a quantitative interpretation of the EPR parameters has not been given so far. In this paper we address ourselves to the first question.

EXPERIMENTAL METHODS

Samples

NaA, KA and CsNaA were prepared by ion exchange of a Linde 4A zeolite in an aqueous solution with an excess of NaCl, KCl and CsCl respectively. 7×10^{-3} kg of each sample were subsequently exchanged overnight in 1 dm^3 of a 4×10^{-4} mole $Cu(NO_3)_2$ solution in order to obtain - on the average - 0.1 Cu^{2+} per unit cell. The samples were repeatedly washed, dried in air at room temperature and stored in a desiccator over a saturated NH_4Cl solution. The contents of Na^+, K^+, Cs^+ and Cu^{2+} in the different samples were determined by atomic absorption spectrometry. The analytical data are given in Table 1. The charge deficit in Table 1 is defined as the difference between the theoretical number of monovalent cations per unit cell (12) and the sum of the cations determined experimentally.

TABLE 1
Number of exchangeable cations per unit cell

Samples	Cu^{2+}	Na^+	K^+	Cs^+	Charge Deficit
CuNaA	0.12	11.67	-	-	0.09
CuKA	0.12	0.05	10.48	-	1.23
CuCsNaA	0.11	5.52	-	2.53	3.73

Pretreatment

The samples were sieved and the fraction 0.25 - 0.50 mm was used to fill U-type quartz reflectometric cells fitted with an EPR sidearm. Dehydration was performed in an O_2-flow at 723 K for 3.6×10^3 s. The rate of temperature change was approximately 100 K/3.6×10^3 s. The CuKA sample was dehydrated at 623 K and at 723 K. The samples were cooled in O_2 and the sample cell was flushed with a He-flow for 1.8×10^3 s before EPR and DRS spectra were taken. O_2-line broadening was achieved by omission of this last step. The crystallinity of each sample after pretreatment was checked by X-ray diffraction with a Seifert-Scintag PAD II instrument.

Spectroscopy

Diffuse reflectance spectra were recorded between 2000 and 210 nm on a Cary 17 spectrometer equipped with a type I reflectance unit. The reference was the "Eastman Kodak White Reflectance Standard". The spectra were computer-processed to obtain plots of the Kubelka-Munck function against wavenumber (ref.8).

Reflectance spectra of NaA, CsNaA and KA, submitted to the same dehydration procedure as their Cu^{2+} analogues, were obtained and processed in the same way and subtracted from the spectra of the corresponding Cu-containing zeolites.

EPR spectra were recorded at 110 K in X-band on a Brucker ER 200D-SRC spectrometer. The spectra were simulated with the SIM14 computer program written by Lozos, Hoffman and Franz of Northwestern University, Evanston (U.S.A.) (ref.9). The two isotopes ^{63}Cu and ^{65}Cu were taken into account with relative abundances of 69 % and 31 % respectively and with a ratio of the nuclear g-factors $g_N(^{65}Cu)/g_N(^{63}Cu) = 1.702$ (ref.10).

RESULTS

Figure 1 shows the experimental and simulated EPR spectra of dehydrated CuNaA, CuCsNaA and CuKA. Table 2 lists the g-values and hyperfine splitting constants derived from the simulated spectra.

TABLE 2

	CuNaA		CuCsNaA		CuKA	
	^{63}Cu	^{65}Cu	^{63}Cu	^{65}Cu	^{63}Cu	^{65}Cu
$g_{\parallel}^{(1)}$	2.3860	2.3860	2.4010	2.4010	2.4050	2.4050
$A_{\parallel}^{(1)}/T$	1.26×10^{-2}	1.35×10^{-2}	1.17×10^{-2}	1.25×10^{-2}	1.17×10^{-2}	1.25×10^{-2}
$g_{\perp}^{(1)}$	2.0640	2.0640	2.0655	2.0655	2.0650	2.0650
$A_{\perp}^{(1)}/T$	2.5×10^{-4}	2.7×10^{-4}	2.0×10^{-4}	2.1×10^{-4}	2.0×10^{-4}	2.1×10^{-4}
$Q^{(1)}/T$	5.0×10^{-4}	5.4×10^{-4}	4.2×10^{-4}	4.5×10^{-4}	4.5×10^{-4}	4.9×10^{-4}
$g_{\parallel}^{(2)}$	-	-	-	-	2.3070	2.3070
$A_{\parallel}^{(2)}/T$	-	-	-	-	1.68×10^{-2}	1.80×10^{-2}
$g_{\perp}^{(2)}$	-	-	-	-	2.0680	2.0680
$A_{\perp}^{(2)}/T$	-	-	-	-	2.5×10^{-4}	2.8×10^{-4}
$Q^{(2)}/T$	-	-	-	-	5.1×10^{-4}	5.5×10^{-4}

The EPR spectra of CuNaA and CuCsNaA consist of only one signal with an effective axial symmetry. Simulation of the spectra was only possible with a Lorentzian lineshape and with a non-zero quadrupole interaction. The spectrum of CuKaA can only satisfactorily be simulated when two different species are taken into account. However X-ray diffraction experiments on this sample indicate that the crystal structure has been partially destroyed, even at the dehydration temperature of 623 K.

O_2 does not change the spectra of CuCsNaA and CuKA, but decreases the intensity of the spectrum of CuNaA by a factor 10. This process is reversible : the original spectrum can be restored by flushing the sample cell overnight with a He-flow.

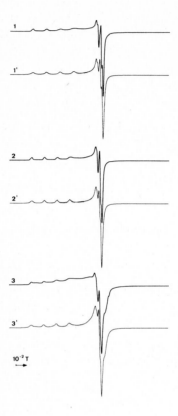

Fig. 1. Experimental and simulated EPR spectra of (1) dehydrated CuNaA, (2) dehydrated CuCsNaA and (3) dehydrated CuKA. The simulated spectra are indicated by a prime.

Fig. 2 shows the reflectance spectrum of dehydrated CuNaA. The insert shows an expansion of the 5000 - 20000 cm^{-1} region. Dehydration of the samples is complete as shown by the absence of the $\nu_3 + \delta$ (5100 cm^{-1}) and 2ν (6800 cm^{-1}) vibrational combination and overtone bands of water. The band near 7200 cm^{-1} belongs to the first vibrational overtone of lattice hydroxyl groups (ref.11). The formation of these hydroxyl groups is caused by proton exchange during the preparation of the samples (see charge deficit in Table 1).

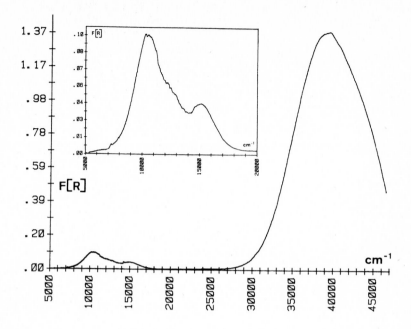

Fig.2. Diffuse reflectance spectrum of dehydrated CuNaA. The insert shows an expansion of the frequency region containing the d-d transitions.

There are three d-d transitions between 10000 cm^{-1} and 16000 cm^{-1} : two separate absorption bands at 10500 cm^{-1} and 15100 cm^{-1} and a shoulder at 12200 cm^{-1}. The ligand-to-metal charge transfer excitations (LMCT) are observable as one broad and intense band with a maximum around 39400 cm^{-1}. Admission of O_2 does not change the spectrum of CuNaA. The spectrum of CuNaCsA has the same general features as that of CuNaA.

The reflectance spectrum of CuKA is more complicated in the region of the d-d transitions with a considerable intensity at the low frequency side of the 10600 cm^{-1} band and a continuous absorption between the two main bands at 10600 cm^{-1} and 15100 cm^{-1} (fig.3).
The maximum of the LMCT band is located at 41000 cm^{-1}. Thus the spectrum of CuKA is the superposition of at least two absorption patterns caused by two types of Cu^{2+} ions in a different crystal environment, in accordance with the EPR results. The frequencies of the d-d transitions and LMCT bands are summarized in Table 3.

46

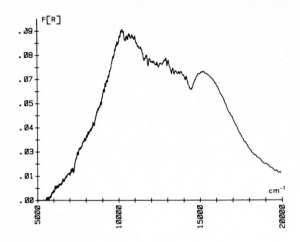

Fig.3. Diffuse reflectance spectrum of dehydrated CuKA in the 5000 - 20000 cm^{-1} region

TABLE 3

d-d transitions and LMCT bands of dehydrated CuNaA, CuCsNaA and CuKA

	CuNaA	CuCsNaA	CuKa
d-d transitions (cm^{-1})	10500 12200 15100	10400 12350 15000	∿8500 10600 ∿13000 15100
LMCT (cm^{-1})	39400	39400	41000

DISCUSSION

This work reports for the first time DRS and EPR spectra for the same dehydrated Cu^{2+} zeolites. The results indicate that with 0.1 Cu^{2+} per unit cell there is only one type of Cu^{2+} in the case of CuNaA and CuCsNaA. The reflectance spectrum has three typical d-d bands (Table 3). Their frequencies are almost independent of the type of co-exchanged cation. The EPR spectra can be simulated by assuming axial symmetry. The g-values and hyperfine splitting constants are slightly but consistently dependent on the type of co-exchanged cation. A good fit of the EPR spectra can only be obtained with inclusion of quadrupole interactions. The main difference with previous EPR results is the much smaller $A_\perp$ values that we obtain. This is - in part - due to the inclusion of the quadrupole interaction.

In the case of CuKA there is obviously more than one Cu^{2+} species present, but the results must be interpreted carefully since a significant structural breakdown occurs even at low exchange levels and a dehydration temperature of 623 K. The supplementary absorption bands in the reflectance spectrum and the second signal (characterized by $g^{(2)}$, $A^{(2)}$ and $Q^{(2)}$ in Table 2) in the EPR spectrum of CuKA are probably due to cupric ions located in places of lattice destruction. Therefore the discussion will be restricted to the three band reflectance spectrum and the corresponding set of EPR parameters $g^{(1)}$, $A^{(1)}$ and $Q^{(1)}$ characteristic of cupric ions in the zeolitic structure.

The three component spectrum of Cu^{2+} can be interpreted in C_{3v} symmetry with the theoretical model of Klier et al. (ref. 4,12,13). Cu^{2+} is coordinated to three O(3) oxygens of the six-rings, connecting α- and β-cages. Table 4 lists the values of the Jahn-Teller coupling constant $\underline{b}$ and the crystal field parameters $\Delta_2 = E(E")-E(E')$ and $\Delta_3 = E(A_1)-E(E')$ calculated from the observed frequencies of the d-d transitions with a spin-orbit coupling parameter of 400 cm^{-1}.

TABLE 4

Jahn-Teller coupling constant $\underline{b}$ and crystal field parameters Δ_2 and Δ_3 for dehydrated CuNaA, CuCsNaA and CuKA

	CuNaA	CuCsNaA	CuKA
Δ_3 (cm^{-1})	14160	13940	13830
Δ_2 (cm^{-1})	10430	10340	10550
$\underline{b}$ (cm^{-1})	58	62	69

The energy difference of the first two d-d transitions (Table 3) increases with increasing $\underline{b}$ (Table 4).

Δ_3 has a significantly higher value for CuNaA than for CuKA or CuCsNaA. Since the value of Δ_2 remains approximately constant for the three samples, the A_1 state is pushed up in energy in the case of CuNaA indicating that the cupric ions are nearly coordinated in the plane of the oxygen six-ring. In dehydrated CuKA and CuCsNaA the cupric ions are displaced from the hexagonal window towards the β-cage due to the electrostatic and steric repulsion from the K^+ and Cs^+ ions in the α-cage. Thus crowding of the α-cages by the large K^+ and Cs^+ ions determines the exact position of Cu^{2+} on the hexagonal window in the β-cage. This crowding of the α-cages is confirmed by X-ray diffraction studies (ref.14,15,16). In CuNaA the Cu^{2+} is at a shorter distance from the O_2 molecules in the supercages than in CuKA and CuCsNaA. Thus line-broadening occurs only for CuNaA. However, since the reflectance spectrum does not change, there is no direct coordination of O_2 to Cu^{2+} and the line-broadening is probably caused by a dipole-dipole interaction. On the basis of the models proposed in this paper we are presently performing ligand field and angular overlap calculations to gain a detailed physical insight into the EPR parameters.

ACKNOWLEDGMENTS

D.P. acknowledges a research grant from the Belgian Government (Geconcerteerde Onderzoeksakties, Ministerie van het Wetenschapsbeleid), which provided also funds for performing this work. R.A.S. acknowledges a permanent research position as Senior Research Associate of the National Fund of Scientific Research (Belgium).

REFERENCES

1 J.D. Mikheikin, G.M. Zhidomirov and V.B. Kazanskii, Russ. Chem. Rev., 5 (1972) 468-483.

2 V.D. Atanasova, V.A. Shvets and V.B. Kazanskii, Russ. Chem. Rev., 3 (1981) 209-219.

3 J. Texter, D.H. Strome, R.G. Herman and K. Klier, J. Phys. Chem., 81 (1977) 333-338.

4 D.H. Strome and K. Klier, in W.H. Flank (Ed.), Adsorption and Ion Exchange with Synthetic Zeolites, A.C.S. Symp. Ser., 135 (1980) 155-176.

5 H.S. Lee and K. Seff, J. Phys. Chem., 85 (1981) 397-405.

6 R.A. Schoonheydt, P. Peigneur and J.B. Uytterhoeven, J.C.S. Faraday I, 74 (1978) 2550-2561.

7 L. Kevan and M. Narayana, in G.D. Stucky and F.G. Dwyer (Eds.), Intrazeolite Chemistry, A.C.S. Symp. Ser., 218 (1983) 283-299.

8 G. Kortüm, Reflectance Spectroscopy, Springer-Verlag, Berlin, 1969, ch.IV, p.111.

9 QCPE program n° 265, Chemistry Department Indiana University, U.S.A.

10 D. Gourier, J. Antoine, D. Vivien, J. Thery, J. Levage and R. Collongues, Phys. Stat. Sol., 41 (1977) 423-429.

11 L.M. Kustov, V.Yu. Borovkov and V.B. Kazansky, J. Catal., 72 (1981) 149-159.

12 K. Klier, P. Hutta and R. Kellerman in J.R. Katzer (Ed.), Molecular Sieves II, ACS Symp. Ser., 40 (1977) 108-119.

13 R. Kellerman and K. Klier, Surface and Defect Properties of Solids (Chem. Soc. London), 4 (1974) 1-33.

14 J.J. Pluth and J.V. Smith, J. Phys. Chem., 83 (1979) 741-749.

15 J.J. Pluth and J.V. Smith, J. Am. Chem. Soc., 102 (1980) 4704-4708.

16 T.B. Vance and K. Seff, J. Phys. Chem., 79 (1975) 2163-2167.

P.A. Jacobs et al. (Editors), *Structure and Reactivity of Modified Zeolites*
© 1984 Elsevier Science Publishers B.V., Amsterdam — Printed in The Netherlands

STUDIES ON THE MODIFIED CLAUS REACTION OVER ALKALINE FAUJASITES BY SIMULTANEOUS INFRARED, KINETICS, AND ESR MEASUREMENTS

H.G. KARGE, Y. ZHANG[1], S. TREVIZAN DE SUAREZ[2], and M. ZIOŁEK[3]
Fritz-Haber-Institut der Max-Planck-Gesellschaft, Faradayweg 4-6,
1000 Berlin 33 (West)(Federal Republic of Germany)

ABSTRACT
        The Claus reaction was investigated in an IR reactor cell using wafers of Me^IX, NaY, and MgY as catalysts. Before, during, and after the reaction, IR spectra were recorded; simultaneously the feed and product streams were analysed by GC. ESR and IR measurements were carried out with the same sample. During the reaction (acidic) OH groups developed on Me^IX, but did not form on NaY. The only reactant species detectable by IR on the working catalyst was chemisorbed SO_2. It was indicated by a newly discovered band around 1240 wavenumbers, and appeared to be an intermediate of the Claus reaction. These species were not correlated to SO_2^- radicals which occurred in very low concentrations on the catalyst surface.

INTRODUCTION
        The modified Claus reaction

$$2H_2S + SO_2 \ \rightleftarrows \ 3/n\ S_n + 2H_2O \tag{1}$$

has attained a significant commercial importance, in particular for sulphur recovery from sour natural gases. On an industrial scale, aluminas or bauxites are being employed to catalyze the Claus reaction. However, zeolites have attracted appreciable interest as model catalysts for this process [1-2]. Some of their properties (electron donor capability, acidity, Si/Al ratio) which may be relevant for the activity in Claus reaction, can be altered in a clear-cut manner *via* modification procedures such as ion exchange, dehydroxylation, dealumination, etc. Activity measurements on

---

[1] On leave from Dalian Institute of Chemical Physics (P.R. of China)
[2] On leave from Universidad del Litoral, Santa Fé (Argentina)
[3] On leave from A. Mickiewicz University, Poznań (Poland)

several X and Y type faujasites have been reported by Dudzik and Ziołek [3-4]. Interaction of the Claus reactants, H_2S and SO_2, with various zeolites was studied mainly by IR [5-9] and ESR spectroscopy [10-12] in static adsorption experiments, but spectroscopic investigations of the working Claus catalyst are still lacking. Therefore, it is almost completely unknown which species are present on the surface of the catalyst on stream. It has been observed, however, that all of the H_2S and SO_2 species detected in static adsorption experiments were removed from the catalyst surface by the competitively adsorbing product H_2O [8, 9]. Thus, it was the aim of our study to search for reactive species occurring on the surface of the working Claus catalyst and, if any were detected, elucidate the relationship between possible reaction intermediates. Since one of the most interesting parameters of a Claus catalyst is its basicity (electron donor capability, see Refs. 10-12) Me^IX (Me^I = Li, Na, K, Rb, Cs) samples were utilized as a first series of systematically modified zeolite catalysts. For the sake of comparison, NaY and MgY were also employed.

EXPERIMENTAL

Materials

NaX (Si/Al = 1.35) and NaY (Si/Al = 2.5) were supplied by Linde Corp. and, at 350 K, five times ion exchanged with 0.1 N solutions of the respective alkaline chlorides. The following samples were obtained (percentage of ion exchange in brackets): LiX (53); KX (84); RbX (55); CsX (64); MgY (73). Helium, hydrogen sulfide (ca. 99 %), and sulphur dioxide (99.97 %) were purchased from Messer Griesheim, Düsseldorf, and used without further purification.

Apparatus and procedures

IR spectra were obtained with Perkin-Elmer spectrometers (models 225 and 580 B, equipped with a data station model 3600). The design and operation of the IR reactor cell has been described elsewhere [13]. For catalyst activation under high vacuum (12h, 10^{-6} Pa), the cell was joined to an all-metal system with a turbo molecular pump. The Claus reaction was usually conducted at 400 K and atmospheric total pressure. The carrier gas (He) contained small partial pressures of H_2S (1.2 kPa) and SO_2 (0.6 kPa) and was passed through the activated zeolite wafer. Mass flow control was achieved by a thermoconductivity controller (Fa. Bronkhorst, models

F-200/201). Conversions were monitored by a Perkin-Elmer gas chromatograph (model Sigma 1) and could be expressed by either H_2S or SO_2 disappearance. A 1.5 m, 1/8 inch Porapak Q/S column was used for separation. During Claus reaction, IR spectra of the working catalyst were taken at short intervals.

RESULTS AND DISCUSSION

Formation of OH groups upon adsorption of SO_2 in the presence of H_2O

Sulphur dioxide was adsorbed from a SO_2/He stream [p(SO_2) = 0.6 kPa] on Me^{I}X zeolites at 400 K. The carrier gas helium contained traces of water which were detected via the $\delta(H_2O)$ band at 1640 cm^{-1}, the band at 3680 cm^{-1} (traces of strongly chemisorbed H_2O) and the band around 3400 cm^{-1} (hydrogen bonded H_2O) when helium was contacted with the zeolite wafer. Upon contact of NaX with the SO_2/He stream, a sharp band appeared at 3650 cm^{-1} and a small one at 3580 cm^{-1}, indicating the formation of structural OH groups. The 3650 cm^{-1} band intensified when the sample was degassed at 400 K (see Fig. 1). A CsX sample behaved similarly, but the intensity of the newly formed OH band was much smaller than with NaX.

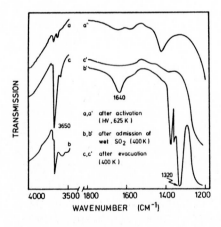

Fig. 1. Formation of acidic OH groups upon the interaction of SO_2 with NaX in the presence of traces of water.

The formation of structural OH groups (bands at 3650 and 3580 cm^{-1}) during the adsorption of SO_2 resembles the result of H_2S interaction with the same zeolites [7, 9]. In contrast to the effect of H_2S, however, adsorption of SO_2 resulted in the formation of OH groups only when small amounts of H_2O were simultaneously present. The 3650 and 3580 cm^{-1} signals did not appear when the carrier gas was thoroughly dried by passing it through a liquid nitrogen trap.

Furthermore, the OH groups generated *via* SO_2/H_2O interaction with the $Me^I X$ zeolites withstood degassing at 400 K whereas under the same conditions most of the OH groups were removed when formed *via* H_2S dissociation [7]. The two reactions leading to formation of OH groups may be visualized by schemes (I) and (II) where $Me^I OZ$ represents an alkaline X type zeolite:

$$Me^I OZ + SO_2 + H_2O \quad \overset{\rightarrow}{\leftarrow} \quad Me^I(HSO_3) + HOZ \qquad\qquad (I)$$

$$Me^I OZ + H_2S \quad \overset{\rightarrow}{\leftarrow} \quad Me^{II}(SH) + HOZ \qquad\qquad (II)$$

Since H_2S is a much weaker acid than H_2SO_3, reaction (II) is more easily reversed than reaction (I). Contrary to the behaviour of X type zeolites, NaY samples failed to develop OH groups (IR band around 3650 cm^{-1}) when contacted with SO_2/H_2O. Similarly, adsorption of SO_2 in the presence of H_2O had little effect in the case of acidic Y type samples such as MgY: the original band of acidic OH groups of MgY (3635 cm^{-1}) was slightly enhanced and a small additional band appeared at 3540 cm^{-1}.

Formation of chemisorbed SO_2 species indicated by a low frequency band

Upon adsorption of sulphur dioxide from a SO_2/He stream on $Me^I X$ zeolites, the well known SO_2 band at ca. 1330 cm^{-1} was observed, similar to those reported in Refs. 5 and 9 for static adsorption experiments. However, the present spectra were significantly improved because much thinner wafers were employed. Thus, an additional band could be detected in the low transmitting region below 1300 cm^{-1}. The investigation of this low frequency band (LFB) around 1240 cm^{-1} was even better facilitated by the use of a computerized Perkin-Elmer 580 B spectrometer which rendered possible accumulating, averaging, and subtracting of spectra. Figure 2 presents the spectra of SO_2 adsorbed on NaX after subtraction of the background spectrum of the adsorbent and the SO_2 gas phase. In the case of the spectra of Fig. 2a the carrier gas was dried prior to contact with the NaX wafer whereas in the experiment of Fig. 2b traces of water were present. Therefore, the base line in Fig. 2b exhibits the $\delta(H_2O)$ peak at 1640 cm^{-1}; this band is negligible in Fig. 2a. One recognizes that in the presence of water (see Fig. 2b) the intensity of the LFB (1240 cm^{-1}) is enhanced at the expense of the 1320 cm^{-1} signal. This seems to indicate that those SO_2 spe-

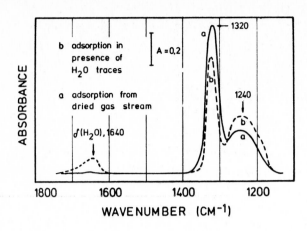

Fig. 2. IR bands
after adsorption of
SO_2 on NaX.

cies which give rise to the band at 1320 cm^{-1} are less strongly held
than the LFB species and, therefore, easily displaced by H_2O molecules. Support for this assignment was provided by desorption experiments.

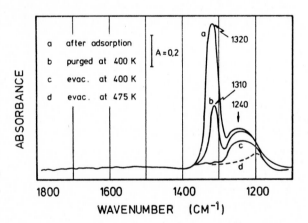

Fig. 3. Desorption of
SO_2 from NaX.

Fig. 3 demonstrates that most of the 1320 cm^{-1} species are removed
by mere purging with helium at 400 K. Evacuation to 10^{-5} Pa completely eliminated the 1320 cm^{-1} band whereas the low frequency
band (LFB) was little affected. It required heating at higher temperatures in high vacuum to desorb the LFB adsorbate. After degassing the sample at 700 K and 10^{-5} Pa, a part of the LFB still remained in the spectrum with the band maximum being shifted to
1190 cm^{-1}.

The overall behaviour of the two SO_2 bands at ca. 1320 and
1240 cm^{-1} is analogous to that of SO_2 adsorbates on alumina [14].
There the higher frequency band of SO_2 (1330 cm^{-1}) also indicates

loosely bound species which are easily displaced by H_2O molecules. Similarly, the low frequency band (at 1070 cm^{-1}) of SO_2 on Al_2O_3 was ascribed to chemisorbed species.

Relationship between chemisorbed SO_2 and SO_2^- radicals

The following experiments were undertaken to clarify the relationship between chemisorbed SO_2, indicated by the low frequency band around 1240 cm^{-1}, and SO_2^- radicals which form upon adsorption on zeolites [10-12]. The cell, which was described in an earlier publication [15], allowed us to conduct IR and ESR investigations with one and the same zeolite/adsorbate sample. After adsorption of 1.2 $mmol \cdot g^{-1}$ SO_2 on NaX at 400 K, the absorbance of the LFB and the spin density, $N(SO_2^-)$, was measured. Subsequently, increasing amounts of SO_2 were desorbed at successively higher temperatures. After adsorption (and also after degassing at room temperature), the total surface concentration of adsorbed SO_2 was greater by three orders of magnitude than the spin density, i.e., less than 0.1 % of the adsorbed SO_2 (N_{ad}) was SO_2^- (see Fig. 4). The total

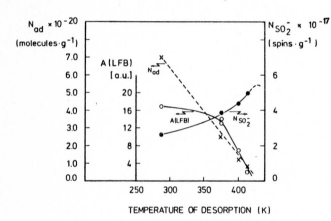

TEMPERATURE OF DESORPTION (K)

Fig. 4. Total amount of adsorbed SO_2, $N(ad)$, absorbance of the low frequency band, A(LFB), and spin density, $N(SO_2^-)$ on NaX/SO_2 as a function of the desorption temperature.

amount of adsorbed SO_2, however, did include not only the LFB species but also (up to 400 K desorption temperature) more loosely bound SO_2 molecules which were indicated by that band at 1320 cm^{-1}. This was at variance with the behaviour of γ-alumina as an adsorbent for SO_2 [14, 15]. With increasing desorption temperatures, the total amount of adsorbed SO_2, N_{ad}, linearly decreased (see Fig. 4). About 90 % of the initially adsorbed SO_2 were already removed after degassing the sample at 425 K and 10^{-6} Pa. The intensity of the LFB, A(LFB), decreased as well, although not in parallel with

N_{ad}, because of the contribution of the 1320 cm^{-1} adsorbate *(vide supra)*. At the same time the spin density $N(SO_2^-)$ slightly increased. Only above ca. 500 K did the radical concentration start to decline. It can be concluded from this pattern that the LFB species are not identical with SO_2^- and no equilibrium exists between chemisorbed SO_2 and SO_2^- radicals. Analogous results have been obtained with SO_2 adsorbed on alumina [15].

Claus reaction

IR spectra of the activated catalysts were scanned prior to reaction and, at short intervals, during the Claus reaction. Additional spectra were obtained after the Claus reaction when the spent catalyst was purged with helium and/or evacuated at reaction temperature.

During the reaction, the OH stretching region was almost completely obscured by broad bands around 3400 cm^{-1} being due to physically adsorbed product water. It also gave rise to the $\delta(H_2O)$

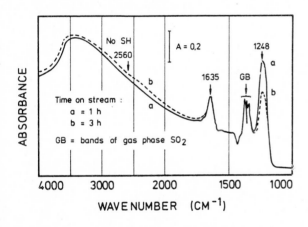

Fig. 5. IR spectrum of a working NaX catalyst during Claus reaction.

signal at 1635 cm^{-1}. This latter band was either nearly constant in intensity or slowly decreased with time on stream because of catalyst decay (sulphur deposition, *vide infra*). When the reaction was completed and the catalyst evacuated at 450 K, an OH band around 3650 cm^{-1} was left in the spectrum. It indicated OH groups which had formed during the reaction because of interaction of $SO_2 + H_2O$ with the zeolite *(vide supra)*. In the case of NaY no such OH band was detected, and the OH band of MgY at 3640 cm^{-1} had almost the same intensity as prior to the Claus reaction.

The low freqency band (LFB) around 1240 cm^{-1}, which had been detected upon adsorption of pure SO_2 *(vide supra)*, also formed during Claus reaction over Me^IX, NaY, and MgY. This was the most important IR result concerning the working Claus catalyst. It is worth noting that, besides the bands of the product H_2O, the LFB of chemisorbed SO_2 was the only signal which indicated reactant species being present on the catalyst on stream. In particular, neither H_2S species (band around 2560 cm^{-1}) nor physically adsorbed SO_2 (1320 cm^{-1}, see Figs. 2 and 3) were detected (compare Fig. 5).

There was no doubt that the LFB indicates SO_2 species being reactive towards H_2S: when the Claus reaction has proceeded for some time and the SO_2 flow was stopped but the stream of H_2S maintained, the LFB immediately vanished (see Fig. 6, spectrum f → g for the example of NaX).

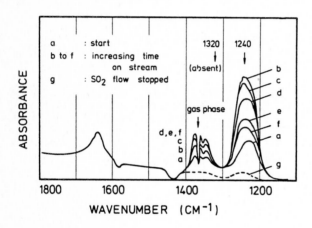

Fig. 6. Variation of the intensity of the LFB at 1240 cm^{-1} with time on stream and after stop of SO_2 flow.

Generally, the intensity of the LFB increased during the first 15 minutes of Claus reaction (at 400 K) and then decreased with time on stream (see Fig. 6, spectra a, b → f). This decrease was most rapid with LiX and NaX (well reproduced), much less pronounced with KX and almost missing with RbX and CsX. A plot of the absorbance, A(LFB), *versus* time on stream is shown for NaX in Fig. 7, curve a.

Simultaneously with scanning the IR spectra, the conversions of H_2S and SO_2 due to Claus reaction were followed by gas chromatography. At the reaction temperature of 400 K, an initial decrease in activity was observed with all the catalysts employed. Within the first hour on stream the conversion usually dropped by about 60 %. Thereafter an almost steady state was maintained. An example is shown in Fig. 7, curve b. The decrease in conversion was

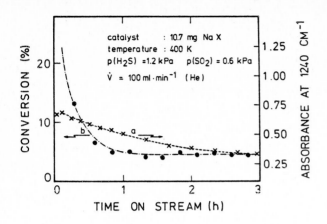

similar to that in LFB absorbance (Fig. 7, curve a). Even though the curves a and b do not coincide, Fig. 7 suggests that a relationship exists between conversion and density of chemisorbed SO_2.

In another experiment using KX as a catalyst, the Claus reaction was interrupted after 3 h by shutting off the SO_2 stream and purging with H_2S/He. The LFB vanished (as in the case of NaX, see Fig. 6) and reappeared after the SO_2 flow had been turned on again. Simultaneously, the steady state conversion of the first period of Claus reaction was almost restored. Thus, the reaction rate did respond to the coverage of the catalyst surface with chemisorbed SO_2.

However, no general correlation was observed between either the intensity of the LFB and the nature of the exchange cation or the density of the LFB species and the steady state conversions over the various catalysts (see Table 1). After 15 min on stream, A(LFB) was similar for NaX through CsX but significantly lower in the case of LiX. Although LiX exhibited a low concentration of chemisorbed SO_2 (band at 1250 cm^{-1}), its activity was, however, equal to that of NaX and KX. In contrast, RbX and CsX showed during Claus reaction high absorbance of the low frequency band, but low conversions. Presently, it cannot be ruled out that SO_2 adsorbed on different zeolite catalysts exhibits different reactivities towards H_2S due to differences in the bonding to the respective adsorption sites. Indeed, slight but non-negligible differences in the wavenumbers of the LFB's were observed (see Table 1). Such variations in reactivity could explain why no simple correlation exists between the rate of Claus reaction and the surface con-

TABLE 1

IR results and steady state conversions

Zeolite	A(OH) [a.u.]	LFB [cm^{-1}]	A(LFB) [a.u.]	Conv. [%] 400 K	450 K
LiX	0.08	1250	0.7	3.7	9.6
NaX	0.06	1242	1.4	3.5	9.4
KX	0.15	1234	1.8	3.9	9.4
RbX	0.13	1238	1.7	1.6	4.3
CsX	0.06	1230	1.3	0.4	-

A(OH):   Absorbance of the OH band around 3650 cm^{-1}, normalized to equal sample thickness, 10 $mg \cdot cm^{-2}$

A(LFB):   Absorbance of the LFB, normalized to 10 $mg \cdot cm^{-2}$, after 15 min.

Conv.:   Conversions during steady state of the Claus reaction, normalized to 10 mg wafer.

centration of the chemisorbed SO_2 when different catalysts were compared. Another explanation may derive from the relatively low reaction temperature (400 K). At low temperatures it could well be that differences in activity of the various catalysts were obscured by mass transfer effects, in particular when the pores of the zeolites were partly blocked by sulphur deposition. In fact, it was observed that at somewhat higher temperatures (450 K) deactivation of the $Me^{I}X$ Claus catalysts was much less severe and the conversion measurements seemed to be more reliable. The sequence in conversions over LiX through CsX, however, was the same as at 400 K reaction temperature. Unfortunately, at higher temperatures it became more difficult to obtain IR spectra of the working catalysts because of sulphur deposition on the windows of the IR reactor cell. More experimental work is needed to clarify the above points.

CONCLUSIONS

A low frequency band (LFB) around 1240 cm^{-1} appears upon SO_2 adsorption as well as during Claus reaction over faujasite-type zeolites, indicating surface species which are reactive towards H_2S. The LFB species appear to be reaction intermediates. They are not, however, precursors of or in equilibrium with SO_2^- radicals, also occurring on zeolites after contact with SO_2. No general correlations emerge between the density of the LFB species, the nature of the exchange cations and the catalytic activity.

ACKNOWLEDGEMENT

The authors are indebted to Prof. Dr. J. H. Block and Drs. A. Gutsze and Z. George for helpful discussions on the manuscript. Financial support of the Senator für Wirtschaft und Verkehr (ERP Sondervermögen) der Stadt Berlin (West) is gratefully acknowledged.

REFERENCES

1   G.T. Kerr and G.C. Johnson, J. Phys. Chem., 64 (1960) 381-382.
2   H. Lee, in M.W. Meier and J.B. Uytterhoeven (Eds.),
    "Molecular Sieves", Advances in Chemistry Series, 121,
    American Chemical Society, Washington, D.C., 1973, 311 pp.
3   Z. Dudzik and M. Ziołek, J. Catalysis, 51 (1978) 345-354.
4   M. Ziołek and Z. Dudzik, Zeolites 1 (1981) 117-121.
5   A.V. Deo, I.G. Dalla Lana, and H.W. Habgood, J. Catalysis,
    21 (1971) 270-281.
6   H. Förster and M. Schuldt, J. Colloid Interface Sci., 52
    (1975) 380-385.
7   H.G. Karge and J. Raskó, J. Colloid Interface Sci., 64 (1978)
    522-532.
8   R.W. Tower, Z.M. George, and H.G. Karge, Preprints of the 6th
    Canadian Symposium on Catalysis, August 19-21, Ottawa,
    Ontario, 1979, pp.87.
9   H.G. Karge and J. Ladebeck, in R. Sersale, C. Colella, and
    R. Aiello (Eds.), Proc. of the 5th Int. Conf. on Zeolites,
    Napoli, Italy, Giannini, Napoli, 1980, 180 pp.
10 Y. Ono, H. Tokunaga, and T. Keii, J. Phys. Chem., 79 (1975)
    752-756.
11 Z. Dudzik and M. Bilska-Ziołek, Bull. Acad. Polon. Sci., 23
    (1975) 699-707.
12 K.C. Khulbe, R.S. Mann, and A. Manoogian, Zeolites 3 (1983)
    360-364.
13 H.G. Karge and K. Klose, Z. Phys. Chem., Neue Folge, 83
    (1973) 92-99.
14 H.G. Karge and I.G. Dalla Lana, J. Phys. Chem., in press.
15 H.G. Karge, S. Trevizan de Suarez, and I.G. Dalla Lana,
    J. Phys. Chem., in press.

P.A. Jacobs et al. (Editors), *Structure and Reactivity of Modified Zeolites*
© 1984 Elsevier Science Publishers B.V., Amsterdam — Printed in The Netherlands

THE STUDY OF BROENSTED AND LEWIS ACIDITY OF DECATIONIZED AND DEALUMINATED ZEOLITES BY IR-DIFFUSE REFLECTANCE SPECTROSCOPY AND QUANTUM CHEMISTRY

V.B. KAZANSKY

Zelinsky Institute of Organic Chemistry, Moscow, U S S R

ABSTRACT

The quantum chemical cluster models of bridged hydroxyl groups in decationized zeolites predict that their acidity is mainly determined by the number of neighbouring Al atoms. This results in the existence of only three main types of hydroxyl groups surrounded by one, two or three Al atoms in the large cavities. This is in agreement with the diffuse reflectance IR study of OH groups in the wide spectral range. This technique permits also to investigate low temperature adsorption of H_2 which was used as a test for Lewis acidic sites formed after dehydroxylation. The conclusion about different nature of these centres in dehydroxylated ZSM-type zeolites and faujasites was made.

INTRODUCTION

The connection of Broensted acidity and catalytic properties of zeolites with the Al content in their framework has been discussed by Beaumont (ref.1), Dempsey (refs.2-3), and more recently by Jacobs (refs.4-5). In (refs.6-8) we considered this problem more quantitatively using results of quantum chemical calculations which were performed for so called cluster models of bridged hydroxyl groups. The summary of this result given below allows to propose a rational systematics of structural OH groups hydrogen forms of various zeolites. It will be compared then with the results of diffuse IR-study of hydrogen groups in zeolites in the wide spectral range which includes in addition to fundamental stretching vibrations their overtones and combined stretching plus bending vibrations. This technique was also applied to the study of low temperature hydrogen adsorption. The use of H_2 molecule as a test for the detection of Lewis acidic sites which were formed in zeolites after their dehydroxylation indicated the different nature of these defects in faujasites and high silica containing zeolites.

CLUSTER QUANTUM CHEMICAL MODELS AND THE NATURE OF STRUCTURAL
HYDROXYL GROUPS IN DECATIONIZED ZEOLITES OF DIFFERENT CHEMICAL
COMPOSITION

Cluster quantum chemical models of hydroxyl groups on the surface of silicagel and in zeolites were developed in our works
(refs.9-11). The broken bonds on the edges of the clusters were
saturated with hydrogenlike univalent pseudoatoms A with variable
Valence Orbital Ionisation Potential (VOIP). Its value was chosen
in such a way that permits the best description of electronic properties and the acidity of hydroxyl groups. For the particular
case of a bridged hydroxylgroup in a zeolite the corresponding
cluster could be represented as follows:

$$\begin{array}{ccc} A_2 & H & A_1 \\ \diagdown O & O & O \diagup \\ A_2 - O - Si & O & Al - O - A_1 \\ \diagup O & O & \\ A_2 & & A_1 \end{array}$$

According to Loewenstein's rule (ref.12) two negatively charged
tetrahedra occupied by aluminum could never be in the neighbouring positions.Therefore the role of pseudoatoms A_1 may be played
only by silicon.On the other hand,A_2 could represent tetrahedra
occupied either by silicon or by aluminum.This results in the following four possible types of structural hydroxyls depending on
the number of surrounding aluminum atoms:

$$\begin{array}{cccc} Si & H & Si \\ \diagdown O & O & O \diagup \\ Si - O - Si & Al - O - Si & (OH\ I) \\ \diagup O & O & \\ Si & Si \end{array} \qquad \begin{array}{cccc} Al & H & Si \\ \diagdown O & O & O \diagup \\ Si - O - Si & Al - O - Si & (OH\ II) \\ \diagup O & O & \\ Si & Si \end{array}$$

$$\begin{array}{cccc} Al & H & Si \\ \diagdown O & O & O \diagup \\ Al - O - Si & Al - O - Si & (OH\ III) \\ \diagup O & O & \\ Si & Si \end{array} \qquad \begin{array}{cccc} Al & H & Si \\ \diagdown O & O & O \diagup \\ Al - O - Si & Al - O - Si & (OH\ IV) \\ \diagup O & O & \\ Al & Si \end{array}$$

The quantum chemical calculations were performed by CNDO method in Boyd-Whithead parametrization which gives the optimal repro-
duction   of the energy of chemical bonds.The VOIP's of A_1 and
A_2 have been found from the condition of the electroneutrality of
the zeolite framework (refs.6,7). For tetrahedra occupied by silicon it was taken equal to 10.7 e.v and for those modeling tetrahedral aluminum 7 e.v. SiO_4 and AlO_4 fragments were assumed to be
regular tetrahedra. Both $Si \diagup^O \diagdown_A$ and $Al \diagup^O \diagdown_A$ angles were taken

equal to 140° and Si-O and Al-O distances equal to 1.63 Å.

The results of quantum chemical calculations are collected in Table 1, where the data for terminal silanol groups are also included.

Table 1

The type of OH group	The number of surrounding Al atoms	Si:Al ratio	q_H	q_{OH}	E_H+, eV
OH I	1	4-7	0.404	-0.466	14.96
OH II	2	3	0.392	-0.481	15.32
OH III	3	1.7	0.384	-0.500	15.60
OH IV	4	1	0.376	-0.507	15.89
SiOH	0	–	0.317	-0.519	17.00

A comparison of presented values leads to the following conclusions:

a) All the considered bridged OH groups are more acidic than the terminal SiOH groups of silicagel since they have lower proton abstraction energies E_H+ and higher positive charges q_H on hydrogen atoms.

b) In agreement with well known experimental facts the acidity of bridged hydroxyl groups grows with increasing Si/Al ratio.

The physical reason of this phenomenon is a low value of VOIP of pseudoatoms simulating the negatively charged aluminium-oxygen tetrahedra. Indeed, the magnitude of 7 eV formally corresponds almost to an alkaline metal or Al metal atom. Therefore, aluminum-oxygen tetrahedra act in a zeolite structure as strong electron donors. This results in the transfer of the electron density to hydroxyl groups and to the decrease of their acidity with the increasing aluminum content.

Let us come back to four types of clusters representing the bridged hydroxyl groups. Each of them has a definite chemical composition. Therefore, the ordered crystal lattice of a zeolite, that consists of the clusters of the only one kind, has both uniform Broensted acidity and the definite silica-to-alumina ratio. For instance, an ideal chain structure which consists only of repeating clusters of the first type has the composition $[SiAlSi_3]_n$ with Si:Al=4. For three dimensional crystal lattice where no aluminum or silicon atoms participate simultaneously in two neighbouring clusters this ratio would be equal to 7.

These figures are certainly approximate and could somewhat change depending on the way of connection of clusters in the real crystal lattice.The clusters could be also separated by additional tetrahedra occupied by silicon.However,it is evident that beyond some critical silica to alumina ratio which is close to the above discussed figures the OH groups of the first type should predominate.Therefore,one might expect the existence of only OH I in high silica containing zeolites.It is also clear that the probability to find only the structural hydroxyls of the first type should increase for the higher silica to alumina ratios.

In a similar way the composition of the structure containing only the clusters of the second type is $[AlSi_3AlSi_3]_n$ with $Si:Al=3$.Such silicon to aluminum ratio is characteristic of Y type faujasites.Therefore one might expect the predominance of OH II in their structure.

For the framework of a zeolite which consists only of the clusters of the third type the composition is $[Al_2Si_2AlSi_3]_n$ with $Si:Al \approx 1.7$.This is just the border between Y and X type faujasites.Therefore for low silica containing Y zeolites and for X zeolites one might expect the presence of both OH II and OH III.The latter should predominate for X type.

Finally,the composition of the clusters of the fourth type is $[Al_3SiAlSi_3]_n$ with $Si:Al=1$.Therefore the hydrogen form of A type zeolite which has such silicon to aluminum ratio should contain structural hydroxyls of the fourth type only.However,this is only a possibility,since the hydrogen form of A zeolite is unstable.

Consequently, the hydrogen forms of zeolites should contain only several definite types of structural hydroxyls. In addition each of them should predominate for distinct crystal structure and chemical composition.The exception are the high silica containing zeolites where one has to expect the presence only of the OH I.

Let us now consider the stereoisomery of structural hydroxyls. For this purpose it is more convenient to take a cluster which contains the silicon atom in its centre.Then for OH I there would exist only one stereoisomer with the acidic proton migrating among three equivalent positions.For the cluster with two aluminum atoms in the first coordination sphere two stereoisomers OH II and OH II' are possible:

```
    Al                        Al                          Al
    O    H                    O    H                  H   O
    Si                        Si                          Si
  O    O—Si              O   O    O—Si            O   O    O—Si
Si         O           Al        O                Al        O
           Si                    Si                         Si
```

OH I   OH II   OH II'

In II' the proton is hydrogen bonded to the neighbouring basic oxygen.Therefore the frequency of its stretching vibrations
should be lower and that of the bending vibrations should be higher than the corresponding frequency for OH II.On the other
hand,the stretching frequency of OH II should be higher than that
for OH I since the latter is more acidic.

For the cluster containing three aluminum atoms there are
still two similar possibilities of OH III and OH III' with localization of the proton between silicon and aluminum or between
two aluminum atoms.In accordance with above quantum chemical calculations the acidic properties of OH III and OH III' should be
somewhat lower than those for OH II and OH II'.This could also
influence their stretching and bending frequencies,however,in a
minor degree than for OH I and OH II.

Thus since the hydrogen form of A type zeolite is unstable one might
expect the existence of the following five types of structural
hydroxyls with decreasing acidic strength in each set: OH I,
OH II, OH III and OH II', OH III'.This conclusion is also true
for amorphous silica alumina catalysts where acidic hydroxyl
groups in principle could be described by similar quantum chemical cluster modesl.

IR SPECTRA AND THE NATURE OF STRUCTURAL HYDROXYLS IN
HYDROGEN FORMS OF ZEOLITES

There are many papers devoted to the study of hydroxyl groups
in zeolites by IR spectroscopy (see for example (ref.13) and references therein).In all cases the conventional transmittance
technique for very thin semitransparent pellets which permits the
measurements of fundamental stretching frequencies in the region
of 3500-3800 cm^{-1} was used.Only in several papers bending OH frequencies at about 1000-1300 cm^{-1} were also studied.The quality
of these spectra is,however,very poor because of the strong absorption of the light by the framework of zeolites.

Recently in our papers (refs.14-17) a new diffuse reflectance technique of IR-measurements was developed,which has more than one order of magnitude higher sensitivity.It permits an easy study of high silica containing zeolites with Si:Al ratios up to several hundreds.In addition,the spectral range could be broadened up to 10000 cm^{-1}.Therefore each type of hydroxyls could be connected with three different frequencies: those of the stretching vibrations,their overtones and with combination of stretching plus bending vibrations.This makes the assignment of different types of hydroxyls more convincing.The resolution of the spectra in the combined region is also considerably improved because the differences in stretching vibrations are added to those of the bending vibrations.Below in addition to the literature data we shall mainly concentrate on the results of these measurements.

IR-spectra of hydroxyl groups in hydrogen forms of high silica containing zeolites

According to the cluster models considered above the bridged hydroxyl groups of only the first type should be characteristic of the zeolites with the Si:Al ratios higher than 4-7. This is in accordance with the published data since only one type of structural hydroxyl groups with stretching frequencies of 3610-3620 cm^{-1} was reported for hydrogen forms of offretite (Si:Al=3.75), omega (Si:Al=4.4), clinoptilolite (Si:Al=5), heulandite (Si:Al=5.43), ferrierite (Si:Al=6) and mordenite (Si:Al=5) (See (refs. 4,13) and references therein). Recently the single band of acidic OH groups at 3610 cm^{-1} was also reported for the hydrogen form of ZSM-5 zeolite (ref. 18).

In (ref.16) a special study of hydrogen forms of mordenite, ZSM-5 and silicalite was carried out in the wide spectral range by the diffuse IR reflectance technique.Despite the Si:Al ratios were changed from 5 to about 125 the positions of the lines in all investigated samples were practically the same (Fig.1). The set of bands with $\omega_{0-1} \approx 3745$ cm^{-1}, $\omega_{\nu+\delta} \approx 4550$ cm^{-1} and $\omega_{0-2} \approx 7325$ cm^{-1} which coincide in their positions with the lines of hydroxyls of silicagel was ascribed to silanol groups.The bands at $\omega_{0-1} \approx 3610$ cm^{-1}, $\omega_{\nu+\delta} \approx 4660$ cm^{-1} and $\omega_{0-2} \approx 7080$ cm^{-1} with the intensities decreasing for lower Al content should be connected with OH I (Fig.1). Their identity for mordenite and ZSM-5 zeolite was also confirmed by similar shifts caused by hydrogen bonding

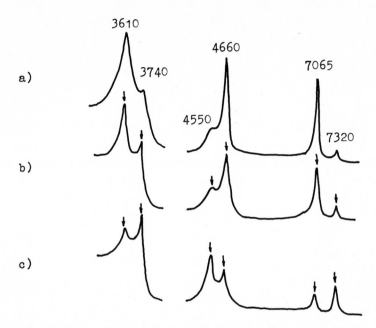

Fig. 1. Diffuse IR spectra of hydrogen forms of high silica zeolites recorded in the wide spectral range. a)Mordenite (Si:Al=5).
b) ZSM-5 (Si:Al=35). c) Silicalite (Si:Al=125).

with adsorbed o-xylene, benzene, dichloromethane and hexane.Thus
the conclusion that after some critical Si:Al ratio the nature of
structural hydroxyl groups in all high silica containing zeolites is identical is in agreement with their IR spectra.Therefore
the unique catalytic properties of high silica zeolites can be
explained either by shape selectivity effects or by the lower
concentration of acidic active sites.

IR-spectra of medium silicon containing zeolites and faujasites

In accordance with above cluster models IR spectra of zeolites
with medium silicon to aluminum ratios contain several bands belonging to different types of structural hydroxyl groups.For instance,two bands were reported for hydrogen forms of erionite
with Si:Al=3.03 (ω_1=3612 cm^{-1}, ω_2=3563 cm^{-1}); stilbite (Si:Al=
=2.6, ω_1=3630 cm^{-1}, ω_2=3650 cm^{-1}); chabasite (Si:Al=2.33,
ω_1=3630 cm^{-1}, ω_2=3540 cm^{-1}) and for faujasites ($3 \geqslant$ Si:Al$\geqslant 1$).
The latter zeolites have been studied in most details (ref.13).
Their IR spectra consist of two lines at about 3650 and 3550 cm^{-1}.
Thus faujasites do not contain any OH I groups.However,in accordance with above mentioned cluster models they are appearing as

the only type of hydroxyle after dealumination resulting in
Si:Al ratio equal to about 30 (ref.19).

On the other hand, there is also some evidence that the lines
3650 and 3550 cm^{-1} are of more complicated nature. Such conclusion was made by Jacobs and Uytterhoeven (ref.20) who used for the
improvement of the resolution of IR spectra the procedure of
their decomposition into several components of the Gaussian shape.

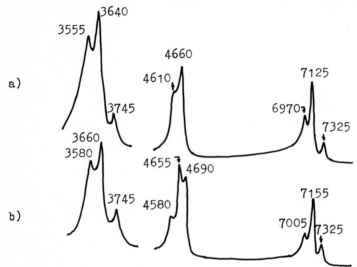

Fig. 2. IR spectra of hydrogen forms of faujasites. (a) Y zeolite. (b) X zeolite.

The diffuse reflectance IR spectrum of Y zeolite with Si:Al
ratio equal to 2.9 and the degree of decationization of about
80 % is shown in Fig. 2 (a). In accordance with the literature it
consists of two bands at 3550 and 3640 cm^{-1} in the fundamental
region. First of them belongs to hydroxyls inside hexagonal prisms
since it is not perturbed by the adsorption of big molecules. On
the contrary, the line at 3640 cm^{-1} that is sensitive to adsorption belongs to OH groups in the large cavities. Our measurements
showed that the lines at 4660 and 7125 cm^{-1} are also sensitive to
adsorption. Therefore we concluded that the OH groups in the large
cavities have the following parameters: $\omega_{0-1}=3646$ cm^{-1}, $\omega_{\nu+\delta}=$
$=4660$ cm^{-1}, $\omega_{0-2}=7125$ cm^{-1} and those inside hexagonal prisms
$\omega_{0-1}=3555$ cm^{-1}, $\omega_{\nu+\delta}=4610$ cm^{-1}, $\omega_{0-2}=6970$ cm^{-1}. Consequently,
the measurements in the broader spectral range did not discover
any new structural OH groups in addition to those already reported before.

The assignment of these hydroxyls to definite types of clusters could be done on the basis of aluminum distribution in Y zeolite that have been reported by Dempsey et al. who studied the single crystal of such zeolite by x-ray analysis (ref.21) (Fig.3). Recently this model was also confirmed by neutron diffraction data (ref.22) and by NMR measurements (ref.23).

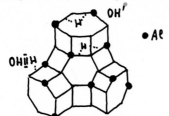

Fig.3.Distribution of Al in the framework of Y zeolite with Si:Al= =2 and the positions of OH groups according to (ref.21).

According to Fig.3 the hydroxyls inside large cavities should be classified as OH II and those inside hexagonal prisms as OH II'.This interpretation is in agreement with the lower stretching and higher bending frequencies of OH II' compared to OH II because of hydrogen bonding to the neighbouring oxygen atoms.

The spectrum of decationized X type zeolite is represented in Fig.2(b).In the fundamental and overtone regions it also consists of two sets of bands which are slightly shifted to the higher wavenumbers compared with positions of the lines in Y zeolite.In the region of combined stretching plus bending vibrations an additional third band that is sensitive to adsorption is resolved. Since the hydroxyle with this combined frequency are absent in the hydrogen form of Y zeolite we ascribe them to OH III in the large cavities.Then the lines which are close in their wavenumbers with the bands observed in Y are connected with OH II and OH II'.The summary of parameters of hydroxyl groups in hydrogen forms of various zeolites is given in Table 2.

TABLE 2

The type of OH group	Si:Al ratio	0-1 cm^{-1}	0-2 cm^{-1}	cm^{-1}	Zeolites,where the corresponding OH groups are observed
SiOH	−	3745	7325	4550	Silicagel,all zeolites
OH I	4-7	3610	7060	4660	High silica zeolites
OH II	3	3640	7125	4660	Erionite,stilbite,cha-
OH II'		3555	6970	4610	basite,Y faujasite
OH III	1.7	3660	7155	4690	X faujasite
OH III'		3580	7005	4580	

The comparison of local and collective approaches in the rational classification of OH-groups in zeolites

Let us compare our classification of OH-groups in zeolites which is based on the local cluster approach with the macroscopic one which was recently proposed by Jacobs in (refs.4,5).According to Jacobs the acidity of a hydroxyl group in a zeolite depends on the mean electronegativity of its framework which acts as a collective donor of electrons.Thus instead of only neighbouring Al atoms the influence of all the alumina in the crystal lattice is taken into account.This results in the continuous spectra of Broensted acidity and of those of the stretching frequencies of OH groups for zeolites with various Si:Al ratios. This model is certainly only another limiting case since it overestimates the influence of collective effects.Our local approach, however,seems to be more realistic for the following reasons:

a) It explains the simultaneous presence of several hydroxyl groups with different acidity in zeolites. On the contrary, the mean electronegativity could be connected with the only one type of hydroxyls.

b) The cluster models permit quantum chemical calculations and the description of acidity and the mechanism of catalytic reactions on the atomic level.

c) The existence of long ranging electronic effects in such insulating substances as zeolites is rather improbable from the general physical point of view.

In other words we consider the mean electronegativity of a zeolite framework only as a secondary factor which is of minor importance.It could explain,for instance,the relatively small differences in OH stretching frequencies for OH II and OH II' in Y and X zeolites or those for OH I frequencies in high and medium silica containing zeolites.However,the acidity of hydroxyl groups is still mainly determined by the number of neighbouring Al atoms.

Both models are leading to quite different conclusions only in the case of high silica containing zeolites.The cluster approach predicts the uniform nature of their Broensted acidity after some critical Si:Al ratio.On the contrary,according to the mean electronegativity principle the acidity should continuously increase for the higher Si:Al ratios.However,this does not agree with our experimental results and with the detection of the acidity of high silica containing zeolites by the heats of NH_3 adsorption (ref.24).

The acidic hydroxyl groups of the amorphous silica alumina catalysts in principle could be also described by similar quantum chemical cluster models.Therefore one might expect that their Broensted acid sites are of the similar nature as in zeolites. This was experimentally confirmed in (ref.25).However,the amount of such acidic centres is about two orders of magnitude less than in hydrogen forms of zeolites since they are formed only as defects because of the partial hydrolysis of the surface.On the contrary,the decationization of zeolites is a stoichiometric reaction.Therefore the number of Broensted acidic sites in their hydrogen forms is comparable with the total aluminum content. Thus the higher catalytic activity of zeolites is rather connected with the amount of the Broensted acidic sites than with their unusual nature.

THE APPLICATION OF THE LOW TEMPERATURE ADSORPTION OF H_2 TO THE STUDY OF LEWIS ACIDIC SITES IN DEHYDROXYLATED ZEOLITES

The adsorption of such test molecules as CO, NH_3, C_5H_5N etc. have been widely applied to the study of Lewis acidity on the surface of oxides.In (refs.17,26,27) we proposed to use for this purpose the low temperature adsorption of molecular hydrogen which has the following advantages:

a) Physically adsorbed and gaseous hydrogen is optically inactive. Therefore it does not create any background in IR spectra.The region of stretching vibrations of H_2 molecules is also free of the absorption bands belonging to the adsorbents.

b) The perturbation of adsorbed hydrogen results in the large red shifts of its stretching frequency (up to ~ 200 cm^{-1}).

c) Since H_2 molecules have a small kinetic diameter all of the surface sites are accessible to their adsorption.

d) Low-temperature adsorption of hydrogen on oxides and zeolites does not cause any irreversible changes of their surface.The amount of the adsorbed hydrogen can be also easily determined by desorption at 300 K.

According to (ref.28) dehydroxylation of H-forms of zeolites may result in formation of two types of electron accepting Lewis sites which are coordinatively unsaturated aluminum atoms and silicon atoms of the lattice:

$$2 \quad \underset{O}{\overset{O}{>}}Si\underset{O}{\overset{O}{<}}\overset{\overset{H}{|}}{O}Al\underset{O}{\overset{O}{<}}O \quad \xrightarrow{-H_2O} \quad \underset{O}{\overset{O}{>}}Si\underset{O}{\overset{O^-}{<}}Al\underset{O}{\overset{O}{<}}O \quad + \quad \underset{O}{\overset{O}{>}}Si^+ \quad Al\underset{O}{\overset{O}{<}} \quad (1)$$

Despite this scheme is widely cited in the literature it is purely speculative and never has been checked experimentally. Recently a serious argument against it was put forward by Kühl (ref.29) who studied dehydroxylated H-forms of faujasites by x-ray fluorescent analysis.He concluded that dehydroxylation mainly results in the dealumination of the zeolite framework leading   to formation of extra-lattice aluminum fragments $^+$AlO which exhibit properties of Lewis acids. The similar conclusion was also made in (ref.30).

Let us discuss in this connection our experimental results on the dehydroxylation of high silica containing zeolites where the probability of the aluminum removal from the framework is the lowest.The diffuse reflectance IR spectra of hydroxyl groups and molecular hydrogen adsorbed at   77 K on the HZSM-5 zeolite pretreated at different temperatures are presented in Fig.4.

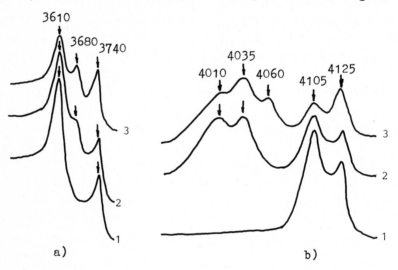

Fig. 4. IR-spectra of OH groups (a) and adsorbed hydrogen (b) for HZSM-5 zeolite recorded at 77 K.Pretreatments are the following: 1 - 770 K,0.1 Pa; 2 - 970 K, 0.1 Pa; 3 - 1270 K, 0.1 Pa.

For the samples evacuated at the temperatures lower than 870K at the very beginning of the dehydroxylation two high frequency bands of adsorbed hydrogen are observed.They belong to H_2 interacting with OH I (4105 cm^{-1}) and with SiOH groups (4125 cm^{-1}).

This assignment is supported by the decreasing of both these lines and those of OH groups upon dehydroxylation at higher tempera-
tures.In addition,the band with $\omega = 4125$ cm^{-1} is observed for H_2
adsorbed on hydroxylated silicagel surface where only silanol
groups are present.

The dehydroxylation of HZSM-5 at 970 K results in appearance
of two new lines of adsorbed hydrogen at 4010 and 4035 cm^{-1}.Af-
ter the further increase of the pretreatment temperature up to
1270 K one more line at 4060 appears.Since the intensities of
all these bands are growing upon dehydroxylation they should be
ascribed to the hydrogen interacting with Lewis acidic sites.
Thus the low temperature adsorption of molecular hydrogen develops three different types of Lewis centres in dehydroxylated
HZSM-5.

Their further assignment could be done in the following way.
The line at 4060 cm^{-1} which appears in the most severe conditions of the pretreatment is likely connected with hydrogen interacting with extra-lattice AlOH fragments which are also observed in the region of their OH stretching vibrations at 3680cm^{-1}
(Fig.4, (ref.30)). Then the bands with comparable intensities at
4010 and 4035 cm^{-1} appearing at the lowest temperature of the pretreatment are connected with trigonal Si and Al atoms which are
formed according to eq. (1).

This interpretation is confirmed by the data on the low temperature hydrogen adsorption on decationized Y-zeolites which are
known to be much less stable than ZSM-5.The rapid temperature increase up to 970 K and calcination at this temperature in air
in "deep bed" conditions followed by evacuation for the complete
dehydroxylation results in the appearance of the band at 4060cm^{-1}
as the predominating one (Fig.5).

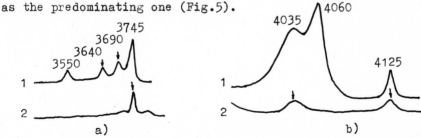

Fig. 5. IR-spectra of OH groups (a) and of adsorbed hydrogen
(b) for HY zeolite pretreated in "deep bed" conditions at 970 K
(I) and for SiO$_2$ pretreated at 970 K.(2).

This is in agreement with the commonly accepted conclusion that such treatment of NH_4Y results in formation of extra-lattice aluminum.

In addition, a weaker band at 4035 cm^{-1} was also observed without any evidences of that at 4010 cm^{-1}. This allows to ascribe this band to hydrogen interacting with trigonal Si atoms since after "deep bed" treatment the amount of trigonal aluminum should considerably decrease. Such assignment is also confirmed by observation of very weak line of hydrogen interacting with trigonal Si at 4035 cm^{-1} after its adsorption on the surface of silicagel dehydroxylated at 970 K (Fig.5).

The different Lewis centres bind molecular hydrogen with different strength. This could be used for estimation of their number from thermal desorption experiments which were carried out in the temperature interval of 77-150 K. It was found that after any pretreatment of HY the amount of lattice Lewis acidic sites never exceeded 0.1-0.2 mmol/g. This is only several per cents of the total aluminum content in this zeolite of about 5 mmol/g. This also supports the conclusion that the dehydroxylation of Y zeolites mainly results in aluminum removal from their framework to extra-lattice positions. It is also clear that such a small amount of trigonal Si or Al could not be detected by x-ray fluorescent analysis in (ref.29).

On the other hand, despite the absolute amount of lattice acid sites in ZSM-5 zeolite is only several times higher than in Y type (the maximum value that was observed in our experiments was about 0.5 mmol/g) it is comparable with the total aluminum content in its framework. Therefore the dehydroxylation of HZSM-5 and other high silica containing zeolites results mainly in formation of trigonal Al and Si atoms in their framework. Thus the scheme (1) is valid for high silica zeolites treated at moderate temperatures and the mechanism of dehydroxylation proposed by Kühl is true for faujasites resulting in the different nature of Lewis acidity of these types of zeolites.

REFERENCES

1 R. Beaumont, D. Bartomeuf, J.Catalysis, 27 (1972) 45.
2 E. Dempsey, J.Catalysis, 39 (1975) 155.
3 E. Dempsey, J.Catalysis, 49 (1977) 115.
4 P.A. Jacobs, W.J. Mortier, Zeolites, 2 (1982) 226.
5 P.A. Jacobs, Catalysis Rev., 24 (1982) 415.
6 V.B. Kazansky, Proc. 4th Natl.Symp.Catal.Ind.Inst.Techn., Bombay, 1978, p.14.

7 I.D. Micheikin, A.N. Lumpov, G.M. Zhidomirov, V.B. Kazansky,
   Kinetika i Kataliz, 19 (1978) 1053.
8 V.B. Kazansky, Kinetika i Kataliz, 23 (1982) 1334.
9 I.D. Micheikin, I.A. Abronin, G.M. Zhidomirov, V.B. Kazansky,
   J.Molec.Catalysis, 3 (1977/78) 435.
10 I.D. Micheikin, I.A. Abronin, G.M. Zhidomirov, V.B. Kazansky,
   Kinetika i Kataliz, 18 (1977) 1580.
11 I.D. Micheikin, I.A. Abronin, A.I. Lumpov, G.M. Zhidomirov,
   Kinetika i Kataliz, 19 (1978) 1053.
12 W Loevenstein, Amer.Mineralog., 35 (1954) 92.
13 J.W. Ward, Zeolite Chemistry and Catalysis, ACS Monograph,
   v.171, Washington D.C., 1976, p. 118.
14 L.M. Kustov, V.Yu. Borovkov, V.B. Kazansky, J.Catalysis, 72
   (1981) 149.
15 L.M. Kustov, V.A. Plachotnik, V.Yu. Borovkov, V.B. Kazansky,
   Kinetika i Kataliz, 23 (1982) 1161.
16 L.M. Kustov, V.Yu. Borovkov, V.B. Kazansky, Zeolites, 3 (1983)
   77.
17 V.B. Kazansky, to be published in Proc. 7th Int.Congr.Catal.,
   Berlin, 1984.
18 J.C. Vedrine, A. Auroux, V. Bolis, P. Desaifve, C. Naccache
   et al., J.Catalysis, 59 (1979) 248.
19 P.A. Jacobs, J.A. Martens, J. Weitkamp, H.K. Beyer, Far.Disc.
   Chem.Soc., 72 (1982) 353.
20 P.A. Jacobs, J.B. Uytterhoeven, J.Chem.Soc., Faraday I, 69
   (1973) 359.
21 E. Dempsey, G.H. Kühl, D.H. Olson, J.Phys.Chem., 73 (1969) 378.
22 Z. Jrzak, S. Vratslav, J. Zajicek, V. Bosacek, J.Catalysis,
   49 (1977) 112.
23 E. Lipmaa, M. Mägi, A. Samoson, M. Tarmak, G. Engelhardt,
   J.Am.Chem.Soc., 103 (1981) 4992.
24 A.L. Klyatchko, G.I. Kapustin, G.O. Glonti, T.R. Brueva,
   A.M. Rubinstein, Proc. VI Seminaire Sovietique-Francaise sur
   la Catalyse, 1983, Moscou, p. 56.
25 V.Yu. Borovkov, A.A. Alekseev, V.B. Kazansky, J.Catalysis,
   80 (1983) 462.
26 L.M. Kustov, A.A. Alekseev, V.Yu. Borovkov, V.B. Kazansky,
   Doklady Akademii Nauk SSSR, 261 (1981) 1374.
27 V.Yu. Borovkov, I.C. Muzyka, V.B. Kazansky, Doklady Akademii
   Nauk SSSR, 265 (1982) 109.
28 J.B. Uytterhoeven, L.G. Cristner, W.K. Hall, J.Phys.Chem.,
   69 (1965) 2117.
29 G.H. Kühl, J.Phys.Chem.Solids, 38 (1977) 1259.
30 Z. Tvaruzkova, V. Bosacek, Chem.Zvesti, 29 (1975) 325.

P.A. Jacobs et al. (Editors), *Structure and Reactivity of Modified Zeolites*
© 1984 Elsevier Science Publishers B.V., Amsterdam — Printed in The Netherlands

MECHANISM OF RADICAL FORMATION IN OLEFIN ADSORPTION ON HIGH-SILICA ZEOLITES AND THE PROBLEM OF THE SYNTHESIS OF AROMATIC STRUCTURES

A.V.Kutcherov and A.A.Slinkin

N.D.Zelinskii Institute of Organic Chemistry, U.S.S.R. Academy of Sciences, Moscow

ABSTRACT

An ESR study on the formation of radicals from various olefins and their thermal transformations on H-mordenite /HM/ was carried out. Various alkenyl and allyl primary radicals are formed at $-78°C$. The fast formation of always the same hydrocarbon radical fragment at $20°C$ is due to isomerization and oligomerization reactions. The centre responsible for the formation of radicals involves not only a redox site but also a Brönsted acid site. This is indicated by the decrease of radical concentration on NH_3 preadsorption. At $300°C$ active centres of HM selectively form radicals with 1,2-dialkylbenzene structure from adsorbed compounds of the most different structures /olefins, n-hexane, CH_3OH/. We suppose that in HM a combination /redox site + acid site/ is localized in a lattice element of defined geometry. We believe that a redox site formation is due to the rupture of a strained Si-O-Si bond in the 4-membered ring of mordenite.

INTRODUCTION

High silica zeolites have proved to be unique catalysts, able to transform various compounds including olefins and CH_3OH into aromatics. Therefore the interest in the investigation of catalytic and physicochemical properties of these contacts is quite understandable.

The purpose of our work was to study the mechanism of the radical formation and the mechanism of the transformation of various hydrocarbons to aromatic compounds on H-mordenite, by the ESR-technique, and to elucidate the role of both redox and acid sites in such processes /ref. 1/.

EXPERIMENTAL

HM samples /SiO_2/Al_2O_3 = 10, decationization α = 92-95%/ were obtained by the thermal decomposition of the NH_4-form of mordenite in air stream at 500-520°C. The ESR spectra were measured with a reflecting spectrometer / ʎ ≈ 3,2 cm/ at temperatures up to 400°C. The ampule with a 100-150 mg sample was placed in the spectrometer cavity, attached to a volumetric adsorption system which permitted evacuation up to 10^{-4} Pa and the inlet of gaseous olefins at pressures from 10^{-1} Pa to 40 kPa. Prior to adsorption, the sample in the spectrometer was heated in air to 400°C and then cooled to 20°C and evacuated for 1 h. The interaction with liquid substances in air was carried out as follows: the sample was heated at 450°C for 1 h, cooled to 20°C and immediately impregnated. The excess of the adsorbed substance was removed by evacuation at 20°C. Then the sample was allowed to come in contact with air and heated to the temperature desired. The kinetics of the radical formation on o-xylene adsorption on the zeolites HM, HNaM / α = 46%/ and HM-18 /SiO_2/ Al_2O_3 = 18, α > 95%/ was studied at 20, 150 and 300°C. The time count was started at 20° from the moment of impregnation and at 150 and 300° from the moment of insertion of the sample into the heated spectrometer cavity.

RESULTS AND DISCUSSION

The interaction of C_2-C_8 olefins with H-mordenite at 20°C is accompanied by the appearance of the same ESR-signal with 11 components /see Fig. 1a/. Cooling of the sample to -196°C leads to a reversible change of the signal /see Fig. 1b/.

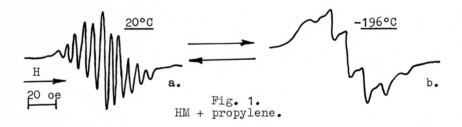

Fig. 1.
HM + propylene.

It is evident that there are two groups of protons in the radical particle: 1/ two equivalent protons with the splitting constant ~8 Oe; 2/ four equivalent protons with ΔH = 16 Oe. The fact that the adsorption of olefins with different structure including C_2H_4 results in the identical ESR spectrum allows to suggest that fast selective formation of the same secondary radical fragment takes place at 20°C /ref. 2/.

The adsorption of various olefins at -78°C leads to the appearance of ESR signals with hfs which differ both in the number of components and the splitting constants. Changes in the hfs of the spectra start only on thawing. Thus, we may assume that at -78°C we find primary radicals. The ESR spectrum of propene at -78°C /see Fig. 2a/ has the simplest hfs, which indicates the formation of radical species with 5 equivalent protons, i.e. the allylic structure has been formed by abstraction of an H-atom from a C_3H_6 molekule. The interaction of isobutene with HM results in the ESR spectrum with 21 equidistant lines and $\Delta H = 5,3$ Oe /see Fig. 2b/, which may be interpreted as a signal from an alkenyl cation-radical containing two groups of protons - 6 equivalent β-protons and 2 α-protons.

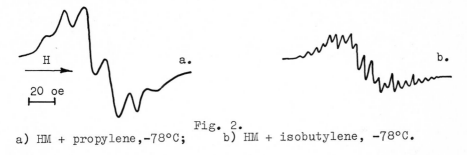

Fig. 2.

a) HM + propylene, -78°C;   b) HM + isobutylene, -78°C.

The fifteen-component spectrum of butene-1 at -78°C is identical with the spectrum given in /ref.3/ which is interpreted as a signal from an alkenyl cation-radical. The form of the ESR signal of pentene-1 at -78°C is identical with Fig. 1a, but its intensity is lower. The absence of a signal in the case of C_2H_4 adsorption at -78°C indicates that an acceptor site in HM cannot activate the parent C_2H_4 molecule as a result of the high ionization energy of ethylene in comparison with substituted olefins.

When samples are heated from -78°C to higher temperatures, the gradual formation of a characteristic signal /see Fig. 1a/ is observed.

It may be summed up that on HM the same radical fragment is formed selectively at 20°C as the result of fast isomerization and oligomerization and that this fragment determines the nature of the ESR spectrum.

It may be suggested that zeolite lattice elements of defined geometry take part in selective formation of radical fragments. Such a structure element seems to involve at least two different sites: an oxidizing site where the radical is formed, and a Brönsted acid site which takes part in chemical transformations. In order to elucidate the role of acid sites the H-D-exchange experiments and experiments with the ammonia preadsorption have been carried out. ESR spectra without hfs are observed on the adsorption of olefins on deuterated HM both at 20° and -78°C, indicating the fast exchange between OD-groups of mordenite and H-atoms of radical particles. Ammonia preadsorption on acid sites results in the linear decrease of the concentration of radicals /see Fig. 3/. Changes in hfs of ESR spectra are not observed, i.e. NH_3 has an influence on the quantity rather than on the structure of radical particles.

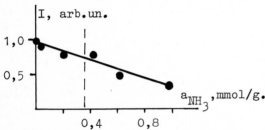

Fig. 3
The influence of preadsorption of NH_3 on the concentration of radicals formed in the interaction of HM with C_3H_6 at 20°C.

In HM the number of the strongest acid sites is equal to ~ 0.35 mmol/g /ref.4/. However complete blocking of these sites does not result in the complete suppression of secondary radical formation and it may be concluded that a combination of the strongest acid and oxidizing properties is not a necessary condition. In HM we suppose the formation of a complex centre of the form:

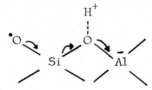

It may be noted that in /ref.5/ the centre of a similar structure was suggested, being formed as a result of homolytic abstraction of an H-atom from a bridged OH-group.

It is necessary to note a sharp difference in the properties of radical particles formed from olefins and aromatic compounds, in their interaction with O_2. Cation-radicals of aromatic compounds are stable on HM in the presence of

O_2, whilst all olefinic radicals disappear quickly and irreversibly after the inlet of oxygen. This fact facilitates the analysis of results observed at the high temperature transformations of various hydrocarbons on H-mordenite.

The heating of HM with adsorbed olefins to 200°C causes a significant complication of hfs, but the signal disappears irreversibly after contact with O_2 indicating that radicals preserve their alkenyl structure. At 250-300°C a new ESR signal with well defined hfs arises which is stable in contact of the

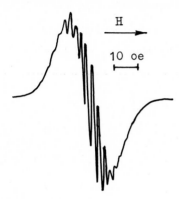

sample with O_2 /see Fig. 4/. Identical signals with hfs are formed as a result of transformations at $\sim 300^{\circ}$C of various compounds such as C_2H_2, allyl alcohol, allyl chloride, butadiene, cycloolefins. Moreover, the ESR signal is the same in the case of such saturated compounds as n-hexane, cyclohexane, methanol. It is important that high temperature transformations are strongly inhibited by ammonia, i.e. they need the participation of Brönsted acid sites.

Fig.4
HM + C_3H_6, 300°C, 20 min

The appearance of an ESR signal with a well defined hfs may be regarded as an evidence of the highly selective formation of radicals with the same defined structure on the oxidizing centres. The comparison of the ESR signal in Fig. 4 with the spectra of various aromatic compounds adsorbed on HM has shown that it is identical with the ESR spectrum of the cation-radical of adsorbed o-xylene. This signal does not change when the sample is heated up to 300°C. At 20°C, the two other xylene isomers, give spectra with hfs that are drastically different from those shown in Fig. 4, but when heating the sample to 200°C they are transformed to the spectrum of adsorbed ortho-xylene. It may be concluded that aromatic radicals selectively formed on HM during the thermal treatment have the 1, 2-dialkylbenzene structure.

As cation-radicals of o-xylene are very stable on HM, the kinetics of their formation at various temperatures allows to elucidate the mechanism of the formation of oxidizing sites /ref.6/. Fig. 5 represents the kinetics of the ESR signal growth upon interaction of various mordenite samples with o-xylene at 20° C.

82

In a 15-200 min. interval the kinetics is
well described by the function $I=f/\sqrt{\tau}/$,
typical for diffusion processes. After
6-8 hours the rate is sharply reduced
and after 2 days a stationary concentration is reached, which does not change
noticeably for 3-5 days. Various mordenite samples essentially differ from
each other by the equilibrium radical
concentration at 20°C. Thus in HM-18
we observe one radical per
~ 100 elemental cells, in HM-10 one /
~ 200 e.c. and in HNaM one /2000 e.c.

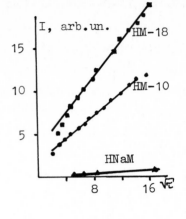

Fig. 5
Kinetics of ESR signal growth

The fact that at 20°C the process is described by the diffusion kinetics is not
surprising: the diffusion of large molecules in HM channels to the redox sites
ought to be the rate determining step. The kinetics of radical concentration
changes on further heating of samples with sorbed o-xylene is not so clear.
The thermal treatment causes a further increase in radical concentrations and
the kinetics of the ESR signal growth is described not by the diffusion equation but by a first order equation. Heating of HM-10 and HNaM samples from
20° to 150°C results in the increase of radical concentration by a factor of
2.1 and 5, respectively. As far as the hfs of ESR signal does not change the
observed process may be treated as an activated "switching on" of the numerous oxidizing sites with the same ionization potential. We may suggest a hypothesis about the activated formation of oxidizing sites on interaction of chemisorbed aromatic molecules with strained Si-O-Si /or Si-O-Al/ bonds of the
zeolite lattice:

$$\begin{bmatrix} \text{"potential"} \\ \text{site} \end{bmatrix} + \begin{bmatrix} \text{chemisorbed} \\ \text{molecule} \end{bmatrix} \xrightarrow{E_{act.}} \begin{bmatrix} \text{redox} \\ \text{site} \end{bmatrix} + \begin{bmatrix} \text{chemisorbed} \\ \text{radical} \end{bmatrix}$$

It is known that there are distorted tetrahedra in the 4-membered rings of
the mordenite lattice. A part of them is located on the wall of the main channel
and has a defined geometrical environment. It is possible to assume that in the
NaM the strained bonds are stabilized by the interaction with Na^{+} cations; in
the H-form the distortion is increased and the rupture is facilitated. Some of

these bonds break upon   thermal decomposition of the NH_4-form. These
strong oxidizing sites are responsible for the nonactivated formation of
various radicals   on HM at low temperatures, however, their concentration is low /1 site per 100-300 e.c./. A great number of the strained bonds in
the lattice is formed by the "potential" redox sites; they may be broken only
on the interaction with a chemisorbed molecule and such a process needs an
activation. This model permits to realize that the difference between HM and
HNaM is due to the stabilizing action of inaccesible Na^+ ions in side holes on
the strained bonds of the lattice. It was found that such explanation may be
extended to the NiHM, MgHM and BaHM samples. In spite of the location of
these cations in sites inaccessible to sorbed aromatic molecules they cause
a sharp decrease of the concentration of strong redox sites which are able to
form radicals at $20^{\circ}C$. However, in these samples as well as in HNaM a considerable number of strong redox sites may be "switched on" by the interaction with various reagents at $200\text{-}300^{\circ}C$.

Thus it may be concluded that Na^+, Ni^{2+}, Mg^{2+} and Ba^{2+} cations stabilize
to some extent /different for various cations/ the mordenite lattice and raise
the activation energy of the redox site formation.

In the analysis of the total mechanism of processes resulting in the formation of alkylaromatic molecules from various types of compounds including saturated ones it is necessary to take into account that on the H-form of high
silica zeolites hydrogenation-dehydrogenation reactions do not proceed. One
may assume that two reactions contribute to the synthesis of aromatic structures: oligomerization resulting in the formation of long hydrocarbon chains
and hydrogenolysis resulting in the removal of short saturated molecules. As
a result of this process an acyclic hydrocarbon fragment with a great degree
of unsaturation may be formed /ref.7/ on the active centre. The cyclization
of such a fragment, for example tetraene, immediately results in the formation
of an alkylaromatic compound.

For comparison, the transformations of some compounds /benzene, xylenes, propylene, methanol/ are studied on zeolites with another structure:
HY, HL and H-rho. In all cases, on high temperature treatment symmetric
ESR-singlets of different intensities appear, but no traces of hfs arise. Thus
the channel geometry and the location of active centres do not provide such

unique conditions for the selective formation of radicals of the type found in mordenite.

It is necessary to emphasize that aromatic structures formed on HM are strongly bound in the zeolite channel and are not released into the gas phase. However, this is not the case with dealuminated mordenites where the number of centres able to form aromatic structures is lowered but where they acquire the ability to desorb the molecules formed and thus to catalyse the reaction. It is important that on dealumination no new "aromatizing" centres appear but that the catalytic process on centres in the mordenite structure itself becomes possible.

Thus the peculiarities of various compound transformations on high-silica zeolites may be due not only to the zeolites ability to form carbonium-ions but also to the presence of special complex centres able to form radical structures. The transformation of these structures may determine the specific properties of high-silica zeolites.

REFERENCES

1 A.V.Kutcherov, A.A.Slinkin, Kinetika i kataliz 23/1982/ 1172-1198; 24 /1983/ 947-959.
2 I.R.Leith, J.Chem.Comm. /1972/ 1282-1283.
3 S.Shih, J.Catal. 36 /1975/ 238-239.
4 A.D.Ruhadze, G.I.Kapustin, T.R.Brueva, A.L.Klyatchko, A.M.Rubinstein, Kinetika i kataliz 22/1981/ 474-479.
5 S.Shih, J.Catal 79 /1983/ 390-395.
6 A.V.Kutcherov, A.A.Slinkin, Kinetika i kataliz, 1984, in press.
7 F.Fajula, F.G.Gault, J.Catal. 68 /1981/ 312-328.

P.A. Jacobs et al. (Editors), *Structure and Reactivity of Modified Zeolites*
© 1984 Elsevier Science Publishers B.V., Amsterdam — Printed in The Netherlands

QUANTUM-CHEMICAL STUDY OF ZEOLITE STRUCTURE STABILITY. COMPARATIVE DISCUSSION OF ZEOLITES AND BOROLITES

A.G. PELMENTSCHIKOV, G.M. ZHIDOMIROV[1] and D.V. KHUROSHVILI, G.V. TSITSISHVILI[2]

[1]Institute of Catalysis, Novosibirsk 630090 (U.S.S.R.)

[2]Institute of Physical and Organic Chemistry, Tbilisi 380086 (U.S.S.R.)

ABSTRACT

The nature of factors responsible for the structural stability of zeolites was studied by MINDO/3 method. An increase in the temperature stability of zeolites with Si/Al = const with a change of counterion in the row H^+, Li^+, Na^+, K^+, Rb^+ was explained in terms of the proposed approach. Structural stabilities of zeolites and borolites (a part of Al atoms in the zeolite framework is substituted by B atoms) were compared.

INTRODUCTION

Thermostability proves to be an important feature of zeolites, determining to the great extent the efficiency of their practical use as catalysts, adsorbents, ion-exchangers, etc. Therefore, determination of the dependence of zeolites thermostability on their composition, structure and way of modification appeared as one of the trends of their experimental (refs.1,2) and theoretical (refs.3,4) investigations. At present, a certain progress has been noticed in the engineering of highly-stable zeolite production (refs.5-8). Meanwhile it is clear, that a better insight into the regularities, responsible for the stability of a zeolite structure, is required for further progress in this field. In this case quantum-chemical calculations may be of some help.

MODEL AND METHOD

The starting point of such investigations will be the choice of a characteristic to be calculated, which could be related to the structural stability of a zeolite. Here two approaches could be pointed out which we   call for short as "integral" and "local" ones. The "integral" approach consists in calculating the dependence of the total binding energy of clusters modelling the key structural elements of a zeolite framework on different variations of their structure and chemical composition. Although some interesting results were obtained in this way (refs.3,4), a few critical remarks concerning the "integral" approach could be made. Particularly, difficulties in correlating total energies of structures of various chemical composition (comparison of energies of clusters with different effective charge (refs.3,4) seems not to be justified) and the problem of taking into account the change in geometry, etc. may be mentioned. Still, the main remark appears to be one stating that the structure stability will be determined primarily by the presence (or absence) of other structural conformation, close in energy, rather than by the overall increase of the binding energy. Calculation of a hypersurface of a rather prolonged cluster is a complicated and a hard-to-realize problem. Therefore, it is expedient to use here qualitative considerations, with an emphasis on the weakest element of the structure. In the systems under consideration such an element is the donor-acceptor bond Al...O, with O being implied to be bound with an counterion. In quantum-chemical calculations the difference in the strength of Si-O and Al...O bonds manifests itself as the difference of the Wiberg indexes of these bonds (ref.9). By calculating the change of the Wiberg index P_{A10} of Al...O bond at various modifications of the zeolite structure, it seems to be possible to follow the character of the change in the zeolite structure stability. This approach is also widely used now (see e.g. (ref.3)).

We believe discussion of the structure stability in terms of its weakest chain (the "local" approach) to be physically quite justified. Yet using only the index P_{A10} for this purpose considerably restricts the possibilities of this approach. In this respect, the calculation analysis of the structural rearrangement itself of the "weak" chain is worthy of notice. In the present communication the development of the "local" approach mentioned is analyzed using the MINDO/3 method. Initially the mechanism and energetics of the donor-acceptor bond formation ($AlO_3 \longrightarrow AlO_4$ transition) were considered for the $Al(OH)_3 + H_2O$ system. In this case, particular consideration was given to the structural rearrangement of AlO_3 fragment: for $(HO)_3Al \ldots OH_2$ (see Fig. 1) $\angle OAlO$ was taken to be 109.5^{O}, for the isolated

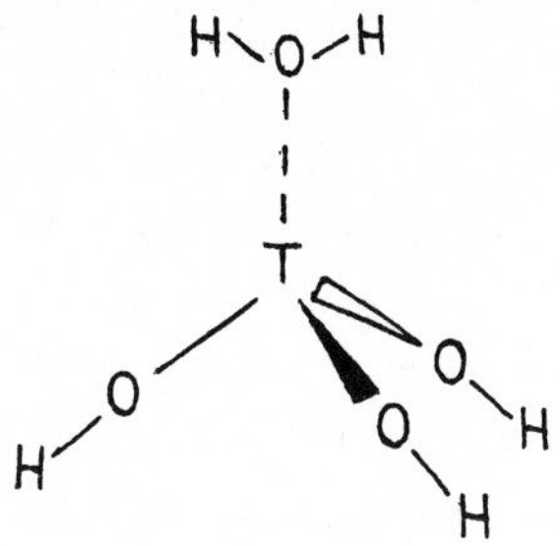

Fig. 1. Complex $(HO)_3T...OH_2$, (T = Al, B).

Al(OH)$_3$ this angle was varied until it reached the minimum energy. Besides, Al-O, Al...O, O-H bonds were varied in calculations:
$\angle$ AlOH = 180° for Al(OH)$_3$ and $\angle$ HOH = 104° for H$_2$O; Al and H$_2$O
atoms lie in the same plane of (HO)$_3$Al...OH$_2$ complex. Similar calculations were made for B(OH)$_3$ + H$_2$O system in order to compare
energetics of AlO$_3 \rightarrow$ AlO$_4$ and BO$_3 \rightarrow$ BO$_4$ transitions. After that
stabilities of these fragments were compared assuming the struc-
tural stability of AlO$_4$ and BO$_4$ in zeolite framework to be determined by interfragmental interactions.

RESULTS AND DISCUSSION

As seen from calculations, a plane structure (see Fig. 2a) is
more preferable for the isolated Al(OH)$_3$, as in this case a minimum energy of repelling between O atoms is achieved. At the same
time, TO$_4$ fragments in oxides (T is a metal atom) are of close to
tetrahedral structure. Therefore, when considering TO$_3 \rightarrow$ TO$_4$
transitions we assume the structural transition of TO$_3$ fragment to occur from the energetically advantageous state in the absence of an extra coordination (e.g. plane state in the case of
AlO$_3$) to the pyramidal one (see Fig. 2b). For Al(OH)$_3$ energy ex-
penditures for such structural transition amount to $E_{str} =$
75 kJ/mol.

A gain in energy upon the formation of an additional bond in
the pyramidal state of Al(OH)$_3$ with H$_2$O molecule is $E_{coor.} =$
126 kJ/mol. Thus, AlO$_4$ fragment in Al(OH)$_3$ + H$_2$O system is energetically advantageous, as $E_{III \rightarrow IV} = E_{str} - E_{coor} = -51$ kJ/mol.

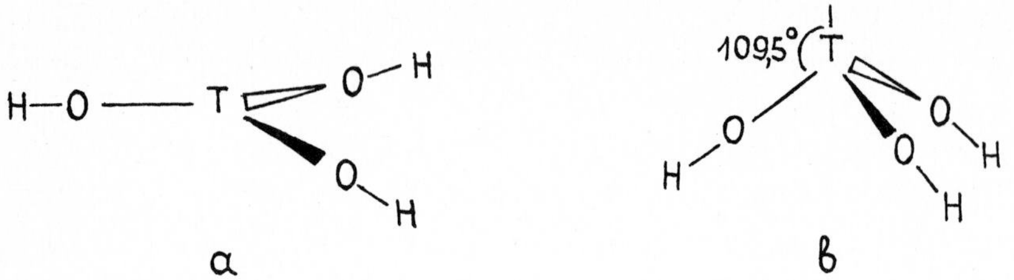

Fig. 2. Plane and pyramidal states of $T(OH)_3$ fragment.

However, most of the coordination bond energy is spent to compensate the energy of structural reconstruction of AlO_3 fragment. Naturally, an increase of electrodonor properties of O atom, participating in the formation of the donor-acceptor extra bond, should increase the stability of AlO_4 fragment. For example, in system $Al(OH)_3 + OH^-$ E_{coor} = 412 kJ/mol, i.e. $E_{III \to IV}$ = -337 kJ/mol (the same bonds as in previous calculations were varied; $\angle AlOH$ = 180° for $Al(OH)_4^-$).

It may be supposed, that the structural stability of TO_4 fragment of a solid is governed to the great extent by interfragmentary interactions. This permits to account for the observed increase in the temperature stability of zeolites with Si/Al = const, occuring upon variation of counterion in the row H^+, Li^+, Na^+, K^+, Rb^+ (refs.10,11). Actually, electron-donor properties of counterions, and consequently, of the related O atoms participating in the formation of the donor-acceptor Al...O bond increase in this sequence. In accord with the previous consideration, this is accompanied by increasing stability of the "weak" chain of the zeolite framework - AlO_4 fragment.

Let us further compare in terms of the adopted model conceptions the stability of AlO_4 and BO_4 fragments in oxide systems. It follows from calculations, that for the isolated $B(OH)_3$, similarly to the case of $Al(OH)_3$ and for the same reasons, the plane state is more preferrable. However, for $B(OH)_3$ E_{str} = 126 kJ/mol, i.e. it considerably exceeds the corresponding value for $Al(OH)_3$, which is an evidence of the difference in covalent radii of B and Al atoms. In this case in $B(OH)_3 + H_2O$ system E_{coor} = 96 kJ/mol will fail to compensate E_{str}, and therefore in this system BO_4

is energetically unadvantageous. Thus, because of the great energy consumption for structural rearrangement needed for the formation of an extra bond BO_3 fragment will demonstrate less tendency to extra coordination than AlO_3. This permits to account for that, in contrast to AlO_3, BO_3 (in the plane state) is encountered in many natural compounds, such as B_2O_3, $B_2O_3SiO_2$, crystalline borosilicates, etc. (refs.12,13). However, in $B(OH)_3$ + OH^- system E_{coor} = 370 kJ/mol, which improves the stability of BO_4, as $E_{III\ IV}$ =-244 kJ/mol. That is, addition of a sufficiently electron-donor O atom into the system will lead to the appearance of BO_4 fragments. It is a familiar fact, that in mixed x Me_2O · (1 - x)B_2O_3 and x Me_2O · (1 - x)$B_2O_3SiO_2$ (Me - an alkali metal, acting as a counterion) BO_4 fragments really exist, i.e. in the relative concentration $N_{BO_4} = \frac{x}{1-x}$ (ref.12).

In recent years catalytic properties of borolites have been extensively investigated. Here a problem concerning the principle possibility of H-form existence in such zeolites and peculiarites of their acid properties are of particular interest (refs.14,15). In a similar manner as in previous calculations we compared energetics of $AlO_3 \longrightarrow AlO_4$ and $BO_3 \longrightarrow BO_4$ transitions in systems $Al(OA)_3$ + $HOSi(OA)_3$ and $B(OA)_3$ + $HOSi(OA)_3$ respectively. Here A is a pseudoatom, with quantum-chemical parameters chosen on condition that the electron structure of the SiO_4 fragment of silica gel is reproduced on a model cluster $Si(OA)_4$ (ref. 16). Bond lengths Al-O, B-O, Al...O, B...O, Si-O and O-H were varied, $\angle$ SiOH = 110°; bridge structures (see Fig. 3) with $\angle$ AlOSi and $\angle$ BOSi = 140° corresponded to AlO_4 and BO_4 fragments. As seen

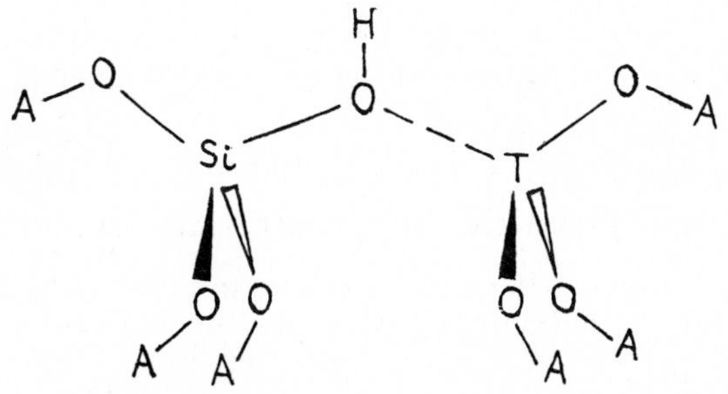

Fig. 3. Model of a bridge structure of zeolite.

from calculations, $E_{III \to IV}$ = -8 and -53 kJ/mol for $BO_3 \to BO_4$ and $AlO_3 \to AlO_4$ transitions, respectively.

From the above reasonings we conclude that, if at all BO_4 fragments exist in H-forms of modified zeolites, they should be much less stable than AlO_4 fragments. (It is appropriate to mention here that in amorphous borosilicates no BO_4 fragments were detected).

The above discussion of zeolite stability brings us to understand the importance of taking into account, when considering adsorption and chemical properties of a solid, possible structural reconstruction of the solid active centers, that is to give up a rigid lattice model. This is most important for consideration of chemisorption properties of Lewis acid centers in alumosilicates (ref.17).

REFERENCES

1   D.W. Breck, Zeolite Molecular Sieves, Wiley, New York, 1974.
2   C.V. McDaniel and P.K. Maher, in J.A. Rabo (Ed.), Zeolite Chemistry and Catalysis, New York, 1976.
3   S. Beran, Zh. Phys. Chem., N.F., 123 (1980) 129-139.
4   S. Beran, P. Jíru and B. Wichterlova, React. Kinet. Catal. Lett., 18 (1981) 51-53.
5   G.T. Kerr, J. Catal., 15 (1969) 200-204.
6   R. Beaumont and D. Barthomeuf, J. Catal., 26 (1972) 218-225.
7   G.T. Kerr, Advan. Chem. Ser., 121 (1973) 219-224.
8   G.T. Kerr, J. Cattanach and E.L. Wu, J. Catal., 13 (1969) 114-116.
9   J. Dubsky, S. Beran and V. Bosacek, J. Mol. Catal., 6 (1979) 321-326.
10 R.M. Barrer and D.A. Langlly, J. Chem. Soc., (1958) 3804-3811.
11 A.S. Berger and A.K. Yakovlev, Zh. Prikl. Khim., 38 (1965) 1240-1246.
12 L.D. Pye, V.D. Fréchette and N.J. Kreidl (Eds.). Borate Glasses. Materials Science Research, v. 12, New York, London, 1978.
13 N.V. Belov, Yu.A. Malinovskii, in N.V. Belov (Ed.), Kristallokhimiya, v. 14, Problemy Kristallokhimii Silikatov, Moskva, 1980, pp. 91.
14 R.M. Barrer and E.F. Freund, J. Chem. Soc., Dalton Trans., 10 (1974) 1049-1053.
15 M. Taramasso, G. Perego and B. Notari. Proceed. of the Fifth Intern. Conf. on Zeolites, Heyden, London, 1980, p. 40
16 A.G. Pelmentschikov, I.D. Mikhejkin and G.M. Zhidomirov, Kin. i Kat., 22 (1981) 1427-1430.
17 A.G. Pelmentschikov, I.N. Senchenya and G.M. Zhidomirov, V.B. Kazanskii, Kin. i Kat., 24 (1983) 233-236.

P.A. Jacobs et al. (Editors), *Structure and Reactivity of Modified Zeolites*
© 1984 Elsevier Science Publishers B.V., Amsterdam — Printed in The Netherlands

CARBOCATION FORMATION IN ZEOLITES. UV-VIS-NIR SPECTROSCOPIC INVESTIGATIONS ON MORDENITES

P. FEJES[1], H. FÖRSTER[2], I. KIRICSI[1] and J. SEEBODE[2]

[1]Applied Chemistry Department, Jozsef Attila University, Beke epület II. emelet, H-6720 Szeged, Hungary

[2]Institute of Physical Chemistry, University of Hamburg, Laufgraben 24, D-2000 Hamburg 13, Federal Republic of Germany

ABSTRACT

Generation of similar enylic carbenium ions from allene, cyclopropane, propene, isopropanol and acetone in H- and NaH-mordenite was proved by u.v.-vis.-spectroscopy. The rate of formation and consecutive oligomerization is markedly enhanced by the zeolite acidity.

INTRODUCTION

It is generally accepted that carbocations act as ionic intermediates in hydrocarbon transformations over solid acids.

Transitions in the ultra-violet and visible range might allow their detection and identification, as has been convincingly demonstrated by Leftin (ref. 1 and 2). Garbowski and Praliaud (ref. 3) have shown that simple unsaturated hydrocarbons are able to form carbenium ions over Y-type zeolites, findings which were recently supplemented by Coudurier et al. (ref. 4).

In a preceding publication the formation of allylic and dienylic carbenium ions from propene and cyclopropane could be proved on H-mordenite (ref. 5). Small amounts of propene and allene prior to adsorption of cyclopropane in Y-zeolites reduced the induction period for isomerization (ref. 6). From this finding Fejes et al. suggested the allylic carbenium ions to be the chain carriers in this reaction. Carrying on these studies, we should like to get some more information on two points of view:

i) how is carbocation formation influenced by modification of the surface acidity, and

ii) are there also other organic compounds of similar structure amenable for carbenium ion generation?

The response should be expected from comparing the carbenium ion formation on H- and NaH-mordenite including compounds with three adjacent carbon atoms, such as allene, isopropanol and acetone in addition to cyclopropane and propene, as

92

we focussed our attention to the formation of allylic carbenium ions. From
these investigations a more detailed picture of hydrocarbon reactions in the
zeolite framework was anticipated.

EXPERIMENTAL

The NH_4-mordenites were prepared from Na-mordenite (Norton) by ion exchange.
The chemical compositions were checked by neutron activation analysis and were
approximately $(NH_4)_7NaAl_8Si_{40}O_{96}$ in case of H-mordenite and $(NH_4)Na_7Al_8Si_{40}O_{96}$
for NaH-mordenite. The decomposition was performed in the optical cell by in
situ heat treatment.

Cyclopropane (99%; Merck), propene (99.98%; Messer Griesheim) and allene
(97%; Matheson) were used without further purification. Isopropanol and acetone
(both Uvasol®; Merck) were stored over dried Linde 4A molecular sieve in order
to remove last traces of water.

The zeolites studied were compressed into self-supporting wafers of about
5 mg/cm^2 thickness. Details on sample evacuation, handling and thermal treatment have been described previously (ref. 5). The spectra were run in transmittance on a Cary 17 spectrometer. All spectra have been subtracted from the
background absorption in order to make this technique more sensitive for detecting small spectral changes, which seems to be exceptionally important in
case of the usually rather broad electronic bands.

RESULTS

Sorption and reaction of allene in H- and NaH-mordenite are fast processes.
At room temperature the formation of different carbenium ions could be pursued
spectroscopically only if the allene pressure was equal to or less than 4 Pa.
Fig. 1 shows the fast development of the bands on H-mordenite at 290 (very
weak), 340 (weak shoulder), 380, 440 and 540 nm.

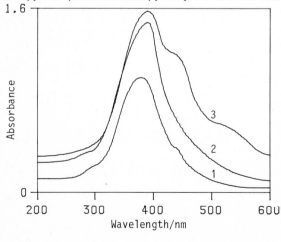

Fig. 1. Transmission electronic spectra (background-cor-
rected) of H-mordenite exposed
to 4 Pa of allene: (1) immediately after allene admission
at room temperature; (2) 18 h
at room temperature; (3) 0.5 h
at 370 K.

On NaH-mordenite the bands were similar to those observed on H-mordenite, but developed more slowly and the main band at 370 nm is less pronounced.

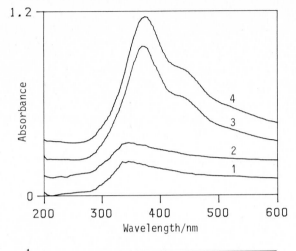

Fig. 2. Transmission electronic spectra of NaH-mordenite loaded with 2660 Pa of propene: (1) after propene admission at room temperature; (2) 3 h at room temperature; (3) 1 h at 390 K; (4) 3 h at 390 K.

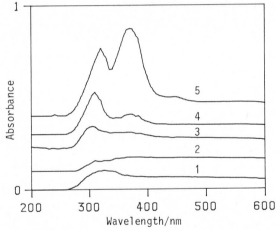

Fig. 3. Transmission electronic spectra of NaH-mordenite exposed to 266 Pa of cyclopropane: (1) after cyclopropane loading at room temperature; (2) 30 min at 370 K; (3) 1 h at 390 K; (4) 1 h at 400 K; (5) 30 min at 430 K.

Upon admission of propene and cyclopropane at room temperature, carbocation formation starts (see Figs. 2 and 3) immediately, but is much slower than with allene under comparable conditions. While in the case of H-mordenite the band positions of cyclopropane and propene are nearly identical, on NaH-mordenite the high-frequency band at 320 nm is more prominent with cyclopropane and undergoes a shift to higher wavelengths during thermal treatment. In general, the formation of the low-frequency species seems to occur more slowly with both cyclopropane and propene.

Sorption of isopropanol on H-mordenite at room temperature led to an instant development of two bands at about 295 and 380 nm with stronger increase of the latter under longer time of contact (see Fig. 4). Upon heating the 380 nm band

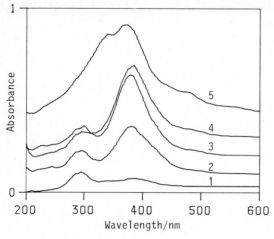

Fig. 4. Transmission electronic spectra of H-mordenite loaded with 133 Pa of isopropanol: (1) after contact with isopropanol at room temperature; (2) 1 h at room temperature; (3) 6 h at room temperature; (4) 18 h at room temperature; (5) 1 h at 370 K.

is shifted by 10 nm to shorter wavelengths and cannot be clearly resolved from an adjacent band of nearly equal intensity at 340 nm, embodying the 295 nm band which is merely perceptible as a shoulder. At the same time new bands emerge at 475 and 555 nm. On NaH-mordenite the behaviour is similar but proceeds more slowly.

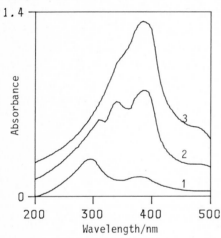

Fig. 5. Transmission electronic spectra of H-mordenite exposed to 133 Pa of acetone: (1) after loading at room temperature; (2) 27 h at room temperature; (3) 1 h at 360 K.

The formation of the carbenium ions is accompanied by the elimination of water from isopropanol, discernible from the appearance of the $2\nu_{1,3}$ band of water at 7090 cm^{-1}. In the n.i.r. spectrum also the overtone of the Si-OH vibration can be seen at 1375 nm, proving the existence of surface hydroxyl groups on NaH-mordenite.

Immediately after adsorption of acetone on H-mordenite at room temperature, absorptions could be detected at 295 and 385 nm, the intensities of which increases with time (see Fig. 5). Later on three new bands appeared at 310, 335 and 480 nm. At longer times there is a general increase of all absorptions, do-

minated by that at 385 nm. In case of NaH-mordenite a similar but slower development of the bands is observed.

DISCUSSION

Sorption of allene, propene, cyclopropane, isopropanol and acetone in mordenites results in surface species which give rise to well-developed bands in
the u.v. and vis. region, the positions of which are summarized in Table 1.
On the basis of the similarity of the spectra obtained, i.e. electronic transitions in the ranges 290-300 nm (usually the most intense bands),440-480 nm and
520-550 nm, it must be concluded that they are caused by surface species of
similar structure. The weak bands at wavelengths higher than 400 nm arise only
after longer times of contact or thermal treatment. The corresponding wavelengths agree fairly well with the values of 320, 396, 473 and 549 nm derived
from the Sorensen equation (ref. 7), which described the band shift of tetramethyl polyenylic cations, as the number of conjugated double bonds increases.
Actually for the tetramethyl allylic ion a band at 305 nm was found in 96% sul-
furic acid (ref. 8). Bearing in mind the influence of the adjacent methyl
groups and solvent interactions, it can be expected that λ_{max} of the allylic
carbenium ion appears rather at lower position. Therefore, the observed bands
are assigned to the $\pi-\pi^*$ transitions of mono-, di-, tri- and tetraenylic carbenium ions (ref. 1,2,9,10). But the exact position of maximum absorption seems
to depend on the interaction between carbocations and the different mordenites.
Also homologous species may be formed by side reactions, explaining the
appearance of some additional bands in the range 300-340 nm. It is far harder
to decide whether the observed surface species are open chains or aromatic in
character, as adsorbed on silica, alumina or zeolites, the two types absorb at
the same wavelengths (ref. 10,11). But from additional infrared studies it must
be concluded that aromatic species are lacking in our case (ref. 12).

Considering the general conclusions drawn from the spectra, the pathways
leading to carbocation formation over zeolitic surfaces can be rationalized in
the following terms.

The most easy way for a propenium ion to be formed in a zeolitic framework
is the interaction of an allene with a Brønsted acid center. Due to its great
reactivity, which leads to instant oligomerization, this ion has not yet been
identified in superacids (ref. 9,13). Also in zeolites this may be responsible
for only weak absorptions in the range 290-340 nm. But with isopropanol and
acetone in H- and NaH-mordenite, bands in this region become more prominent and
can be assigned to the allyl (propenyl) carbenium ion in accord with an interpretation of Garbowski et al. (ref. 14). Once this ion has been formed, the
polyenylic species can originate from consecutive additions of allene molecules.

The main difference between the propene transformation in H- compared to NaH-mordenite is that, owing to the lower number of Brønsted acid sites in the latter, a less pronounced production of dienyl and polyenyl species takes place. Sorption of cyclopropane in NaH-mordenite gives spectra similar to propene, but nevertheless the surface transformations are slower. Possible mechanisms for allylic carbenium ion formation in acidic zeolites have been given elsewhere (ref. 5). The origin of a dienylic species at 370-375 nm could be explained as follows:

$$CH_2\text{-}CH\text{-}CH_2^+ + CH_3\text{-}CH=CH_2 \rightarrow CH_3\text{-}\overset{\overset{\displaystyle CH_3}{|}}{CH}\text{-}CH\text{-}CH\text{-}CH_2^+$$

$$\xrightarrow{-H^+} CH_3\text{-}\overset{\overset{\displaystyle CH_3}{|}}{C}=CH\text{-}CH=CH_2 \quad \xrightarrow{-H^-} \quad CH_2\text{-}\overset{\overset{\displaystyle CH_3}{|}}{C}\text{-}CH\text{-}CH\text{-}CH_2^+$$

The hydride ion abstraction could be performed by a propyl or an allylic ion or with lower probability on a Lewis acid site.

Dehydration of isopropanol over zeolites is well documented in the literature (ref. 15-19). It is held unanimously that the reaction is catalyzed by protons with respect to both propene and ether formation. As shown in Fig. 4, the sorption of isopropanol leads to the development of bands at 300 and 375 nm, increasing in intensity with time. The effect of heat treatment was similar. It is believed that the precursor of allylic and polyenylic carbenium ions is propene, formed by dehydration of the alcohol. This is proved by the overtone band of water in the n.i.r. region arising parallel to the u.v. bands at 300 and 375 nm.

Adsorption of acetone has been studied on oxides and zeolites (ref. 20 and 21). Depending on the acidity and the type of the acid sites (Brønsted or Lewis) ketones of low molecular weight are bound to the surface as carbonyl and carboxylate structures. Chang and coworkers (ref. 22) consider that the main products of the transformations of acetone are diacetone and mesityl oxide, together with hydrocarbons containing more than six carbon atoms. These are presumed to be oligomeric species formed from propene. In the case of mordenites the restricted dimensions of the pore channels might not favour the production of bulky molecules such as diacetone and mesityl oxide. Polyenyl cation formation can be visualized via propene and its oligomers instead. Adsorption of acetone on H-mordenite and NaH-mordenite leads to the development of absorption bands at 310 and 380 nm, in agreement with previous suppositions. The introductory step of allylic ion formation involves dehydration of acetone:

$$CH_3-\underset{O}{\overset{\|}{C}}-CH_3 + H^+ \rightarrow CH_3-\underset{OH}{\overset{+}{\underset{|}{C}}}-CH_3 \rightarrow \overset{+}{CH_2}-\underline{CH-CH_2} + H_2O$$

The results obtained with different substrates, all able to form alkenium cations of identical or similar structure upon interaction with H-type mordenites, convincingly demonstrate the validity of the sequence of ion-forming capability (ref. 23), originally proposed for aequeous sulphuric acid solutions:
dienes > alcohols > paraffins. In our experiments, propenium ion formation was extremely fast with allene and slowed down in the order:
allene >> propene > cyclopropane > isopropanol > acetone.

The experiments shed some light on the role of Brønsted-acidity of mordenites in carbocation formation: upon higher acidity, the development of absorption bands characteristic of propenium etc. ions needed a shorter time of contact and/or lower temperature.

Table 1. - Electronic transitions observed upon sorption and heat treatment of different adsorbates on zeolites H-M and NaH-M. Values in brackets indicate either very weak absorptions or such of questionable presence

adsorbate	adsorbent	wavelength/nm				
allene	H-M	(290-295)	(340)	380-390	440	540
allene	NaH-M	(290-300)	(345)	350-370	430-440	520
propene	H-M	(280-290)	325-335	375		
propene	NaH-M	(290-300)	335-345	370-375	435-450	
cyclopropane	H-M	(280-290)	325	370-385		
cyclopropane	NaH-M	(280-300)	310-320	370-375	450	
isopropanol	H-M	295-300	340	370-385	475	555
isopropanol	NaH-M	295-310	330	370-380	(455)	555
acetone	H-M	295 310	335	380-390	480	
acetone	NaH-M	295 310		375-380	470	555

This work requires further refinement in two respects: there is no direct proof yet as to the hydride ion-abstraction ability of Lewis acid centers and, apart from a few suggestions of kinetic origin (ref. 24), it remains to be seen whether the carbocations thus produced can play the role of ionic chain carriers in hydrocarbon transformations.

ACKNOWLEDGEMENT
We thank M. Hoffmockel for experimental assistance. The financial support of the Hungarian Academy of Sciences, the Deutsche Forschungsgemeinschaft and the Fonds der Chemischen Industrie is gratefully acknowledged.

REFERENCES

1 H.P. Leftin and E. Hermann, 3rd Int.Conf.Catal., Amsterdam, 1965, Vol. 2, pp. 1064-1090
2 H.P. Leftin and M.C. Hobson Jr., Adv. Catal., 14 (1963)115-201
3 E.D. Garbowski and H. Praliaud, J. Chim. Phys. Phys.-Chim. Biol., 76 (1979) 687-692
4 G. Coudurier, T. Decamp and H. Praliaud, J.C.S., Faraday I, 78 (1982) 2261-2676
5 H. Förster, S. Franke und J. Seebode, J.C.S., Faraday I, 79 (1983) 373-382
6 P. Fejes, I. Hannus, I. Kiricsi and K. Varga, Acta Phys. Chem. Szeged, 24 (1978) 119-130
7 T.S. Sorensen in Carbonium Ions, ed. G. Olah and P. Schleyer, Wiley-Interscience, New York, 1970, Vol. 2, pp. 807-835
8 T.S. Sorensen, J. Am. Chem. Soc., 87 (1965) 5075
9 N.C. Deno, J. Am. Chem. Soc. 85 (1963) 2291-2995, 2995-2997, 2998-3000
10 A.V. Kiselev and V.I. Lygin, Infrared Spectra of Surface Compounds, Wiley, London, 1975
11 H.P. Leftin and W.K. Hall, Proc.Int.Conf.Catal., Paris, Vol. 1 (1961) pp. 1353-1372
12 H. Förster and J. Seebode, to be published
13 G.A. Olah and M.B. Comisarow, J.Am.Chem.Soc., 86 (1964) 5682
14 E. Garbowski, J.-P. Candy and M. Primet, J.C.S., Faraday I, 79 (1983) 835-844
15 R. Maggiore, L. Solario, C. Crisafulli and G. Schembari, Annali di Chimica, (1981) 697-705
16 B. Skundric and M. Spanic, Z. Phys. Chem. N.F.,125 (1981) 99-106
17 J. Sedlacek, Coll. Czech. Chem. Commun., 46 (1981) 2466-2478
18 P.A. Jacobs, Cat. Rev.-Sci. Engn., 24 (1982) 415-440
19 H. Knözinger, in S. Patai (Ed.), The Chemistry of Hydroxyl Groups, Wiley, London, 1971, pp. 641-718
20 G. Senkyr and H. Noller, J.C.S., Faraday I, 70 (1974) 997-1004
21 J.A. Lercher, H. Noller and G. Ritter, J.C.S., Faraday I, 77 (1981) 621-628
22 C.D. Chang and A.J. Silvestri, J. Catal., 47 (1977) 249-259
23 J. Rosenbaum and M.C.R. Symons, J. Chem. Soc. (1961) 1-7
24 M.L. Poutsma in J.A. Rabo (Ed.), Zeolite Chemistry and Catalysis, American Chemical Society, Washington, D.C., 1976, ACS Monograph 171, pp. 437-528

P.A. Jacobs et al. (Editors), *Structure and Reactivity of Modified Zeolites*
© 1984 Elsevier Science Publishers B.V., Amsterdam — Printed in The Netherlands

QUANTUM CHEMICAL STUDY OF THE CHARACTERISTICS OF MOLECULES INTERACTING WITH ZEOLITES

S. BERAN

The J. Heyrovský Institute of Physical Chemistry and Electrochemistry, Czechoslovak Academy of Sciences, CS-121 38 Prague 2 (Czechoslovakia)

ABSTRACT

The CNDO/2 method was used to study the physical characteristics of interaction complexes of carbon monoxide and water molecules with the Li, Na, Mg, Ca and Al cations situated in the S_{II} cationic positions of faujasites, modelled by $T_6O_6(OH)_{12}$ clusters (T=Si or Al). Calculation demonstrate that formation of such complexes brings about transfer of electron density from the molecule to the cationic centre and, particularly, polarization of the molecule. Both these effects then result in a strengthening of the C-O bond of CO and in a weakening of the O-H bonds of water, depending on the cation type. Simultaneously, it was found that correlation can be drawn between the calculated bond strength of interacting molecules and the stretching vibrational frequencies observed for adsorption complexes of these molecules.

INTRODUCTION

One of the major factors which has a marked effect on the physical, chemical and catalytical properties of zeolites is the type of cation compensating the negative charge on the zeolitic skeleton refs. (1-5). The differences in the properties of zeolites containing various cation can be estimated by investigating the interactions of the zeolite active sites with molecules refs. (1-5). Various molecules and experimental techniques may then be employed to determine active and, for interacting molecules, accesible centers, as well as the ways in which these centers act on the interacting molecules.

Carbon monoxide and water represent molecules which have recently been extensively used refs. (2-4) for this type of study and which were found to be capable of forming stable interaction complexes with active sites in faujasites.

IR measurements of the frequencies of CO molecules adsorbed on various cationic forms of faujasites have revealed the presence of two or three types of interaction complexes of CO with the zeolite. The vibrational bands of interacting CO molecules at highest frequencies are attributed to their specific interactions with cations situated in the zeolite cationic positions, while the bands lying at lower frequencies seem to be connected with the nonspecific interactions of CO with the zeolite framework. For faujasites with Na, Mg, Ca and Al cations, the frequency of the highest observed vibrational band of CO was found to attain values of 2172, 2197, 2213 and 2225 cm^{-1}, respectively, compared to the value of 2143 cm^{-1} observed for the isolated CO molecule (refs. 2-4).

Similarly, the interaction of water with the alkaline or alkaline earth cations in faujasites is known to give rise to interaction complexes (refs. 2,5). Formation of such complexes is again connected with a shift in the water molecule vibrational frequences, depending on the cation type. The asymmetric stretching vibrational frequency of the water molecule forming a complex with the Li,Na, Mg and Ca cations in faujasites was observed (refs. 2,5) to attain values of 3714,3694, 3642 and 3640 cm^{-1} respectively, compared to a value of 3756 cm^{-1} found for the isolated molecule.

One of the ways of obtaining additional information on the type and nature of interaction complexes of molecules with active sites of zeolites is calculations on model complexes of these molecules with zeolites cluster models. For this reason the CNDO/2 method is employed to calculate the physical characteristics of the complexes of the CO and water molecules with the cations situated in the S_{II} cationic position of faujasites.

MODEL AND METHOD USED

When studying a solid phase by quantum chemical methods, a model of the solid phase – a cluster– must be used, consisting of a limited number of atoms. Such models do not accurately describe all the features of the real solid phase (e.g.,for an ionic crystal they do not include the overall electrostatic field of the crystal and generally there are problems with termination of the cluster model) and are, to a certain degree, approximative. On the other hand, the results of a number works (refs. 6-21) employing such models for the solid phase and using semiempirical methods of quantum chemistry have demonstrated that this approach is capable of yielding a good deal of qualitative information on the properties of the solid phase, which are frequently difficult to obtain by other techniques.

The cationic forms of faujasites were modelled by the $Si_5AlO_6(OH)_{12}Li$, $Si_5AlO_6(OH)_{12}Na$, $Si_4Al_2O_6(OH)_{12}Mg$, $Si_4Al_2O_6(OH)_{12}Ca$ and $Si_3Al_3O_6(OH)_{12}Al$ clusters, depiciting the sixfold zeolite window facing the large cavity with a cation located in the S_{II} cation position that have already been used successfully for calculations of the physicochemical properties of the corresponding cationic forms of faujasites (refs. 12-17). The cluster geometry was taken from X-ray data (refs. 22-25) and was not adapted from different lengths of the Si-O bonds (ref. 12). The models were terminated by hydrogen atoms lying on the lines determined by the terminal Si-O and Al-O bonds at a distance 1.08×10^{-10}m from the terminal O atom.

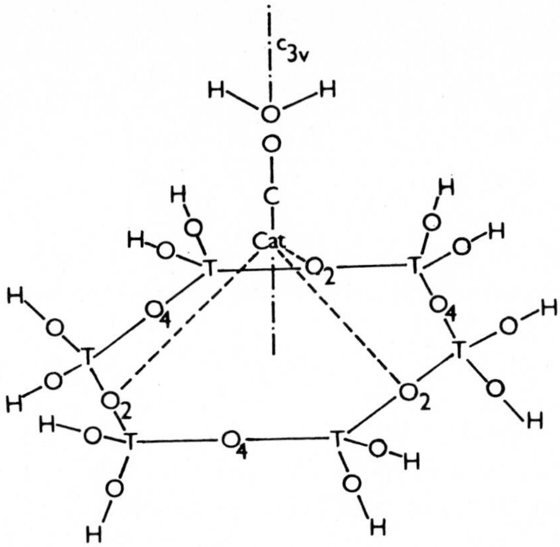

Fig. 1. Depiction of the type of CO and H$_2$O interactions with the cationic zeolite centre and the cluster model used in calculations.

When investigating interaction complexes of a CO molecule with the cation situated in the S$_{II}$ position, the molecule was located on the threefold zeolite axis and its C atom lay close to the cation (cf. Fig. 1) in a distance r, corresponding to the sum of the covalent radii of the cation and of the C atom (cf. Table 1). It should, however, be noted that the distances estimated in such a way are somewhat lower than the equilibrium distances obtained by optimization (ref. 13).

Similarly with interaction complexes of a water molecule with cations in the S$_{II}$ position, the O atom of water was located on the c$_{3v}$ zeolite axis which halved the HOH angle. The distance r between the cation and the O atom (situated close to the cation) corresponded to the sum of the covalent radii of the cation and of the O atom (cf. Fig. 1).

The lengths of the C-O and O-H bonds in CO and water have values of 1.13x10^{-10} and 0.95x10^{-10}m, respectively, and the HOH angle was 105°.

The calculations were carried out by the standard version of the CNDO/2 method (refs. 26,27) with s, p basis sets for the Si and Al atoms (ref. 12). The changes in the properties of the CO and water molecules forming interaction complexes with individual cations in the zeolites were characterized in terms of the charge densities on the atoms and the C-O and O-H Wiberg bond orders (ref. 28) which are known to correlate with the stretching vibrational frequencies (refs. 13,26). Calculations on the effect of the electrostatic field of the cation (modelled by a positive point charge corresponding to the cation charge) on the molecular characteristics were performed with a modified version (refs. 19,20) of the CNDO/2 method whose Hamiltonian contained a contribution describing the interaction of the electrons and nuclei of the molecule with this point charge.

RESULTS AND DISCUSSION

As shown by our previous calculations (refs. 13, 17-19, 21), a molecule interacting with a zeolitic active centre is influenced by this centre in two ways. Firstly, the zeolitic centre is capable of accepting (or donating) electron density from (or to) the molecule and, secondly, the molecule is polarized by the electrostatic field of the zeolite, determined particularly by the atoms forming the active centre with which the molecule is interacting. In the interactions of a molecule with cations in faujasites, the properties of the cation were demonstrated to play a basic role (refs. 13,17-19,21).

The electron-acceptor properties of the cationic zeolite centre (which were found to be more important than its electron-donor ability) may be roughly estimated from the values of the energies of the lowest unoccupied molecular orbital (E_{LUMO}) calculated for the cluster models with individual cations. For the cations studied here, these lowest unoccupied molecular orbitals were found to be located particularly on the cation and the E_{LUMO} energy of the Li, Na, Mg, Ca and Al cations was calculated (refs. 12,14-17) and found equal -1.8, -1.9, -3.1, -2.3 and -5.7 eV, respectively. This indicates that the electron-acceptor capability of the Li-, Na- and Ca- cationic centres of faujasites is roughly the same; it is somewhat higher for the Mg cationic site and substantially higher for the Al site.

The degree of polarization of a molecule interacting with a cationic zeolite centre is determined by both the cation charge and the distance between the cation and the molecule. The charge densities on the Li, Na, Mg, Ca and Al cations located in the S_{II} cation position of faujasites were calculated (refs. 14-17) to have values of 0.05, 0.30, 1.0, 1.3 and 1.0, respectively. The estimated distances between the cation and the molecule are listed in Tables 1 and 2 together with the electrostatic field of the cation (q_c/r^2) in a distance of r. The values of this electrostatic field for individual cations have the following order: Na < Ca < Mg < Al for the CO molecule and Li < Na < Ca < Mg for the water molecule (cf. Tables 1 and 2).

Carbon monoxide

In a recent communication (ref. 13), we demonstrated that the transfer of electron density from the CO molecule, (which takes place for molecules interacting with electron acceptor cationic centres), leads to a strengthening of the C-O bond and, therefore, to a shift of the vibration to higher frequencies.

The same effects brings about polarization of the CO molecule leading to the $C^- -O^+$ dipole moment. Interactions of the CO molecule with positively charged cations through its C atom apparently result in this type of polarization and thus should cause a strengthening of the C-O bond. The degree of strengthening of the C-O bond and, consequently, also the magnitude of the shift in the vibrational frequency are then a result of the extent of both these effects - the molecule polarization and electron density transfer.

Calculations on the complexes of the CO molecule with the cations in the zeolite clusters revealed that a transfer of electron density from the CO molecule to the cationic zeolite centre occurs (particularly from the σ^* molecular orbital). The amount of electron density transferred corresponds roughly to the electron-acceptor ability of the individual centres characterized by their E_{LUMO} value (cf. Table 1). Therefore, the magnitude of the electron density transferred from CO to the zeolite calculated for individual cationic sites should result in the following order of the strength of the C-O bond: Al > Mg > > Na > Ca, as it is apparent from the q_{tot} values in Table 1.

TABLE 1

Distances between the C atom and the cation, r (in 10^{-10} m), the electrostatic field of the cation, q_c/r (in arbitrary units), the CNDO/2 charges on the C and O atoms, q, the total charge on CO, q_{tot}, the Wiberg bond orders, p_{C-O}, and the vibrational frequency, ν_{C-O}, (in cm^{-1}) for interaction complexes of CO with cations in faujasites[a].

Type of complex	r	q_c/r	q_C	q_O	q_{tot}	p_{C-O}	ν_{CO}[b]
CO	–	–	0.042	-0.042	0.000	2.615	2143
Na-CO	2.31	0.056	0.139	0.025	0.164	2.640	2172
			(0.009)	(-0.009)		(2.650)	
Ca-CO	2.51	0.206	0.101	0.028	0.130	2.652	2197
			(-0.079)	(0.079)		(2.692)	
Mg-CO	2.07	0.233	0.168	0.082	0.240	2.675	2213
			(-0.084)	(0.084)		(2.707)	
Al-CO	1.95	0.263	0.187	0.104	0.291	2.679	2225
			(-0.100)	(0.100)		(2.716)	

[a] Values in parentheses describe the effect of a point charge of the individual cations on the CO molecule located at a distance of r (ref. 18).
[b] (refs. 2,3).

The polarization of the CO molecule interacting with the cation (not including transfer of electron density) can be estimated from calculations on the CO interactions with the point charges corresponding to the calculated charge den-

sities on the individual cations located in faujasites. The results of such
calculations (given in brackets in Table 1) have revealed that the extent of CO
polarization by individual cations is proportional to the electrostatic field
formed by these cations and has the following order: Al $>$ Mg $>$ Ca $>$ Na.

Information on the changes in the strength of the C-O bond of the CO molecule as a result of its interactions with the cations in faujasites is provided
by the Wiberg bond orders, p_{C-O}, found for interaction complexes (cf. Table 1).
If these bond order values are compared quantitatively, it is found that the
strength of the C-O bond for individual cations exhibits the following order:
Al $>$ Mg $>$ Ca $>$ Na. Simultaneously, it emerges that the bond orders calculated for
the zeolite cluster models (and, therefore, including the effect of electron
transfer) and those estimated from the CO interactions with the point charges
exhibit the same trend. This, in agreement with our previous calculations (ref.
13), confirms the conclusion that the electrostatic field of the zeolite centre
plays a decisive role in perturbation of the CO molecule interacting with the
centre and, consequently, in changes in the strength of the C-O bond.

When comparing the C-O bond orders calculated for individual complexes of CO
with the cationic sites with the values of the vibrational frequencies observed
for the CO interactions with the cations in faujasites, quite good correlation
is obtained (cf. Fig. 2). This result indicates that the model used for theoretical description of this problem, as well as experimental assignment of vibrational bands are reasonable.

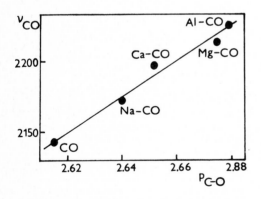

Fig. 2. Correlation between the vibrational frequencies ν_{CO} (cm^{-1}), of CO interacting with the cationic centres of faujasites and the Wiberg bond orders calculated for corresponding interaction complexes of CO, p_{C-O}.

Water

Similarly as for CO, interaction of a water molecule with a cationic site in
faujasites results in transfer of electron density from the nonbonding molecular orbital of water to the zeolite, as well as in polarization of the O-H bonds
leading to an increase in the O$^-$-H$^+$ dipole moments. In this case, however, both
these effects bring out a weakening of the O-H bonds.

It follows from the values of E_{LUMO} calculated for the Li, Na, Mg and Ca

TABLE 2

Distances between the cation and O, r(in 10^{-10}m), the electrostatic field of the cation, q_c/r (in arbitrary units), the charges on O and H, q,the total charge on water, q_{tot}, the Wiberg bond orders,p_{O-H}, and the asymmetric stretching frequency, ν_{water}(cm^{-1}), for complexes of H_2O with cations in faujasites[a]

Type of complex	r	q_c/r	q_O	q_H	q_{tot}	p_{O-H}	ν_{H_2O}
water	–	–	-0.288	0.144	0.000	0.979	3756
Li-OH$_2$	2.11	0.011	-0.242 (-0.306)	0.175 (0.153)	0.108	0.965 (0.974)	3714
Na-OH$_2$	2.27	0.050	-0.270 (-0.327)	0.178 (0.163)	0.086	0.965 (0.972)	3694
Ca-OH$_2$	2.47	0.213	-0.292 (-0.400)	0.181 (0.200)	0.070	0.957 (0.964)	3640
Mg-OH$_2$	2.03	0.243	-0.280 (-0.400)	0.210 (0.400)	0.140	0.948 (0.964)	3640

[a]Values in parentheses describe the effect of a point charge of the individual cations on the water molecule located at a distance r (ref. 18).

cations in faujasites that the electron-acceptor ability of the Mg cation is substantially higher compared with the abilities of the remaining studied cations which are roughly the same. Therefore, the amount of electron density accepted by the Mg cation should be larger than the density received by the remaining cationic centres. The calculated amounts of electron density transferred from water to the zeolite realy confirms this assumption (cf. q_{tot} in Table 2). The transfer of electrons calculated for the Li, Na and Ca cations, however, decreases from Li to Ca, while the E_{LUMO} indicate rather the opposite trend.

On the other hand, the electrostatic field of the individual cations differ significantly (cf. q_c/r^2 in Table 2) exhibiting the following order: Mg > Ca > Na > >Li. Polarization of the water molecule by the charges located on these cations was found to have the same trend as the magnitude of the electrostatic field (cf. q_O and q_H in brackets in Table 2).

The influence of both these effects on the characteristics of water interacting with the cationic centres is apparent from the values of q_O, q_H and p_{O-H} given in Table 2. The values of the p_{O-H} bond orders calculated for the complexes of water with the cations in zeolites correlate with the bond orders estimated from the water interaction with the charges on cations and decrease with increasing electrostatic field of the cation. These results, therefore, again indicate that the electrostatic field of the cation play a major role in perturbation of a molecule interacting with a cationic centre, although transfer of electrons from the molecule also contributes to this perturbation. At the same time, a correlation can be drawn between the O-H bond orders calculated

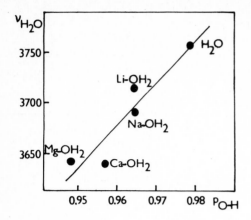

Fig. 3. Correlation between the asymmetric stretching vibrational frequency, ν_{H_2O} (cm^{-1}), of water interacting with the cationic centres of faujasites and the Wiberg bond orders calculated for corresponding interaction complexes of water, p_{O-H}.

for the water molecule forming a complex with the zeolite cation and the asymmetric stretching frequencies of water adsorbed on the corresponding cationic forms of faujasites (cf. Fig. 3).

REFERENCES

1 P.A. Jacobs, Carboniogenic Activity of Zeolites, Elsevier, New York, 1977.
2 J.W. Ward, ACS Monograph, No. 171 (1976) 118.
3 C.L. Angell and P.J. Schaffer, J. Phys. Chem., 70 (1966) 1413.
4 V. Bosáček, D. Brechlerová and M. Křivánek, Adsorption of Hydrocarbons in Microporous Adsorbents, Eberswalde, Berlin, 1982, p. 26.
5 C.L. Angell and P.J. Schaffer, J. Phys. Chem., 69 (1965) 3463.
6 G.V. Gibbs, E.P. Meagher, N.D. Newton and D.K. Swanson, Struct. Bonding Cryst., 1 (1981) 195.
7 I.D. Mikheikin, I.A. Abronin, G.M. Zhidomirov and V.B. Kazansky, J. Mol. Catal., 3 (1978) 435.
8 J. Sauer, P. Hobza and R. Zahradník, J. Phys. Chem., 84 (1980) 3318.
9 W.J. Mortier and P.J. Geerlings, J. Phys. Chem., 84 (1980) 1982.
10 V.I. Lygin and V.V. Smolikov, Zh. Fiz. Khim., 49 (1975) 1526.
11 E.C. Hass, P.G. Mezey and P.J. Plath, J. Mol. Structure, 77 (1981) 389.
12 S. Beran and J. Dubský, 83 (1979) 2538.
13 S. Beran, J. Phys. Chem., 87 (1983) 55.
14 S. Beran, J. Phys. Chem. Solids, 43 (1982) 221.
15 S. Beran, P. Jírů and B. Wichterlová, J. Phys. Chem., 85 (1981) 1951.
16 S. Beran, J. Phys. Chem., 86 (1982) 111.
17 S. Beran, Z. Phys. Chem., N. F., 130 (1982) 81.
18 S. Beran, Z. Phys. Chem., N. F., 134 (1983) 93.
19 S. Beran, Chem. Phys. Lett., 91 (1982) 86.
20 S. Beran, Zh. Fiz. Khim., 57 (1983) 1178.
21 S. Beran, P. Jírů and L. Kubelková, J. Mol. Catal., 12 (1981) 341.
22 H. Herden, W.-D. Einicke, R. Schöllner, W.J. Mortier, L.R. Gellens and J.B. Uytterhoeven, Zeolites, 2 (1982) 131.
23 G.R.Eulenberger, D.P.Schoemaker and J.G.Keil, J.Phys.Chem.,71 (1967) 1812.
24 W.J.Mortier, H.J.Bosmas and J.B.Uytterhoeven, J.Phys.Chem.,76 (1972) 650.
25 J.J. Pluth and J.V. Smith, Mat. Res. Bull., 7 (1972) 1311.
26 J.A. Pople and D.L. Beveridge, Approximate Molecular Orbital Theory, McGraw-Hill, New York, 1970.
27 D.P. Santry and G.A. Segal, J. Chem. Phys., 44 (1966) S3289.
28 K. Wiberg, Tetrahedron, 24 (1968) 1083.

P.A. Jacobs et al. (Editors), *Structure and Reactivity of Modified Zeolites*
© 1984 Elsevier Science Publishers B.V., Amsterdam — Printed in The Netherlands

NEW ROUTES IN ZEOLITE SYNTHESIS

H.LECHERT

Institute of Physical Chemistry, University of Hamburg,
Laufgraben 24 , 2000 Hamburg 13 (F.R.of Germany)

ABSTRACT

The control of zeolite crystallization by a systematic separation of the nucleation and the growth process is investigated in detail for the example of Y-
zeolites. In connection with the results of these studies the influence of organic cations and the templating effect for the formation of different new zeolites is discussed.

INTRODUCTION

Looking into the literature and especially into the patent literature dealing
with the synthesis of zeolites one may obtain the impression that this field is
more an art than a field of research with straightforward methods allowing clear
questions on the respective systems and getting well interpretable answers.

Nevertheless in the recent years many efforts were made to improve the understanding of the processes of the zeolite crystallization and getting in this
way the possibility of a better control of the crystallization itself and the
properties of the final product, which has to fulfill the demands of more and
more sophisticated applications e.g. in the petroleum processing industry.
An excellent review about the chemistry of the zeolites has been given recently
by Barrer (ref.1).

The purpose of this article shall be to outline the most important principles
during the single steps of the zeolite crystallization and the consequences
which can be drawn for the control of this process to obtain improved conditions
and products.

Because Y-type zeolites are still the most important in industrial application, most of the mentioned problems will be discussed for the crystallization of
these zeolites.

Furthermore an attempt shall be made to discuss the methods of the addition
of organic cations which have been applied in recent years for the synthesis of
new zeolite materials in the scope of the guidelines obtained from the mentioned principles.

THE PROCEDURES OF ZEOLITE SYNTHESIS

Zeolites are usually crystallized from alkaline aqueous gels at temperatures between about 70°C and 300°C. The composition of the reaction mixture is suitably defined by a set of molar ratios

$$SiO_2/Al_2O_3 \; , \; H_2O/SiO_2, \; OH^-/SiO_2 \; , \; M^+/SiO_2$$

where M^+ represents in the most cases the Na-ion, but may also stand for other alkali- , alkaline earth- or ammonium-ions. In recent years the preparation of special zeolites has been carried out with addition of quaternary ammonium salts, amines or other polar organic substances. In these cases the definition of the batch composition has to be completed by the ratio

$$R_4N^+/SiO_2$$

The zeolite framework which crystallizes from a given batch is determined mainly by its compostion and its temperature. Strong influences of the cation M^+ in the solution can be observed especially in the case of the organic components mentioned above. These components can be regarded often as socalled templates around which the structural elements of the zeolite can be formed in the reaction mixture.

Experiments covering a broad range of compositions are usually summarized in the well-known crystallization diagrams (refs.1 to 3). An example of such a diagram is given in Fig. 1 for the system

$$Na_2O \; Al_2O_3 \; SiO_2 \; H_2O$$

at different molar ratios of the components. In this diagram the total alkali concentration c_{Na_2O} is plotted on the ordinate and the SiO_2/Al_2O_3-ratio on the abszissa. The temperature and the total concentration of Al_2O_3 and SiO_2 are kept constant. The different phases crystallizing at different batch compositions at a given temperature of 90°C are marked as fields. Diagrams of this kind have been suggested first by Zdanov (ref.2).

An interesting fact which can be seen from this diagram is that the content of alkali has a much stronger influence on the kind of the crystallizing zeolite than the SiO_2/Al_2O_3-ratio.

Looking into the patent literature it can be seen that the boarders between the phases in Fig.1 may be shifted by measures like e.g. an aging of the gel or special ways of mixing the substances, indicating that the zeolite crystallization is a complicated process which is decisively controlled by nucleation phenomena.

In the first kinetic studies done by Zdanov (ref.4) and Meise and Schwochow (ref.5) on zeolite A and by Culfaz and Sand (ref.6) on Mordenite an autocatalytic behaviour has been clearly proved.

The autocatalytic behaviour and the importance of the nucleation allows, now, an improvement of the control of the crystallization process by applying

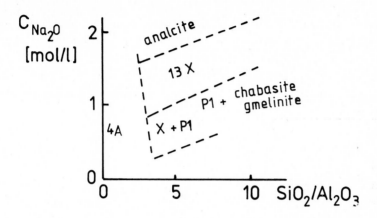

Fig. 1. Crystallization diagram for the system $Na_2O\ Al_2O_3\ SiO_2\ H_2O$ for a temperature of 90°C and $(Al_2O_3 + SiO_2)/H_2O = 0.0005$.

suitable seeds.

Next to important advantages in the technical zeolite production which can be made less sensititve against seeding of unwanted species by impurities, as a more principal consequence the nucleation and the growth of the crystallization process can be separated as has been shown by Kacirek and Lechert (ref.7).

THE KINETICS OF THE ZEOLITE GROWTH AND THE SYNTHESIS OF ZEOLITE Y

Fig.1 shows that under the given conditions no zeolite Y can be observed. Studying carefully the region of the X-zeolite it furthermore may be noticed that from batches in which the composition has been adjusted by mixing solutions of NaOH, $NaAlO_2$, and Na-waterglass without any special precautions the Si/Al-ratio in the crystallized products of NaX increases with decreasing alkalinity from about 1 at the upper boarder to about 1.4 at the lower boarder only weakly dependent on the SiO_2/Al_2O_3-ratio of the batch. Below the boarder the zeolite NaP1 crystallizes.

Adding to batches with the compositions

$$NaAlO_2\ (3-10)SiO_2\ m\ Na_2O\ 400\ H_2O \qquad 0.33\ \leq\ m\ \leq\ 1.33$$

from which only pure NaP1-zeolite can be obtained, seeds of NaX then products of faujasite in the Y-region are obtained with Si/Al-ratios between 1.7 and

3.2 (see ref.7).

Since, in the indicated region only nucleation of NaP1 zeolite occurs the growth of the seeds reflect the kinetics of the Y-zeolite growth.

For a theoretical description the general assumption can be made that the increase of the volume V of the crystals is proportional to the free surface S of the seeds where k is the rate constant of the growth process.

$$\frac{dV}{dt} = k\ S(x) \qquad (1)$$

Calculating the fraction x of the formula units of faujasite the differential equation

$$\frac{dx}{dt} = 3\ k\ x^{2/3}\ x_o^{1/3}\ \bar{r}_o^{-1} \qquad (2)$$

can be obtained from which follows by integration:

$$x = x_o\ (\ 1 + kt/\bar{r}_o\ )^3 \qquad (3)$$

where x_o is the number of formula units of seeds added to the gel and $\bar{r}_o$ the average radius. k has the dimension length/time and describes the linear growth of the crystals. As has been proved in (ref.7) Eq.3 describes very well the growth of the crystals. For samples with Si/Al-ratios between 1.65 and 2.39 in the final product, k has values between $9.0\ 10^{-2}$ and $1.45\ 10^{-2}$ μm/h.

Table 1 shows some examples for the crystallization of Y-zeolites with high Si/Al-ratio in dependence on different excess alkalinities (Na-Al) given by $(NaOH) - (NaAlO_2)$ in the batch.

TABLE 1

Dependence of the Si/Al-ratio of zeolite Y grown from batches with different excess alkalinity (Na-Al) using seeds of zeolite X with $\bar{r}_o$ = 0.9 μm.

Composition of the batch			Composition of the product				Crystallisation time
$\frac{SiO_2}{Al_2O_3}$	$\frac{(Na-Al)}{SiO_2}$	$\frac{H_2O}{Al_2O_3}$	% Y	% NaP1	amorph.	Si/Al	hours
20	0.76	780	100	–	–	2.78	96
20	0.79	780	96	4	–	2.92	144
20	0.72	780	100	–	–	2.97	187
20	0.71	780	95	–	5	3.00	312
30	0.70	780	100	–	–	3.10	690
30	0.695	780	100	–	–	3.14	624
30	0.69	780	98	–	2	3.24	695

The data of the Table 1 show that the composition of the final zeolite expressed by the Si/Al-ratio is sensititvely dependent on the excess alkalinity
(Na-Al) in the batch and can be easily controlled by this parameter. For the
preparation of Y-zeolites with lower Si/Al-ratio it is suitable to decrease
also the SiO_2/Al_2O_3-ratio in the batch.

As a result of a large number of experiments it can be observed that for
the compositions given above pure faujasites can be obtained if the ratio x_o/r_o^3
is larger that a value of about 2 μm^{-3}. Under these conditions the nucleation
of NaP1 is slower than the growth of the X-nuclei added to the batch.

In an experiment in which the seeds have been added to one of the initial
components before the formation of the gel and heating to the crystallization
temperature, no reaction occurs. The seeds are only effective if they are added
to the ready gel after mixing the components. This experiment shows that the
crystallization occurs not by a transformation of the gel which should have
been favoured in the first experiment but only after a dissolution of the gel
via the solution phase.

From these results the conclusion can be drawn that the solution phase can
be reduced without any change of the crystallization conditions if the alkali
excess in the solution is adjusted to be constant. Experiments have shown that
the solution phase can be reduced to about 1/5. For the technical application
this has the advantage that for the production of the same quantity of zeolite
only vessels of much smaller size may be used which increases the space-time-
yield and decreases the energy demand of the process.

Another parameter which can be controlled is the particle size. Because in
the course of crystallization no nucleation of Y-zeolite takes place different
amounts of seed crystals after complete crystallization of the material must
lead to different crystal sizes of the product. The average volumes of the
crystals must have the inverese ratios of the number of seed crystals.

Fig.2 shows the particle size distributions of a batch with

$SiO_2/Al_2O_3 = 14$, $(Na-Al)/SiO_2 = 0.73$ and $H_2O/Al_2O_3 = 800$,

the seeds had an average radius of 0.23 μm. The amount of seeds was varied from
0.044% to 0.44% to 4.4%.

The expected particle volumes were 100:10:1.The observed ratios obtained
from the evaluation of raster scan micrographs from which also the data in Fig.2
have been obtained were 70:10:1. The lower value of the first figure is due
to the fact that in the batch with the large crystals because of the long
crystallisation time already a nucleation of NaP1 had taken place which led
to a content of about 30% NaP1 in the final product.

As can be seen already from Table 1 the crystallization time increases rapidly with the Si/Al-ratio in the final product. Careful studies of the crystallization kinetics of substances with different Si/Al-ratio showed that the empi-

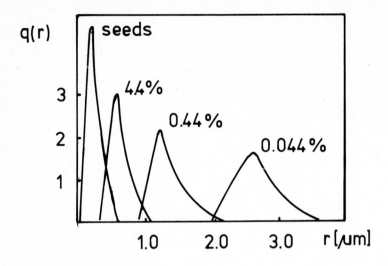

Fig. 2. Distribution density curves of the radii of NaY particles grown from one batch (see text) with different quantities of seed crystals indicated beside the curves.

rical relation Eq. 4 between this ratio and the rate constants k can be obtained if the rate constant is taken in relation to the silica content of the solution phase.

$$\log(k/(SiO_2)_{sol}) = \text{const } (Si/Al) \qquad (4)$$

The constant adopts a value of about 10 for Si/Al-ratios below 2.5 and is equal to about 20 for Si/Al above 2.6 . It can be seen that the value of the rate constant k of the crystallization varies in the range of Si/Al-ratios of 1.1 to 3.4 about four orders of magnitude.

From Eq. 2 follows that for a given k the rate of the crystallization is increased if seeds with smaller average radii are used. The lower limit which can be used by a system of crystallizing zeolite can be obtained from a crystallization curve of a batch without seeds. In Fig. 3 the crystallization curves of a batch of Y-type zeolite with and without seeds are compared. Deviating from the batches used for the experiments described until now silica sol has been used as a silica source to allow a nucleation of Y-zeolite as it is described in the patent literature.

Apparently, after an induction period the batch without seeds shows a much more rapid crystallization which is due to the fact that the system uses structures as seed which are much smaller than the seeds of the radius 0.25 μm applied for the other experiment. An estimation of the average radius of these struc-

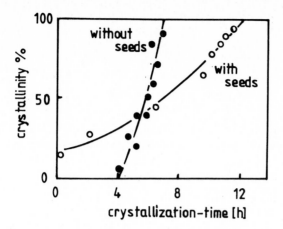

Fig. 3. Comparison of two crystallization curves of a batch of Y-zeolite with
and without seeds.

tures obtained by a fit of the crystallization curve after the induction period
give values of about 0.02 to 0.03 μm.

The crystallization of such a system can be described by an equation of the
following kind

$$x = x_0' \left(1 + k(t-t')/\bar{r}_{opt} \right)^3 \qquad (5)$$

where x_0' depends on the special conditions determining the number of the nuclei
in the reaction mixture, t' is the induction period and $\bar{r}_{opt}$ is the average
radius of the optimal nuclei mentioned above. The question is now, how it is
possible to obtain seeds with this optimal size. This problem is in principle
solved in a patent by Elliot and Mc Daniel (ref.9) where as seeding agent a
gel is applied with a rather high SiO_2/Al_2O_3-ratio and a high alkalinity. Typical compositions of gels active in seeding the formation of Y-type zeolites
are

$$NaAlO_2 \ (6-12) \ SiO_2 \ (6-12) \ Na_2O \ 160 \ H_2O$$

In (ref.9) the aging time of the gel plays an important role. Kinetic studies
by Hsu et al. (ref.10) with batches seeded by these gels after different aging
times showed that the crystallisation rate is identical with the rate obtained
for the batch without seeds, only the induction period is shortened in dependence on the aging time as it is demonstrated in Fig.4.

It can be seen that the induction period decreases with increasing aging
time and remains then constant for a long time which is important for the
practical application .

The mechanism of the activity of the gel may be understood in the following
way: The high alkalinity causes a high supersaturation of silicate and alumi-

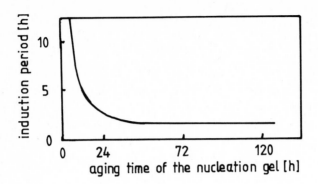

Fig. 4. Correlation between the aging time of the nucleation gel and the induction period of the crystallization process.

nate and the formation of a large number of nuclei. The growth of these nuclei
proceeds until the aluminum in the gel is exhausted which is the case already
for very small particles because of the high SiO_2/Al_2O_3-ratio in the batch.
The system remains then stable. In the X-ray diffraction pattern these nuclei
show no crystallinity. The beginning of the formation of faujasite like species can be detected by ^{23}Na-broadline NMR.

Similar effects may be caused by a low temperature aging of the whole reaction mixture as has been mentioned by Rollman (ref.11) for the synthesis of
Y-zeolites.

Gels of this kind have been found also for the synthesis of zeolite L and
zeolite Omega by Hsu et al. (ref.12).

The kinetics of the nucleation has been studied e.g. by Kacirek (ref.13)
and by Zdanov (ref.14) who calculated the rate of the formation of the nuclei
via the rate of crystallization and the particle size distribution of a final
product by extrapolating back to the radius zero of the respective particles.
Generally it seems to be obvious that the surface of the gel plays an important
role for the nucleation process.

ZEOLITE CRYSTALLIZATION IN THE PRESENCE OF ORGANIC ADDITIVES AND TEMPLATION

Beside the control of the zeolite crystallization via an adjustment of the
nucleation process or the growth conditions, in recent time the addition of
organic substances, especially of tetraalkyl-ammonium ions has become an important tool for the initiation of the formation of new zeolite structures and
the synthesis of more common zeolites with unusual compositions.

An example where all these aspects can be demonstrated and different concepts of the activity of these substances discussed, is the tetramethyl-
ammonium ion (TMA).

Dramatic changes of the composition of zeolites grown in the presence of
TMA have been observed first by Barrer and Denny (ref.15) and by Kerr (ref.16).
for the example of the A-zeolite . This zeolite crystallizes from a system with
only Na-ions, demonstrated in Fig.1, strictly with Si/Al = 1. Adding TMA-ions,
samples with higher Si/Al-ratio can be obtained. This was carefully examined
by Kacirek and Lechert (ref.17) using a solution 1.8 molar solution of TMA-
silicate which conains a high amount of double fourmembered rings (ref.18) and
mixing this solution with a gel of Si/Al = 1. If seeds of A-zeolite are added
samples with A-structure, the socalled N-A-zeolites can be obtained up to Si/Al-
ratios of 2.6.

In this case two concepts may be discussed. At first the TMA-ion provides
the reaction mixture with building units of the final product which favours
the formation of this product kinetically. Besides a templation effect may be
present, forming sodalite units around the TMA-ions, which fit exactly into
these cages.

Another idea of the increase of the Si/Al-ratio of a given zeolite in the
presence of TMA-ions is that the space which is needed to incorporate these
large ions into a given structure is too small compared e.g. with the Na-ions.
If the formation of the structure is favoured by the conditions in the batch
during the crystallization the higher Si/Al-ratios will be favoured because
of electrostatic reasons. This concept seems to be realized e.g. in the.formation
of Sodalite as has been reported in a paper by Baerlocher and Meier (ref.19).
If Sodalite is synthesized in a system containing only Na-ions it crystallizes
in a framework composition $Na_3Al_3Si_3O_{12}$.Adding TMA-ions, the composition changes
to $(CH_3)_4N$ $AlSi_5O_{12}$ and the TMA-ions are incorporated into the sodalite cages.

Going to high silica contents of the gel and lower alkalinities to starting
compositions

$$SiO_2/Al_2O_3 = 15 - 25 \text{ , } H_2O/SiO_2 = 15 - 25 \text{ and } OH^-/SiO_2 = 0.6 - 1.0$$

which are with Na-ions typical for the crystallization of Y-zeolite and with
K-ions typical for the crystallization of L-zeolite, after the replacement of
15 - 20 % of the Na-ions by TMA-ions zeolite Omega (ref.20) is formed and in
the system with the K-ions Offretite (ref.21) crystallizes.

In the final product the TMA-ions are situated inside the Gmelinite-cages
which are common to both structures of Omega and Offretite and it may be suspected that the formation of these structures is initiated by the formation
of the Gmelinite subunits by a templation effect of the TMA.

As another interesting fact should be mentioned that the presence of K-ions

causes the formation of Offretite which has beside the Gmelinite cages Cancrinite cages as building unit in contrast to the Omega zeolite which consists
of a special arrangement of only Gmelinite cages. Similar effects are observed
in the system without TMA-ions where the presence of the K-ion forces the formation of L-zeolite consisting of columns of Cancrinite cage instead of Y-
zeolite in the pure Na-system with the sodalite cage as building unit.

One of the most important inventions in the field of the application of
organic substances for the dirigation of the zeolite crystallization is the
formation of the zeolites of the ZSM 5 family which are formed preferably in
the presence of tetraalkylammonium ions with longer hydrocarbon chains (ref.22).
The ZSM 5 itself is formed in the presence of the tetrapropylammonium ion (TPA)
in batches with a wide range of alkalinities and SiO_2/Al_2O_3-ratios. The patent
literature covers the range:
$SiO_2/Al_2O_3 = 5 - \infty$, $OH^-/SiO_2 = 10^{-10} - 1.0$, $R_4N^+/SiO_2 = 0.01 - 2.0$,
$H_2O/SiO_2 = 0.7 - 3000$, $M^+/SiO_2 = 0.3 - 3.0$
M^+ may be an alkali- an alkaline earth or an ammonium ion. The reaction temperatures are described between 40°C and 210°C with crystallization times between
200 days and some hours.

The importance of this zeolite family lies above all in the special channel
dimensions causing unusual selectivity effects in hydrocarbon processing (see
e.g. ref. 23) and in the high Si/Al-ratios causing high stability and a high
acidity of the Broenstedt sites in the H-form.

It has been proved by Howden (ref.24) by thermoanalytical measurements and
by Boxhoorn et al. (ref.25) by ^{13}C-MAS-NMR measurements that the TPA -ion is
localized in the intersections of the channels of the ZSM 5 structure. Further
it has been shown that the TPA-ion is needed preferably in the nucleation step
of the crystallization process (ref.26). In this case the organic cation is
obviously used as a template around which the channel intersections are formed.
According to Howden (ref.24) these intersections are filled with the template
to about 70 %.

The templating function of the organic component is obvious also from the
observation that even anionic detergents like n-dodecylbenzene-sulphonate are
active in templating the formation of ZSM 5 as it has been observed by Hagiwara
et al. (ref.27). As exchangable counterions usually the ions described as
M^+ above have to be regarded.

An extended range of the patent literature is devoted to the replacement
of the expensive alkyl ammonium salts by less expensive substances like e.g.
amines where especially the hexamethylendiamin has been used with good success
(ref.28). A series of other experiments has been carried out with alkoholes
as organic component, where dioles have proved to be favourable in the ammonium-

system. This system has the advantage that for the preparation of the H-form
no exchange of the sodium ions is necessary.

In connection with these attempts in the recent patent literature a puzzling
variety of substances can be found which have been grown in a broad range of
conditions with different templates. These templates are in the most cases
quaternary ammonium salts or amines with different structures. For SiO_2/Al_2O_3-
ratios above about 20 mostly substances crystallize which have to be related
to the ZSM 5 family. For lower SiO_2/Al_2O_3-ratios zeolites similar to Chabazite,
Ferrierite or Mordenite are obtained. These assignments are due to similarities
in the X-ray pattern and to sorption experiments which are used for a determination of the pore diameter and the pore volume.

In a systematic study of the crystallization kinetics of ZSM 5 by Rollman
and Vayosik (ref.29,30) the influence of different concentrations of the single
components has been investigated. For crystallization times not sufficient
for complete crystallization can be seen that for different OH^-/SiO_2-ratios
a distinct maximum in the rate of crystallization exists which is dependent
on the other conditions of the experiment and has to be determined in each
special case. Own experiments have shown that for low values of the mentioned
ratio big crystals can be obtained demonstrating that the nucleation is slow
and the crystallization occurs primarily by the growth of the nuclei. At high
OH^--concentrations the nucleation dominates and small crystals are obtained.

Looking at the ratio TPA^+/SiO_2 it can be seen that an appreciable crystallization generally begins above values of about 0.03. Near about 0.1 a further
increase of the TPA-content has only weak influence. The concentration of 0.03
corresponds with about 3-4 TPA-ions in a unit cell of the ZSM 5 structure. This
suggests that for the nucleation process this concentration seems to be a lower
limit.

Systematic studies of the function of the templates are very rare. Remarkable in this connection is a paper by Daniels, Kerr, and Rollman (ref.31) in
which polymers of the type

$$x = 10 - 60$$

have been added to a batch with $SiO_2/Al_2O_3 = 30$, $H_2O/SiO_2 = 20$, $OH^-/SiO_2 = 1.2$
$Na^+/SiO_2 = 1.2$ from which Y and P zeolite crystallizes.

For n = 4 at 90°C in 3 - 15 days already at $N^+/SiO_2 = 0.01$ instead of Y
and P Gmelinite crystallizes.with stacking faults similar to the Chabasite
structure. Increasing this ratio to about 0.25 pure Gmelinite is obtained. Increasing it to 0.45, no crystallization occurs. Only polymers with n = 4, 5,
and 6 are effective in initiating the formation of Gmelinite. The length of the

repetition unit of the polymer is in this case nearly equal to the lattice constant of the Gmelinite. For higher n again Y and P crystallizes.

Summarizing, it may be expected that this special field of zeolite synthesis will proceed in the discovery of more specific templates for the synthesis of zeolites of economical interest. For these zeolites certainly some development will take place in the optimization of the crystallization processes. An excellent review about the role of organic substances in the synthesis of zeolites has been published by Lok et al. (ref. 32).

Efficient new principles in constructing new frameworks which may be predicted theoretically are for the present not evident. So, the discovery of new zeolite frameworks will be certainly for some further time a matter of accident.

On the other hand in the recent time important progress has been made with new analytical methods for the investigation of the precursors of the zeolite nucleation and the zeolite growth in the solution phases and in the distribution of the different components in the crystal. From a more detailed knowledge of the structure of the solutions as well as the ready crystallized material conclusions for a better control of the nucleation phenomena can be expected.

METHODS FOR THE CHARACTERIZATION OF SILICATE AND ALUMINATE SPECIES IN SOLUTIONS

In the preceding section it has been mentioned that one of the possible control effects of the NA-zeolite growth was that the TMA-silicate provided double four membered rings in the system.

Effects of this kind should be even more important for the nucleation step which controls the formation of the zeolite species as has been demonstrated in connection with the synthesis of the Y-zeolite.

For the investigation of different silicate species in solution in earlier studies paper chromatography (see. ref. 18) and the trimethyl-silylesters of the respective silicic acids have been used which have been separated by gas-liquid or by gel-permeation chromatography (refs.33-36).

These methods have the disadvantage that they don't work in the equilibrium state and are relatively slow. Generally, in the alkaline solutions which are used for the zeolite synthesis the establishment of the equilibria between the different silicate and possibly alumosilicate species are comparatively fast. For the analysis by the indicated methods the solutions are acidified. In the acid solutions the equilibration is much slower and the different operations necessary for the paper chromatography or the preparation of the trimethyl-silyl esters can be carried out without disturbing the composition too much so that it can be hoped to get a true picture of the respective equilibrium mixture.

More direct information can be obtained by suitable spectroscopic methods which can be applied nondestructively to the equilibrium system.

Two methods which have proved to be applicable in this field are the laser Raman spectroscopy and the high resolution NMR spectroscopy of the nuclei ^{29}Si and ^{27}Al.

The first Raman experiments have been carried out by Mc Nicol et al. (ref. 37) who observed no special changes in the liquid phase during the crystallization. In contrast to this Angell and Flank (ref. 38) and Guth et al. (refs. 39,40) and Roozeboom et al. (ref. 41) found direct evidence of aluminosilicate complexes in solution. The latter authors explained their experiments by the existence of the species I of the following scheme and postulated a series of further aluminosilicates. As an example the five membered ring is shown as species II in the scheme..

The species I has been postulated also by Derouane et al. (ref. 42) as the ion transported in silica rich gels during the synthesis of ZSM 5.

A method which has become very helpful in the analysis of silicate solutions is the high resolution ^{29}Si-resonance. The problems of the assignment of the different peaks of the spectra and the most important literature is summarized in an article by Harris et al. (ref. 43). E.g., in a solution of potassium silicate 0.65 molar in Si and KOH/SiO$_2$ = 1:1 eleven different silicate species could be detected by very careful decoupling experiments of enriched ^{29}Si-silicate. The structures of these silicates are schematically represented in Fig.5.

A comparison of a Na-silicate solution and a solution containing TPA-ions has been carried out by Cavell et al. (ref. 44). The authors found that the monomer was the most abundant ion in each case. At the addition of TPA-ions an appreciable condensation could be observed, where especially the dimer, the cyclic trimer and species with unstrained silicons with two neighbours as they are present e.g. in chains, can be detected.

The ^{27}Al-resonance is above all suited to decide whether the aluminum is present in a solution in tetrahedral or octahedral coordination. An example

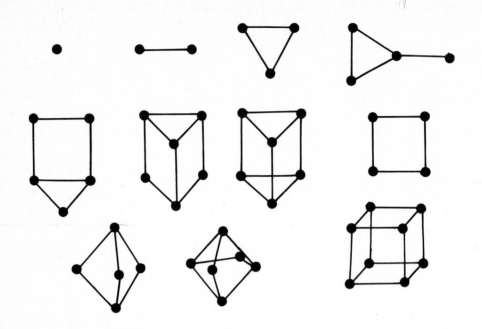

Fig. 5. Structures of silicate ions detected in a potassium silicate solution
by Harris et al. (ref. 43). The points represent a SiO_4-unit.
The connecting edges belong to shared oxygen ions.

can be found in a paper of Derouane et al. (ref. 45) who identified a TPA-
aluminosilicate complex by ^{27}Al-resonance in an aluminum-containing sodium silicate solution with TPA-ions.

With special techniques intermediate states before the detectability of a
crystalline phase can be studied by high-resolution NMR of ^{29}Si and ^{27}Al in
the solid state as has been demonstrated e.g. by Engelhardt et al. (ref. 46)
and Thomas et al. (ref. 47).

From a systematic application of these methods on crystallizing gels it
can be expected that in the near future many of the mechanisms of nucleation
and growth of zeolites will become more transparent.

THE DEALUMINIZATION

For the application of zeolites the Si/Al-ratio of the alumino-silicate
lattice is one of the most important parameters, because it determines the
stability of these substances and the acidity of their catalytical acitve
sites.

Among the substances of technical importance the ZSM 5 is the only zeolite
in which the Si/Al-ratio can be varied to substances with essentially no aluminum already in the synthesis. For the other zeolites among which the Y-zeolite

and the Mordenite are the most important especially for the preparation of catalysts, other means for dealuminization have to be used. One of the methods which has been widely discussed in the literature is the controlled removal of the aluminum by complexing agents like e.g. EDTA or in the case of Mordenite directly by hydrochloric acid. For Y-zeolites Si/Al-ratios of about 12 and for Mordenite of about 60 can be obtained by this method without an appreciable loss of crystallinity (refs. 48,49).

A great importance has the preparation of socalled "ultrastable" zeolites which can be obtained by a special steam treatment at elevated temperatures. This steam treatment causes a migration of a part of the aluminum from lattice positions into cation positions from which it can be easily removed. After another heat treatment the structure shrinks under stabilization. The conditions of the ultrastabilization have been discussed in detail by Mc Daniel and Maher (ref. 50).

In the recent years chemical methods have been developed for the replacement of the aluminum in the aluminosilicate lattice by silicon.

The most important of these methods has been described by Beyer et al. (ref. 51,52). The dehydrated zeolite samples are exposed to a nitrogen stream which is saturated with $SiCl_4$ at about 400°C for some time. Then the temperature is increased. At 470°C the reaction begins and the aluminum is more or less completely released as $AlCl_3$. The reaction is exothermic and must be carried out under careful controlled conditions, otherwise the crystal structure will break down. With this method almost aluminum free Y-zeolites can be obtained (ref. 53). Si/Al-ratios between 20 and 100 are easily to prepare. The method fails until now for Mordenite and L-zeolite.

Another method has been described by Breck and Skeels (ref. 54) in which the exchange of the aluminum is made by treating the zeolite with a solution of Na_3SiF_6.

NEW ZEOLITE TYPES

Beside the aluminosilicate zeolite in the recent years some other zeolite systems have been described which shall be mentioned in short.

The substitution of aluminum against boron has been studied by Taramasso et al. (ref. 55), and Klotz and Ely (ref. 56). Marosi et al. (ref. 57) and Hinnenkamp et al. (ref. 58) . Kouwenhoven et al. (ref.59) and Rubin et al. (ref. 60) have studied ferrosilicates and Klotz (ref. 61) has reported on chromiumsilicates.

The most promisable new class of zeolites has been found in the $AlPO_4$ system by Wilson et al.(see e.g. ref. 62). Under application of a large number of templating agents a series of new zeolites has been obtained among which the structures of sodalite and erionite-offretite can be found.

For the present cannot be seen which of these new developed zeolite systems will become of practical importance. For the transition metal systems new aspects in catalytic activity may be expected. For the other types special effects connected with the pore dimensions and possibly the acidity will become interesting.

ACKNOWLEDGEMENTS

The author thanks the "Fonds der Chemischen Industrie" for the support of his work.

REFERENCES

1 R.M. Barrer, Hydrothermal Chemistry of Zeolites, Academic Press, London, New York, 1982, 360 pp.
2 S.P.Zdanov, in Molecular Sieves, Society of the Chemical Industry, London, 1968, p.68.68.
3 D.W. Breck, Zeolite Molecular Sieves, John Wiley and Sons, New York, 1974 771 pp.
4 S.P. Zdanov, Adv.Chem.Series, 101 (1971) 20.
5 F.E. Schwochow and F. Meise, Adv.Chem.Series, 121 (1973) 169.
6 A. Culfaz and L.B. Sand, Adv.Chem.Series, 121 (1973) 209
7 H. Kacirek and H. Lechert, J.Phys.Chem., 79 (1975) 1589.
8 H. Kacirek and H. Lechert, J.Phys.Chem., 80 (1976) 1291.
9 C.H. Elliott Jr. and C.V. Mc Daniel, U.S.Pat 3,639,099.
10 Hsu Rhu-Reng and Zhang Jianmin, Chemical Journ. of Chinese Universities, 3 (1982) 437.
11 L.D. Rollmann, Proceedings of the NATO ASI on Zeolites and their Application, Lisbon 1983, Martinus Nijhoff Publishers, in preparation.
12 Hsu Rureng , personal communication.
13 H. Kacirek, Dissertation University of Hamburg, 1974.
14 S.P. Zdanov and N.N. Samulevich, in L.C.V. Rees (Ed.) , Proceedings of the 5th International Conference on Zeolites, Heyden, London, 1980, pp. 75-84. 15 R.M. Barrer and P.J.Denny, J.Chem.Soc., (1961) 971.
16 G.T. Kerr, Inorg.Chem. 5 (1966) 1537.
17 H. Kacirek and H. Lechert, ACS Symp.Series, 40 (1977) 244.
18 W. Wieker, Neuere Entwicklungen der anorganischen Chemie, VEB Verlag der Wissenschaften, Berlin, 1974.
19 C. Baerlocher and W.M. Meier, Helv.Chim.Acta, 52 (1969) 1853.
20 Union Carbide Corp. , Brit.Pat., 1,178,186 (1972).
21 E.E. Jenkins, U.S.Pat., 3,578,398 (1971).
22 R.J. Argauer and G.R. Landolt, U.S.Pat. 3,702,886 (1972). R.W. Grose and E.M. Flanigen, U.S.Pat. 4,061,724 (1977).
23 E.G. Derouane, Catalysis by Zeolites, Studies in Surface Science and Catalysis 5, Elsevier , Amsterdam, 1980, pp.5-18.
24 M.G. Howden, CSIR Report 413, Counc.for Scientific and Industrial Research, Pretoria, 1982,
25 G. Boxhoorn, R.A. van Santen, W.A. van Erp, G.R. Bays, R. Huis and D. Clague, J.Chem.Soc. , Chem. Comm., (1982) 264.
26 E.G. Derouane, S. Detremmerie, Z. Gabelica and N. Blom, Appl.Catal., 1 (1981) 201.
27 H. Hagiwara, Y. Kiyozumi, M. Kurita, T. Sato, H. Shimada, K. Suzuki, S. Shin, A. Nishijima and N. Todo, Chem. Letters (1981) 1653.
28 L. Marosi and H.Schlimper, Europ.Pat.Appl. 0 041 621 (1981).
29 L.D. Rollmann and E.W. Valyosik, Europ.Pat.Appl. 0 021 674
30 L.D. Rollmann and E.W. Valyosik, Europ.Pat.Appl. 0 021 675.

31 R.H. Daniels, G.T. Kerr and L.D. Rollmann, J.Am.Chem.Soc. 100 (1978) 3097.
32 B.M. Lok, T.R. Cannan and C.A. Messina, Zeolites, 3 (1983) 282.
33 L.S. Dent-Glasser and E.E. Lachowski, J.Chem.Soc., Dalton Trans.,(1980) 399.
34 L.S. Dent-Glasser and E.E. Lachowski, J.Chem.Soc., Dalton Trans.,(1980) 393.
35 L.S. Dent-Glasser and D.N. Smith, J.Chem.Soc.,Chem.Commun. (1980) 727.
36 T. Shimono, T. Isobe and T. Tarutani, J.Chromatogr. 205 (1981) 49.
37 B.D. Mc Nicol, G.T. Pott and K.R. Loos J.Phys.Chem. 76 (1972) 3388.
38 C.L. Angell and W.H.Flank, ACS Symposium Series 40 (1977) 194.
39 J.L. Guth, P. Caullet, P. Jacques and R.Wey, Bull.Soc.Chim.France 3-4 (1980) 121.
40 J.L.Guth, P. Caullet and R. Wey, in L.V.C. Rees (Ed.) Proceedings of the 5th International Conference on Zeolites, Heyden, London, 1980, p. 30.
41 F. Roozeboom, H.E. Robson and Shirley S. Chan, Zeolites, 3 (1983) 321.
42 E.G. Derouane, S. Detremmerie, Z. Gabelica and N. Blom, Appl.Catal., 1 (1981) 201.
43 R.K. Harris, C.T.G. Knight and W.E. Hull, J.Am.Chem.Soc. 103 (1981) 1577.
44 K.J. Cavell, A.F. Masters and K.G. Wilshier, Zeolites, 2 (1982) 244.
45 see ref.42.
46 G. Engelhardt, B. Fahlke, M. Mägi and E. Lippmaa, Zeolites, 3 (1983) 292.
47 J.M. Thomas, C.A. Fyfe, S. Ramdas, J. Klinowski and G.C. Gobbi, J.Phys.Chem. 86 (1982) 3061.
48 G.T.Kerr, A.W. Chester and D.H. Olson Acta.Phys.Chem.Hung 24 (1978) 169.
49 R.W. Olsson and L.D. Rollmann, Inorg.Chem. 16 (1977) 651.
50 C.V. Mc Daniel and P.K. Maher, in J.A. Rabo, Zeolite Chemistry and Cataly-sis, ACS Monograph 117, p. 285
51 H.K. Beyer and I. Belenykaja, Catalysis by Zeolites, Studies in Surface Science and Catalysis 5, Elsevier, Amsterdam, 1980, p. 203.
52 P. Fejes, I. Kiricsi, I. Hannus, A. Kiss and G. Schobel, React.Kinet.Catal. Letters, 14 (1980) 481.
53 J. Klinowski, J.M. Thomas, M. Audier, S. Vasudevan, C.A. Fyfe and J.S. Hart-mann, J.Chem.Soc., Chem.Commun., (1981) 570.
54 D.W. Breck and D.A. Skeels, presented at the 6th International Conference on Zeolites, Reno 1983.
55 M. Taramasso, G. Perego and B. Natari, Fr.Pat. 2 478 063 (1981).
56 M.R. Klotz and S.R. Ely, U.S.Pat. 4 285 919 (1981).
57 L. Marosi, J. Stabenow and M. Schwarzmann, Ger.Offen. 2 909 929 (1980).
58 J.A. Hinnenkamp and V.V. Walatka, U.K.Pat.Appl. 2 062 603 (1981).
59 H.W. Kouwenhoven and W.H. Stork, U.S.Pat. 4 208 305 (1980).
60 M.K. Rubin, C.J. Plank and E.J. Rosinski, Eur.Pat.Appl. 0 013 630 (1980).
61 M.R. Klotz , U.S.Pat. 4 299 808 (1981).
62 S.T.Wilson, B.M. Lok, C.A. Messina, T.R. Cannan and E.M. Flanigan in G.D. Stucky. F.G. Dwyer (Eds.) Intrazeolite Chemistry, ACS Symposium Series 218 Washington 1983, p.79.

STUDY OF MORDENITE CRYSTALLIZATION III : FACTORS GOVERNING MORDENITE SYNTHESIS*

P. BODART, J. B.NAGY, E.G. DEROUANE[1] and Z. GABELICA

Facultés Universitaires de Namur, Laboratoire de Catalyse

Rue de Bruxelles, 61 - B-5000 NAMUR (Belgium)

[1]Present address : Mobil Research and Development Corporation

Central Research Division, P.O. Box 1025,

Princeton, NJ 08540 (U.S.A.)

ABSTRACT

Mordenite is synthesized from different aluminosilicate gels in sealed pyrex tubes at 165°C. Its crystallization domain is established for these conditions and the coexisting phases are identified. Differences with published data are explained assuming partial dissolution of the silica-rich pyrex glass. Morphologies and composition of the mordenite crystallites are related to the molar ratios of the reactants. In addition, the composition range of mordenite and the structure of coexisting phases (analcite, gismondine, silica polymorphs...) give information about the nature of some probable mordenite precursor species : single four-membered rings and single six-membered rings.

INTRODUCTION

It has been recognized that the initial composition of the system $mNa_2.nAl_2O_3$ $.pSiO_2.qH_2O$ influences the nature of zeolitic materials that crystallize from these gels, their composition and their morphology (ref. 2 and 3).

We have recently shown that mordenite can be synthesized in high yield, hydrothermally, from such a Na-Al-Si hydrogel at 165°C (ref. 1). The aim of this work is to define the composition domain where mordenite crystallizes, and to show the influence of the ingredient ratios (with constant water content) on the composition and morphology of the crystalline mordenite grown under these particular conditions.

SYNTHESIS PROCEDURE

The various hydrogels are obtained by mixing aqueous sodium silicate (Merck, art. 5621), silicagel (Davison, grade 950), sodium aluminate (Riedel de Haën, art. 13404), sodium hydroxide (Riedel de Haën, art. 06203), γ-alumina (Merck, art. 1095) and deionized water (ref. 1).

* Part II, see ref. 1

Typical composition ranges, expressed as oxide molar ratios,are the following :

$1.04 \leqslant Na_2O/Al_2O_3 \leqslant 20.5$

$2.95 \leqslant SiO_2/Al_2O_3 \leqslant 77.9$

$H_2O/(Na_2O + Al_2O_3 + SiO_2 + H_2O) \simeq 0.9$

The gel, sealed in pyrex tubes, is heated at 165°C in static conditions, under autogeneous pressure,for 4 days. After cooling at room temperature, the solid products are filtered, washed with cold water and dried at 120°C for about 12 h.

CHARACTERIZATION OF THE SOLID PRODUCTS

X-ray powder diffraction patterns, recorded on a Philips PW-1349/30 diffractometer using the Cu-K$_\alpha$ radiation,are used for the identification of the crystalline phases (ref. 2). Crystal morphologies are determined by scanning electron microscopy (Jeol JSM 35). Associated energy dispersive X-ray analysis (EDX) gives the global composition (Si, Na, Al) of the samples, and, in particular, that of individual single zeolite particles. The procedure was described previously (ref. 4).

RESULTS AND DISCUSSION

Gel composition ranges producing pure mordenite (optimal domain) or mordenite and coexisting phases (extended domain) are shown on figure 1. It appears that, in our conditions, pure mordenite can be obtained from hydrogels richer in Al than those reported in the literature (Table 1)

TABLE 1 : Comparison of the domain for mordenite crystallization defined from our experiments (165°C, 4 days),with other literature data

	Mordenite synthesis		
	in pyrex tubes (this work)		in stainless steel
	optimal domain	extended domain	autoclaves (ref.5)*
Na_2O/SiO_2	0.19-0.37	0.12-0.50	0.07-0.30
SiO_2/Al_2O_3	7-24	6.8 -99	10-50
Na_2O/Al_2O_3	2-7	1.4 -29	1.2 -15

* : composition producing pure mordenite, T<200°C, with no limitation in crystallization time

This can be explained by assuming a partial dissolution of the pyrex walls of the autoclaves, which essentially results in a release of silica species with a subsequent consumption of OH^- ions. Moreover, for some particular compositions of the starting gel (4.6 $Na_2O.Al_2O_3.15.4\ SiO_2.\ 197\ H_2O$), pure mordenite is obtained from synthesis run in pyrex tubes, while pure analcite crystallizes in stainless steel autoclaves (fig. 2).

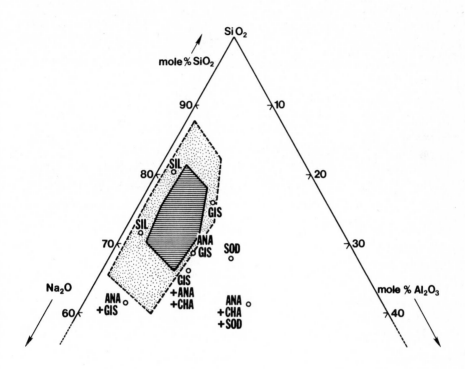

Fig. 1. Domain for mordenite crystallization in the $Na_2O-Al_2O_3-SiO_2-H_2O$ system (H_2O content ≈ 90-95 mole %) : 165°C, 4 days, pyrex tubes.

gel composition producing pure mordenite (optimal domain)
gel composition producing both mordenite and coexisting phases (extended domain); ANA, GIS, CHA, SOD and SIL refer to analcite, gismondine, chabazite, sodalite and silica coexisting phases respectively.

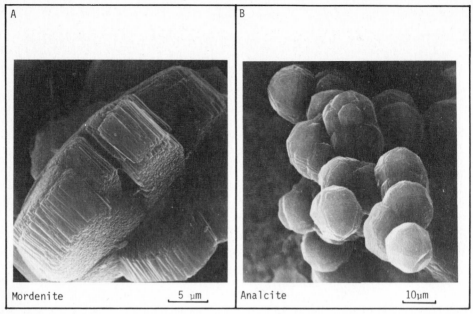

Fig. 2. Scanning electron micrographs of mordenite (A) and analcite (B) crystals obtained from the same (4.6 $Na_2.Al_2O_3.15,4\ SiO_2.197\ H_2O$) gel heated at 150°C in pyrex tubes and in stainless steel autoclaves respectively.

In the extended domain, the following coexisting phases are identified : a silica polymorph appearing in a silica-rich gel, analcite and gismondine in alu- minic and alkaline media. In our conditions, chabazite and sodalite have never been found associated with mordenite, although they crystallize under conditions very similar to those for gismondine or analcite.

Scanning electron micrographs (Fig. 3) show typical crystal morphologies of mordenite. Usually, aggregates of parallel acicular crystals are observed (Fig. 3b and 3c). Increasing alkalinity produces large prismatic crystals elongated and twinned along the c-axis (Fig. 3a). Highly siliceous gels yield tiny flat crystallites (Fig. 3d).

The evolution of the mordenite Al-content with the gel composition, expressed as $\left[\dfrac{OH^-}{H_2O} \cdot \dfrac{SiO_2}{Al_2O_3}\right]$ ratio, is shown in Fig. 4. OH^- concentration is calculated as- suming that one Na_2O molecule produces 2 OH^- groups and that each Al_2O_3 consumes 2 OH^- to give 2 AlO_2^- species (ref. 6).

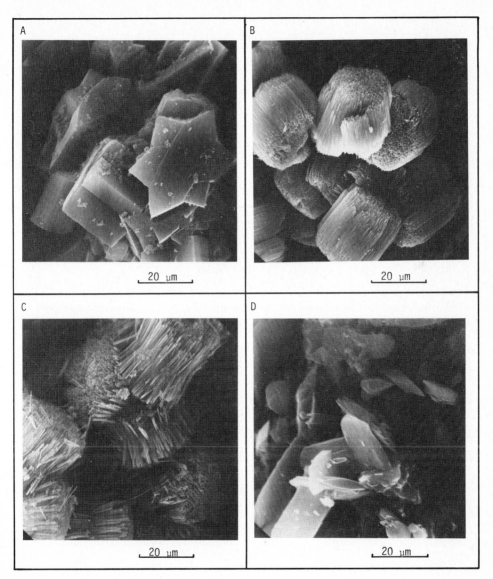

Fig. 3. Typical micrographs of mordenite crystallized at 165°C in pyrex tubes
from various gel compositions :
(A) 3.5 $Na_2O.Al_2O_3.11.3\ SiO_2.249\ H_2O$
(B) 2.4.$Na_2O.Al_2O_3.11.0\ SiO_2.219\ H_2O$
(C) 1.95 $Na_2O.11_2O_3.8.1\ SiO_2.175\ H_2O$
(D) 2.4 $Na_2O.Al_2O_3.19.0\ SiO_2.353\ H_2O$

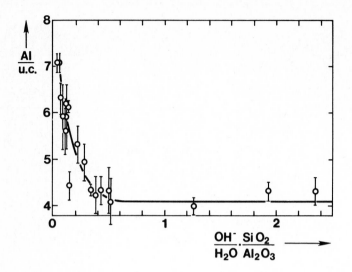

Fig. 4. Evolution of the Al-content (Al per unit-cell) of mordenite, with the initial gel composition (pyrex tubes, 165°C)

Alkaline media produce silica-rich mordenite which contains 4 Al atoms per unit cell. This is partly due to a better solubilization of the hydrogel and to a more rapid dissolution of the pyrex walls at high pH. High SiO_2/Al_2O_3 ratios in the initial gel also lead to silica rich mordenite. Its Al-content per unit cell always ranges between 4 and 8 atoms. According to previous structural studies, this corresponds to a mordenite having either 0 or 2 Al atoms in the single four-membered rings of its structure (ref. 7 to 10). Consequently, the structure of mordenite can be considered as an assemblage of silicic six-membered ring sheets linked by single four-membered rings having 0 or 2 Al-atoms. (ref. 10). Moreover, all the different coexisting phases detected have common structural units with mordenite : silicic six-membered ring sheets for silica polymorphs (ref. 11), six- or four-membered rings for the Al-rich gismondine, analcite, chabazite and sodalite (ref. 7). As the structure of mordenite consists of six-membered rings without Al-atoms and of four-membered rings with 0 or 2 Al-atoms, the following structures are proposed for the precursor species and

or $Si_6O_{18}H_{12}$

Similar species have also been detected in silicate solutions (ref. 12 to 14).

CONCLUSIONS

Our study allowed us to define the domain in which mordenite crystallizes from $Na_2O.Al_2O_3.SiO_2.H_2O$ hydrogels, in pyrex autoclaves, at 165°C. The morphology of the crystallites so obtained is found to be dependent on the gel composition. Alkaline or silica-rich media favour the formation of Si-richer mordenite. The composition of mordenite as well as the structure of the coexisting phases suggest that species similar to simple structural units such as six- or four-membered rings are involved in the beginning of the crystallization process of mordenite.

ACKNOWLEDGMENTS

P. Bodart thanks IRSIA-IWONL for financial support.

REFERENCES

1 P. Bodart, J.B.Nagy, E.G. Derouane, Z. Gabelica, A. Gourque, S. Maroie and J. Verbist, Clays, Clay Miner., submitted for publication.
2 R.M. Barrer, "Hydrothermal Chemistry of Zeolites", Academic, London, 1982, 360 pp.
3 D.W. Breck, "Zeolite Molecular Sieves", Wiley, New York, 1974, 771 pp.
4 Z. Gabelica, N. Blom and E.G. Derouane, Appl. Catal., 5(1983) 227.
5 L.B. Sand, U.S. Pat. 3,436,174 (1969).
6 L.D. Rollmann and E. W. Valyocsik Eur. Pat. 21,674 (1981).
7. W.M. Meier and D.H. Olson, "Atlas of Zeolite Structure Types", Pub. Structure Commission of International Zeolites Association, 1978, distrib. Polycrystal Book Service, Pittsburgh, p.69.
8. G. Debras, J. B.Nagy, Z. Gabelica, P. Bodart and P.A. Jacobs, Chem. Lett. (1983) 199.
9 W.M. Meier in L.B. Sand and F.A. Mumpton (Eds.) Natural Zeolites, Occurence, Properties, Use, Pergamon, Oxford, 1976, p.53.
10 E.G. Derouane and J.G. Fripiat, Proc. 6th Int. Conf. Zeolites, Reno, 1983, in press.
11 C. Frondel, "Dana's System of Mineralogy, volume III, Silica Materials", Wiley, New York, 1962, 334 pp.

12 R.K. Harris, C.T.G. Knight and W.E. Hull, J. Am. Chem. Soc. 103 (1981) 1577.
13 R.K. Harris and C.T.G. Knight, J. Molec. Struct. 78 (1982) 273.
14 G. Engelhardt, D. Hoebbel, M. Tarmak, A. Samoson and E. Lippmaa, Z. Anorg. Allg. Chem., 484 (1980) 22.

P.A. Jacobs et al. (Editors), *Structure and Reactivity of Modified Zeolites*
© 1984 Elsevier Science Publishers B.V., Amsterdam — Printed in The Netherlands

STRUCTURAL PECULIARITIES AND STABILISATION PHENOMENA OF ALUMINIUM DEFICIENT MORDENITES

H.K. Beyer, I.M. Belenykaja, I.W. Mishin[*] and G. Borbely
Central Research Institute for Chemistry, Hungarian Academy of Sciences, Pusztaszeri ut 59-67, 1025 Budapest (Hungary)

ABSTRACT

Samples of synthetic sodium mordenite dealuminated to different degrees by acid-leaching were steamed and, after steaming, again treated with acids. Thermal analysis, adsorption measurements, X-ray diffractometry and lattice vibration infrared spectrophotometry indicate isolated vacant lattice sites (partly dehydroxylated hydroxyl nests) in the acid-leached samples. The chemical nature of these vacancies is discussed. During steaming a secondary medium-size pore system is formed, and the isolated vacancies disappear. Dealumination by acid-leaching results in a considerable decrease of thermal stability. Lattice stabilisation is only caused by hydrothermal reactions. Acid-leaching after steaming practically does not affect the framework of the steamed samples.

INTRODUCTION

The extraction of tetracoordinated aluminium from the lattice of a zeolite (clinoptilolite) by leaching with acids was first reported by Barrer and Makki (ref.1) in 1964. This attempt has to be regarded as the first chemical modification of a zeolite framework. Beginning with 1967 Belenykaja et al. (ref.2) published a series of papers concerning the dealumination of mordenites by acid treatments. It was reproted that not more than 75 % of the lattice aluminium can be removed and that high dealumination degrees are accompanied by partial lattice destruction. Kranich et al. (ref.3) succeeded in reducing the aluminium content of mordenite essentially to zero by thermal treatment of H-mordenite at 970 K followed by leaching with boiling mineral acids. A similar dealumination process includes the hydrothermal treatment of H-mordenite at about 850 K and the aluminium removal by boiling acids. After 9 steaming-leaching cycles a Si/Al ratio of about 50 has been obtained (ref.4).

The present paper deals with the nature of vacant lattice sites formed by dealumination of mordenite with acids and with structural

---

[*] On leave from Zelinsky Institute of Organic Chemistry, Academy of Sciences of the SSSR, Moscow.

changes caused by thermal dehydroxylation and hydrothermal reactions, respectively.

EXPERIMENTAL

The starting material was a synthetic large-pore Na-mordenite $Na_7[Al_7Si_{41}O_{96}]$ without binder supplied by VEB Elektrochemisches Kombinat Bitterfeld. It was treated with 6 N HNO_3 for 4 hours at 298 (sample 1) and 373 K (samples 2-5), respectively. The acid treatment was repeated once again in case of sample 1, 3 and 4 and 3-times in case of sample 5. No silicon was extracted by these procedures. A portion of each acid-leached sample was treated in a water steam atmosphere (about 10^5 Pa) for 2 hours at 873 K. Aluminium was extracted from portions of the steamed samples by refluxing with 6 N HNO_3 for 2 hours (designed as steamed/leached). Table 1 contains the chemical composition of the prepared samples.

TABLE 1
Chemical composition of the prepared dealuminated mordenites.

| Sample | acid-leached | | | | steamed/leached · | | | |
| | mmole·g^{-1} | | | Si/Al | mmole·g^{-1} | | | Si/Al |
	Na_2O	Al_2O_3	SiO_2		Na_2O	Al_2O_3	SiO_2	
1	0.002	0.627	15.58	12	0.006	0.079	16.53	105
2	0.003	0.438	15.89	18	0.006	0.094	16.43	87
3	0.010	0.300	16.14	27	0.002	0.067	16.55	124
4	0.004	0.252	16.24	32	0.008	0.077	16.53	107
5	0.004	0.235	16.26	35	0.003	0.055	16.57	151

The thermoanalytical curves were recorded with a Derivatograph (MOM, Budapest). Care was taken to maintain "shallow bed" conditions during the experiments. The nitrogen adsorption isotherms were measured with a reproducibility better than 0.5 % using the usual volumetric apparatus. X-ray diffraction pattern were taken on a Phillips diffractometer PW 1130/00 with Si powder as inner standard. The intensity of the (150) reflexion was chosen as a measure of the crystallinity because it proved to be nearly unaffected by the different sample treatments. The mid-infrared transmission spectra were recorded with a Nicolet 7119 FT-IR spectrophotometer using the KBr pellet technique.

RESULTS

Thermoanalytical investigations

The thermogravimetric (TG) curves of all mordenite samples dealuminated with acids show a high dehydration step overlapping at
about 430 K with a second weight loss step covering a broad temperature range up till about 1000 K (fig. 1). The second step may be
due to dehydroxylation reactions. Steaming of the samples results
in a strong decrease of both TG steps.

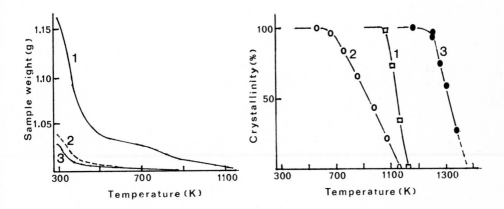

Fig. 1. TG curves of mordenite
(sample 2) after leaching with
acid (1), steaming (2) and
steaming/leaching (3).

Fig. 2. Thermal stability of Na-
mordenite (1), sample 5 after
acid-leaching (2) and after
steaming/leaching (3).

The dealumination of mordenite by acids results in a decrease of
thermal stability (fig. 2). The crystallinity starts to decrease at
about 650 K. This process is accompanied by an alteration of the
lattice symmetry from orthorhombic to triclinic as evident from
the splitting of all except the (002) and (020) reflexions after
treatment at higher temperatures. However, the low crystallinity
of the heat-treated samples does not allow to determine the exact
parameters of this cell with lowest symmetry. On the other hand,
steaming of the dealuminated samples causes an enormous increase
in thermal stability (fig. 2). The lattice degradation of neither
the steamed samples nor the original Na-mordenite does not result
in lattice symmetry alterations.

Adsorption measurements

Nitrogen adsorption isotherms of dealuminated mordenites with
Si/Al ratios up till 18 (samples 1 and 2) show the near rectilinear

shape characteristic for the filling of micropores (fig. 3). At
higher dealumination degrees (samples 3 and 4) the decrease of the
amount adsorbed in micropores at low relative pressures and the increase of the amount adsorbed at higher pressures point to a second
medium-size pore system located either in the zeolite crystals or
in an amorphous phase formed by partial lattice destruction.

The adsorption isotherms of the steamed and steamed/leached
samples indicate in each case a considerable decrease of the amount
of zeolitic phase and a corresponding increase of the medium-size
pore volume (fig. 3). From the hysteresis loop the pore size distribution was calculated having its maximum at 4 nm. Acid leaching
of the steamed samples has no or - at lower dealumination degrees -
only a minor effect on the amount adsorbed provided that this value
is related to the amount of SiO_2 in the sample.

X-ray diffraction

Unit cell parameters of the samples studied are summarized in
table 2. The unit cell contraction caused by dealumination is anisotropic as already described by Olsson and Rollmann (ref. 5).
This behaviour has been described to the preferential removal of Al
from site 1 and 2 where 75 % of the lattice aluminium is located.

Hydrothermal treatment of the dealuminated mordenites results in
an increase of the cell volume though the original value of the
starting material is not reached. This unit cell dilation is due,
first of all, to the increase of the lattice constant b.

The intensities of the X-ray diffraction peaks and, hence, the
structure factors of the originale Na-mordenite, the acid-leached
samples and the steamed/leached products show characteristic differences. Acid-leaching causes a moderate increase of the (110)
and (220) peaks while the following steaming results in a 2-3 fold
increase of the (110), (020), (200) and (111) reflexions. The intensities of some higher indexed reflexions like (202) and (350)
decrease somewhat on dealumination and steaming.

Acid leaching of the steamed samples has no further influence
on the unit cell parameters and the structure factors.

Mid-infrared spectrophotometry

As already reported (ref. 6), dealumination of mordenites by
acids shifts the framework vibration bands in the mid-infrared
spectra towards higher wavenumbers (fig. 4). The intensity of all
bands between 600 and about 800 cm^{-1} decreases strongly. At

TABLE 2

Unit cell parameters of dealuminated mordenites

sample	lattice constants, nm			unit cell volume, nm^3
	a	b	c	
sodium mordenite	1.815	2.052	0.754	2.807
1 acid-leached	1.815	2.022	0.747	2.743
steamed/leached	1.808	2.027	0.747	2.737
3 acid-leached	1.812	2.022	0.745	2.737
steamed/leached	1.815	2.033	0.748	2.760
5 acid-leached	1.808	2.018	0.746	2.721
steamed/leached	1.814	2.029	0.748	2.754

variance with Barthomeuf et al. (ref.6) our spectra provide no evidence for the assignement of a 710 cm^{-1} band to isolated $AlO_{4/2}$ tetraeders and for the formation of a new band at 820 cm^{-1}. The latter is already present in the spectrum of the original sample as a shoulder with about the same intensity while the former decreases for all samples in about the same extent in spite of considerable differences in the residual aluminium content. However, a new broad band appears at 935 cm^{-1}.

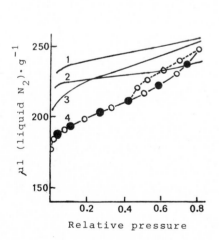

Fig. 3. N_2 adsorption isotherms at 78 K. 1 = sample 1; 2 = sample 2; 3 = sample 3; 4 = sample 4 after steaming (o) and steaming/leaching (●). The dashed line indicates the desorption branch.

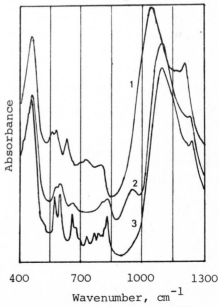

Fig. 4. Infrared spectra of Na-mordenite (1), acid-leached mordenite - sample 5 (2) and sample 5 after steaming (3).

Steaming of the dealuminated mordenites remarkably improves the sharpness and resolution of all bands indicating a higher degree of ordering in steamed samples. Steaming results also in the complete disappearence of the band at 935 cm^{-1} and does not cause further shifts of the band frequencies. Extraction of residual aluminium after steaming does not change the spectra of the steamed samples.

DISCUSSION

The extraction of framework Al by acids has to result in isolated vacant sites on places originally occupied by the removed aluminium. For stoichiometric reasons "nests" constiting of 4 hydroxyl groups (structure I in the reaction scheme) has been proposed (ref.1). Later the existence of such nests was questioned because the hydroxyl content of dealuminated mordenites does not correspond to structure I. The TG results reported in this paper indicate that after dehydration at about 430 K most of the nests are at least partially dehydroxylated (structure II or III). Alternatively, a direct (ref.7,8) or an indirect formation of new Si-O-Si linkages (structure V) involving hydroxyl nests as intermediate (ref.9) was assumed. These suggestions generally accepted to describe the thermal dehydroxylation implies the transformation of part of the 5-membered rings of the mordenite structure to 4-membered ones that must result in strong lattice distortions and, hence, effecting the instability of the lattice. Basing on the conception that opened rings increase the flexibility of the lattice, structure III was suggested to be existent in partially dehydroxylated mordenites (ref.10). However, also structure II has to be taken into consideration. A second band of coordinatively bound pyridine at 1462 cm^{-1} found by Karge (ref.11) in the spectra of acid-leached mordenites besides the common one at 1452 cm^{-1} could be possibly ascribed to coordinatively unsaturated charged silicon atoms acting as Lewis sites (structure II).

h.t. = hydrothermal

Lattice vacancies created by aluminium removal from the framework of HY zeolite by hydrothermal treatment (ref.12) are occupied by migrating silicon and oxygen atoms (ref.13). This process results in "cummulated vacant sites", i.e. in the degradation of larger regions of the crystal. In such a way a secondary medium-size pore system is formed (ref.14). It is evident that under hydrothermal conditions reactions of such type may proceed also in the lattice of mordenites dealuminated by acids.

The decrease of thermal stability by extraction of framework aluminium and the opposite effect of steaming are in accordance with the conceptions outlined in the reaction scheme. With increasing concentrations of isolated vacant sites the structure should become more and more instable. On the other hand, the extreme high stability is fully understandable if structure IV is formed. Consequently, dealumination of mordenites (and generally of zeolites) resulting in isolated vacant lattice sites does not cause an increase of thermal stability as often claimed in the literature (e.g. ref.7,9). It seems to be evident that in all these cases hydrothermal conditions were not strictly excluded during the heating process.

The lattice symmetry change to triclinic upon heat treatment of acid-leached mordenites points also to essential structural differences between acid-leached and steamed mordenites. The anisotropic contraction of the unit cell during acid-leaching and its less pronounced anisotropic dilation during steaming supports the conception of isolated vacancies because it is evident that the distance between the two T-atoms T1 and T2 decreases in the sequence

$$
\begin{array}{ccc}
\text{O}\diagdown\text{T1}\diagdown\text{Al}\diagdown\text{T2}\diagup\text{O} & \longrightarrow & \text{O}\diagdown\text{T1}\diagdown\text{Si}\diagdown\text{T2}\diagup\text{O} \longrightarrow \text{O}\diagdown\text{T1}\cdots\text{H}\cdots\text{O}\cdots\text{T2}
\end{array}
$$

From the near rectilinear shape of the N_2 adsorption isotherms of dealuminated mordenites with Si/Al ratios up till about 18 it can be concluded that the micro- and macropore structure of the crystals are intact, i.e. only isolated vacant lattice sites could be formed and complete lattice degradation of larger regions of the crystals should be excluded. At higher dealumination degrees, however, a medium-size pore system also observed by Wolf and John (ref.9) is created. Steaming is always accompanied by the formation of a medium-size pore system. The adsorption isotherms alone do not allow to distinguish between medium-size pores inside the mordenite crystals (bidispersal pore system) or in a separate amorphous phase. However, the absence of the broad peak in the $2\Theta = 20{-}30$ range of the diffractograms characteristic for amorphous SiO_2 indicates that the medium-size pores are located in the mordenite crystals themselves and, indirectly, that migrating silicon and oygen atoms fill up the isolated lattice vacancies. Acid-leaching of steamed dealuminated mordenites has practically no effect on the amount of adsorbed N_2 provided that this value is related to the amount of SiO_2 in the samples. Consequently, the Al removed by steaming could not be localized inside the pore system and it should form a second phase of oxide-hydroxide species outside the crystals.

The infrared results are also in accordance with the conceptions concerning the outlined dealumination-dehydroxylation mechanism. The band intensity decrease in the 600-800 cm^{-1} region may be due to the decrease of the number of intact rings. The band at 935 cm^{-1} present only in the spectra of mordenites dealuminated with acids may be ascribed either to vibrations of "open rings" or to the bending vibration of nest hydroxyl groups. The highly ordered structure of steamed samples reflected by the high resolution of the spectra is also in accordance with the elimination of the isolated vacant lattice sites and the uniformity of the tetraedral lattice elements in structure IV.

REFERENCES

1 R.M. Barrer and M.B. Makki, Can. J. Chem., 42 (1964) 1481-1487.
2 I.M. Belenykaja, M.M. Dubinin and I.I. Krishtofori, Izv. Akad. Nauk SSSR, Ser. Khim., 1967, 2164-2171; 1968, 2184-2190.
3 W.L. Kranich, Y.H. Ma, L.B. Sand, A.H. Weiss and I. Zwiebel, Adv, Chem. Ser., 101 (1971) 502-512.
4 N.Y. Chen and F.A. Smith, Inorg. Chem., 15 (1976) 295-297.
5 R.W. Olsson and L.D. Rollmann, Inorg. Chem., 16 (1977) 651-654.
6 B.H. Ha, J. Guidot and D. Barthomeuf, JCS Faraday I, 75 (1979) 1245-1253.
7 I.V. Mishin, A.L. Klayachko-Gurvich, B.I. Shikunov, G.A. Ashavskaya, G.I. Kapustin and A.M. Rubinstein, Izv. Akad. Nauk SSSR, Ser. Khim., 1973, 1346-1348.
8 I.M. Belenykaja, M.M. Dubinin and I.I. Krishtofori, Izv. Akad. Nauk SSSR, Ser. Khim., 1971, 2635-2640.
9 F. Wolf and H. John, Chem. Techn., 25 (1973) 736-739.
10 H.K. Beyer, I.M. Belenykaja, M.M. Dubinin and F. Hange, Izv. Akad. Nauk SSSR, Ser. Khim., 1982, 1457-1463.
11 H.G. Karge, Z. physik. Chem. Neue Folge, 122 (1980) 103-116.
12 G.T. Kerr, J. Phys. Chem., 71 (1967) 4155-4156.
13 J. Scherzer and J.L. Bass, J. Catalysis, 28 (1973) 101-115.
14 U. Lohse, H. Stach, H. Thamm, W. Schirmer, A.A. Isirikyan, N.I. Regent and M.M. Dubinin, Z. anorg. allg. Chem., 460 (1980) 179-190.

P.A. Jacobs et al. (Editors), *Structure and Reactivity of Modified Zeolites*
© 1984 Elsevier Science Publishers B.V., Amsterdam — Printed in The Netherlands

ON CATALYTIC ACTIVITY OF SYNTHETIC OFFRETITE IN CONVERSION REACTIONS OF METHANOL AND O-, M-, P-XYLENES

G.V.TSITSISHVILI, TS.M.RAMISHVILI, M.K.CHARKVIANI

Institute of Physical and Organic Chemistry, Academy of Sciences
of the Georgian SSR, Jykia 5, Tbilisi, 380086, U.S.S.R.

ABSTRACT

Catalytic properties of synthetic TMA-offretite and its ammonium deri-
vates have been investigated. TMA-offretite has been synthesized from the
natural raw material by a modified method.

It is shown that hydrogenous and decationated forms of offretite are selective catalysts for DME /dimethyl ether/ formation in methanol dehydration.
Decationated offretites activate also the isomerization of xylenes.

INTRODUCTION

Offretite is rare in the nature /1/. Its synthetic analogue /2-5/ is a Na,
K, TMA /tetramethylammonium/ zeolite. It has a channel system forming
a three-dimensional porous structure. Offretite is stable up to 965 $^\circ$C /1/.
The large TMA-cations /d=6.94 $\overset{\circ}{A}$/ in offretite are situated chiefly in gmelinite cages and partly in the large channels, the potassium cations /d=2.66 $\overset{\circ}{A}$/
at the windows of gmelinite cages and also in cancrinite cages. By removal
of the TMA-cations from TMA-offretite by heating the zeolite in air at
450-500 $^\circ$C /4,6/ or by exchange of potassium cations for ammonium cations
/7,8/ the geometrical accessibility of gmelinite cages and large channels for
the reacting molecules is increased; besides, in thermally processed TMA-
-offretite and NH$_4$-offretite structural hydroxyl groups of different locations
and acidic strength /4,8/ are formed. The number and strength of these acid
sites depend chiefly on the content of potassium ions in the zeolite and on
the temperature of preliminary processing /8/. A decrease of ion content in
H-offretite increases the number and strength of acid sites. Samples containing a small amount of potassium ions have a crystalline structure and re-

present solid strong acids the strength of which may be compared with that of H_2SO_4 dissociated to 90 % /8/. Specific for H-offretite is, first of all, its high acidity /8/, which significantly exceeds that of faujasites and mordenite, as well as the possibility to change its reaction selectivity by regulating the number of potassium ions in offretite /9/.

In spite of these peculiarities, catalytic properties of offretites are not widely studied. They have been investigated in the reactions of hydrocarbon cracking /4, 8, 10/ and also, the role of coke formation in the offretite porous structure during the methanol conversion into hydrocarbons has been elucidated /11/.

The catalytic activity of synthetic offretite, obtained by the exchange of TMA for protons, in n-hexane cracking is comparable to that of synthetic erionite and NH_4-faujasite and is significantly higher than that of ZK-5 and H-mordenite /4,12/. It is active in the cracking of 2-methylpentane as well /10/.

METHODS

Preparation of Catalysts

Catalytic properties of synthetic TMA-offretite and its ammonium derivatives have been investigated. The initial TMA-offretite was synthesized by the method /5/, modified in the Laboratory of Physical Chemistry of the Institute of Physical and Organic Chemistry, Academy of Sciences of the Georgian S.S.R., and had the following chemical formula:

$0.48K_2O \cdot 0.33Na_2O \cdot 0.15/TMA/_2O \cdot 0.14CaO \cdot 0.08MgO \cdot 1.00Al_2O_3 \cdot 5.94SiO_2 \cdot 3.91H_2O$

Adsorption capacity of TMA-offretite is given in Table 1.

TABLE 1

Adsorption capacity of TMA-Offretite

Sample	Adsorption capacity for water, mmol/g /cm^3/cm^3/, P/P_s =0.4; 20 $^{\circ}$C
TMA-offretite without preheating	7.80 /0.14/
TMA-offretite preheated at 520 $^{\circ}$C	9.00 /0.16/

Ammonium forms of offretite with different degree of ion exchange were obtained by prolonged calcination of TMA-offretite in air at 550 $^\circ$C and subsequent exchange with 2N NH_4Cl or NH_4NO_3 for 4 hours at 70 $^\circ$C, followed by the washing off of Cl^- and NO_3^- ions. Characteristics of the investigated samples are given in Table 2.

TABLE 2

Characteristics of the investigated samples

N	Sample	Reagent	Degree of cation exchange for ammonium ion, α, %
1.	TMA-offr.	-	-
2.	NH_4-offr.	NH_4Cl	47.1
3.	NH_4-offr.	NH_4Cl	70.5
4.	NH_4-offr.	NH_4NO_3	45.5
5.	NH_4-offr.	NH_4NO_3	72.3

Before carrying out the reaction, ammonium forms were preheated in a flow of pure oxygen for 2h at 300 $^\circ$C /H-offr.300/ and 450 $^\circ$C /H-offr.450/, TMA--offretite was heated for 15 h at 450 $^\circ$C /TMA-offr.450/ and at 200 $^\circ$C /TMA-offr.200/. Subsequently the catalysts were passed by a stream of helium for 2 h, and the temperature was reduced to the needed one in a helium flow.

Materials, Apparatus, Analysis

Chromatographically pure methanol and o-,m-,p-xylenes have been used in the experiment. Fresh samples of catalysts were used. The catalytic activity of offretites has been determined by the pulse method with helium as carrier gas.

The analysis of reaction products was carried out with a Chromograph ЛХМ-8МД . Methanol conversion products were freezed for 20 min in a quartz loop immersed into liquid nitrogen, and then analysed. Xylene isomerization products were analyzed without preliminary freezing.

Determination of the sample crystallinity was carried out by the method of i.r.-spectroscopy; i.r.-spectra of the catalysts have been recorded with a Zeiss Model UR-20 spectrometer. The investigated samples preserved their crystalline structure after the reaction.

Methanol conversion was investigated in the temperature range from $113°$ to $448\,°C$, helium flow rate - 12-60ml/min. TMA-offr.200, TMA-offr. 450, and H-offretite, prepared from the sample 4 were used /Table 2/. Catalyst mass was 0.05 g; methanol dose- 1.6μl. The experiments were performed under conditions allowing to neglect the external diffusion.

Xylene conversion was studied in the range of 250-600 $°C$, at helium flow rate 25 ml/min, xylene dose - 2μl, catalyst mass - 0.06 g. TMA-offr. 200, TMA-offr.450 and H-offr., obtained from samples 2-5 /Table 2/ have been studied. The use of various chromatographic columns, filled with Pora pak T at 120 $°C$, zeolite KX at 35 $°C$ and 180 $°C$, and silica gel "ASK" /Sojuzreactiv, USSR/ at 35 $°$ and 105 $°C$, allowed us to separate completely both the products of methanol dehydration and dehydrogenation: dimethyl ether /DME/, formaldehyde, water, unreacted methanol, the hydrocarbon components of the reaction mixture, hydrogen and carbon monoxide. Hydrogen was determined only qualitatively.

Isomerization products were separated at 70 $°C$ on a 3 m column, containing Chromaton N-AW with the stationary phase /13%/ benton-245, DMODA /dimethyldioctadecylammonium/ - vermiculit and vaseline oil.

Catalytic activity in methanol conversion was evaluated by the apparent constant of the first order reaction rate /13/ $K_{ef} = /F_o/273Rm/\ln/1-x/$, where F_o is the helium flow rate related to the pressure at the entrance of the reactor /14/, ml.sec^{-1}, m- the catalyst mass, g, R - the gas constant, x - the conversion degree, k_{ef} - the reaction rate constant, sec^{-1}. Activity in the xylene conversion was estimated by the conversion degree of xylene into isomerization products. Isomerization selectivity was determined as the mole ratio of the formed isomers to the mole sum of all reaction products, beside unreacted xylene.

CONVERSION OF METHANOL

TMA-offr.200, as it was expected, turned out to be inactive in methanol conversion; on other catalysts - TMA-offr.450, H-offr.300 and H-offr.450 - methanol mainly was dehydrated to DME. Formation of CO and CH_4 traces, and a significant amount of hydrocarbon gases $/C_2-C_4/$ was observed on H-offr.450 and TMA-offr.450 above 243 $^{\circ}$C and 350°C, resp. Besides, a strong coke formation on all catalysts - offretites was observed, mainly above 300 $^{\circ}$C. In spite of this, offretites showed an activity which was stable with time and was maintained for nearly 50 hours of the measurement.

The temperature dependence of activity in methanol dehydration exhibits an extreme /Fig. 1/. Calculated values of E_{app} for methanol dehydration on TMA-offr.450, H-offr.450 and H-offr.300 are nearly equal and are varying in the range from 11.8 to 9 kcal/mole. The temperatures 210 $^{\circ}$C, 243 $^{\circ}$C and 350 $^{\circ}$C for H-offr.450, H-offr.300 and TMA-offr.450, respectively, correspond to the maximum on the respective curves of the temperature dependence of activity.

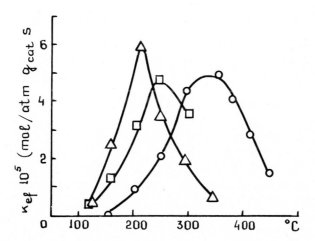

Fig. 1. Temperature dependence of k_{ef} of methanol dehydration to DME on TMA-offr.450 / O /, H-offr.450 /Δ / and on H-offr.300 /□ /.

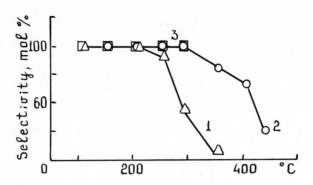

Fig. 2. Temperature dependence of the selectivity of DME formation on H-offr.300 /□ /, H-offr.450 /△ / and on TMA-offr.450 / o /.

Selectivity of DME formation in case of H-offr.450 and TMA-offr.450 /Fig. 2, curves 1 and 2/ begins to decline approximately in this region, and hydrocarbon components appear in the reaction mixture. They contain particularly propylene and propane /12 and 5 mass %, resp., at 295 $^{\circ}$ on H-offr. 450/, as well as small amounts of n-butane and trans-butene-2. The formation of formaldehyde is observed only at lower loads on the catalyst /F_o/m=240 ml/ /min.g/. About 4 mass % of formaldehyde is formed on H-offr.450 above 295 $^{\circ}$C. Trace amounts are observed above 448 $^{\circ}$C on TMA-offr.450. A high yield of DME, about 40 mol%, at F_{He} = 40ml/min /with 100% selectivity/ on H-offr.450 at 210 $^{\circ}$C and with 85% selectivity on TMA-offr.450 at 356 $^{\circ}$C is obtained.

According to k_{ef} values of methanol dehydration reactions for the ascending branches of the curves of the temperature dependence of activity /Fig.1/, the investigated zeolites are arranged in the following order:

H-offr.450 > H-offr.300 > TMA-offr.450 > TMA-offr.200 /inactive/.

The inactivity of TMA-offr.200 is explained by the inaccessibility of zeolite channels for CH_3OH molecules, as the channels are blocked by large TMA cations. As it is known, a prolonged calcination of TMA-offretite is needed for their removal /4,6/. The activity in methanol dehydration is increased with the dehydroxylation of H-offr.300 /Fig. 1/; this is, apparently, associated with a symbatic increase of the number and strength of Lewis acid sites in H-offr. 450 /7,8/, and the lower activity of TMA-offr.450 may be explained by the existence of weak Lewis acid sites in these geolites /6/.

ISOMERIZATION OF XYLENES

Ammonium forms of offretite /samples 2-5, Table 2/, previously dehydroxylated and deammoniated at 450°C /H-offr.450/, turned out to be high - temperature catalysts for the isomerization of o-, m- and p-xylenes; intramolecular isomerization is activated by them only above 400-450°C. Isomerization is accompanied also by dealkylation reactions /benzene and toluene are formed/ and by the formation of deposit products. In Table 3 some of these data are given.

With the increase of cation exchange in offretite from 45.5% to 72.3% the yield of isomerization products in o-xylene isomerization increases, whereas that of m-xylene isomerization decreases and the degree of p-xylene isomerization either remains constant /samples 4,5; Table 2,3/ or increases /samples 2,3; Table 2,3/.

Conversion of m- and o-xylenes on H-offretites is of certain practical interest. In the presence of H-offr. with α= 72.3%, obtained by the exchange with NH_4NO_3, the yield of m-xylene is 54.4mol% /Table 3/ at 558°C, significantly exceeding the equlibrium value. The isomerization selectivity is 78.5%.

TABLE 3

Isomerization of o-, m-, p-xylenes on H-offretites

Catalyst	Tempe-rature, $^\circ$C	Yield of isomeri-zation products, mol%		Dealkyla-tion, mol% /related to toluene/	Isomerization selectivity, mol%

		Isomerization of o-xylene			
		p-	m-		
H-offr./4/*	458	1.8	12.2	-	100.0
	564	7.2	38.4	3.6	92.7
H-offr./5/	450	4.4	19.8	3.6	87.1
	558	7.4	54.4	16.8/toluene+ benzene/	78.8
H-offr./2/	458	10.2	28.4	-	100.0
	564	7.8	20.6	0.6	97.9
H-offr./3/	396	8.6	51.6	9.8	86.0
	570	8.4	49.8	10.2/toluene+ benzene/	84.7

		Isomerization of m-xylene			
		p-	o-		
H-offr./4/	600	13.4	2.8	3.4	82.6
H-offr./5 /	558	11.2	0.8	6.2	67.2
	600	11.2	0.8	6.4	67.0
H-offr./2/	458	15.4	4.4	1.2	94.2
	564	16.6	8.6	4.4	85.1
H-offr./3/	458	14.0	2.4	8.2	66.7
	563	11.4	3.4	11.0/toluene+ benzene/	57.3

		Isomerization of p-xylene			
		m-	o-		
H-offr./4/	600	60.0	0.7	2.6	95.9
H-offr./5/	458	49.2	0.0	7.8	86.3
	550	48.6	1.4	8.6	85.3
H-offr./2/	450	29.4	2.6	2.6	92.5
	568	43.2	5.0	4.0	92.3
H-offr./3/	569	62.8	2.8	8.4	88.6
	600	57.2	2.4	7.2	89.1

* Figures in brackets correspond to the numbers of samples given in Table 2.

The p-xylene yield at 564°C is 72% of the equilibrium value and the selectivity 85.1% in the conversion of m-xylene on H-offr. with α = 47.1% obtained by exchange with NH_4Cl.

It should be noted that isomerization reactions of xylenes on H-offr. proceed without hydrogen and high pressure and are characterized by high selectivity without trimethylbenzene formation.

TMA-offr.200 and TMA-offr.450 are inactive in xylene isomerization. Formation of deposit products has been observed on the latter. The obtained data, apparently, allow to suppose that, possibly, Lewis acid sites take part in the isomerization of xylenes on H-offretites, as it is known that H-offr. heated at 450-600°C is a strong aprotonic acid /7,8/.

REFERENCES

1 D.Breck, Zeolite Molecular Sieves, "Mir", Moscow, 1976, 196pp.
2 M.K.Rubin, German Patent P 1806154, 6, 16.10.1968.
3 R.Aiello, R.M.Barrer, J.A.Davies, J.S.Kerr, Trans.Faraday Soc. 66 /1970/ 1610-1617.
4 T.E.J.R.Whyte, E.L.Wu, G.T.Kerr, P.B.Venuto, J.Catalysis, 20 /1971/ 88-96.
5 E.E.Jenkins, U.S. Patent 3578398, 11.05.1971.
6 E.L.Wu, T.E.Whyte, P.B.Venuto, J.Catalysis, 21 /1971/ 384-393.
7 C.Mirodatos, A.Abou-Kais, J.C.Vedrine, D.Barthomeuf, J.Chem. Soc., Faraday Trans., I, 74/1978/ 1786-1795.
8 C.Mirodatos, D.Barthomeuf, J.Catalysis, 57/1979/ 136-146.
9 C.Mirodatos, R.Beaumont, D.Barthomeuf, Acad.Sci.Paris, Ser.C., 281/1975/ 959-961.
10 N.Y.Chen, Proceedings 5th Int.Congr.Catalysis, North-Holland, Amsterdam, v.2, 1973, pp.1343.
11 P.Dejaifve, A.Auroux, P.C.Gravelle, J.C.Vedrine, J.Catalysis, 70/1981/ 123-136.
12 P.B.Weisz, J.N.Miale, J.Catalysis, 4/1964/ 527-531.
13 D.W.Bassset, H.W.Habgood, J.Phys.Chem., 64/1960/ 769-773.
14 V.V.Ushchenko, T.V.Antipina, Zh.Phys.Khim., 2/1969/ 540-541.

P.A. Jacobs et al. (Editors), *Structure and Reactivity of Modified Zeolites* 151
© 1984 Elsevier Science Publishers B.V., Amsterdam — Printed in The Netherlands

SYNTHESIS AND STUDY OF PROPERTIES OF ZSM-II TYPE SILICALITES OF I-VIII GROUP ELEMENTS

K.G. IONE, L.A. VOSTRIKOVA, A.V. PETROVA and V.M. MASTIKHIN

Institute of Catalysis, Novosibirsk 630090 (U.S.S.R.)

ABSTRACT

ZSM-II type silicalites have been prepared by hydrothermal synthesis in the presence of salts of I-VIII group elements and the activity and selectivity of these zeolites in methanol conversion have been examined.

It has been found that Be^{2+}, B^{3+}, Al^{3+}, Ga^{3+} cations are fixed in the silicon-oxygen framework with a tetrahedral coordination and are catalytically active in methanol conversion to hydrocarbons. Cr^{3+}, V^{4+}, Mo^{6+}, Mn^{2+}, Cu^{2+} cations do not enter the structure of the framework and have an octahedral coordination; silicalites containing these cations are found to be practically inert in methanol conversion.

The catalytic function of zeolites in reactions following the acid-base mechanism is related to the donor-acceptor interactions in the field of the aluminium atom which substitutes isomorphously a silicon atom in the silicon-oxygen framework of zeolites. Such interactions produce Brönsted acid centers (BAC) of the type $(Si-O-Al)^{-}H^{+}$ or $3(Si-O-Al)^{-}Al^{3+}OH^{-}...H^{+}$. Attempts have been made (ref.1) to change the activity and selectivity of a zeolite by introducing polyvalent cations Me^{n+} via ion exchange, which results in the change of the concentration of hydroxyl groups. In this case the composition of the anion part of the active center remains unchanged.

Earlier we made attempts to determine the state (ref.2,3) and the catalytic properties (ref.4) of non-alumosilicate zeolites with Ga^{3+}, Cu^{2+}, Fe^{3+} and B^{3+} introduced into the framework. In this investigation ZSM-II type silicalites with $SiO_2/Al_2O_3 \approx 1400$ have been prepared by hydrothermal synthesis in the presence of salts of I-VIII group elements and the activity and selectivity of these zeolites in methanol conversion have been examined.

EXPERIMENTAL

Zeolites were prepared by hydrothermal treatment of 30% aqueous silica sol in the presence of tetrabutylammonium bromide, one of the salts of I-VIII group cations and sodium hydroxide for 1-30 days at 415-475 K. No source of aluminium was added into the initial mixture. Samples decationated up to the Na_2O content of 0.1-0.5 wt. % with a crystallization degree (Re,%) no less than 80-90% were used in catalytic tests. Re,% was determined by comparing the X-ray patterns of the product and a silicalite (So) in the θ region 10-15°. The silicalite used as a reference was synthesized by the above described procedure but without addition of inorganic salts. The effective surface area (S, m^2/g) of the zeolite was determined from the value of argon adsorption at the temperature of liquid nitrogen. ^{27}Al, ^{11}B, ^9Be, ^{71}Ga, ^{29}Si NMR spectra were registered using a "Bruker-CXP-300" spectrometer. ESR spectra were taken using a JES-3BX spectrometer in X- and Q- regions at temperatures of 77 and 300 K.

The catalytic activity was determined in a pulse regime using a 2 cm^3 reactor with a vibro-fluidized catalyst bed. The conditions employed were: composition of the reaction mixture 12-15% vol. of methanol in helium, contact time 5-12 s, reaction temperatures 653, 683, 723 and 773 K.

RESULTS

In the whole range of temperatures and times of crystallization studied ($NaOH/SiO_2$ = 0.1-0.2) the main crystalline component of a solid product was a phase with the structure of ZSM-II. The effective surface area (S) depends almost linearly on the quantity of the zeolite phase in the residue obtained (Re,%). This indicates that under the synthesis conditions both amorphous and crystalline non-zeolite phases of the final product are nonporous. At the same $NaOH/SiO_2$ ratio and crystallization time the dependence of the Re% value on the temperature of crystallization exhibits an extreme in the temperature range from 413 to 475 K (see Fig. 1a).

As the temperature of crystallization rises up to 443 K and 475 K, the content of an amorphous residue, α-quartz and cristobalite increases in the final product. Moreover, under the same experimental conditions the amount of the zeolite phase in the residue increases as the quantity of polyvalent cations Me^{n+} in the initial silica gel decreases (see Fig. 1b).

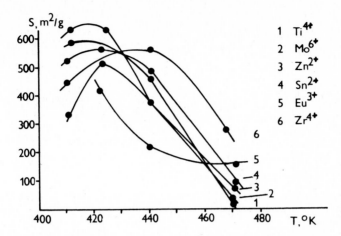

Fig. 1a. Effect of the temperature of crystallization on the content of the silicalite phase with ZSM-II structure in the products of crystallization.

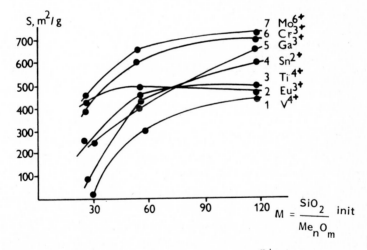

Fig. 1b. Effect of the content of Me^{n+} cations on the content of the silicalite phase with ZSM-II structure in the products of crystallization.

Of special interest is the fact that the value of Re,% depends on the nature of the polyvalent cation introduced. For example, the highest values of S and Re have been obtained for the product with Mo^{6+}, Cr^{3+}, Zr^{4+}, Sn^{2+} additives and the lowest values for the product with In^{3+}, V^{4+} and Eu^{3+} salts. The SiO_2/Me_nO_m ratio

in a solid product has been shown to be much lower   for sili-
calites with Mo^{6+}, Eu^{3+}, Tb^{3+}, Sb^{3+}, Se^{6+} and In^{3+} additives than in
the initial silica gel(see Fig. 2) .

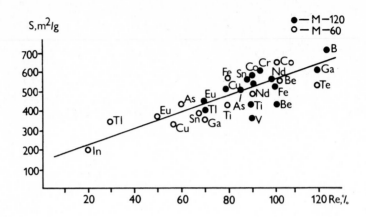

Fig. 2. Dependence of Re (%) and S(m^2/g)on the nature
of the element introduced.

The state of polyvalent cations in silicalite. In the NMR
spectra of silicalites synthesized in the presence of Be^{2+}, B^{3+},
Al^{3+}, Ga^{3+} salts the most intensive signals can be attributed to
the cations fixed in a tetrahedral oxygen surrounding.

TABLE 1
Conditions of NMR spectra registration. H_o = 7.04 mT.

Cation	Reference	Chemical shift in the NMR spectrum of silicalite, ppm
Al^{3+}	1 M $AlCl_3$	59
Be^{2+}	1 M $NA_2Be(OH)_4$	-5.9
Ga^{3+}	1 M $Ga(NO_3)_3$	+155
B^{3+}	1 M H_3BO_3	-24-25

Cr^{3+}, V^{4+}, Mo^{6+} and Cu^{2+} in the zeolites are fixed in the
octahedral oxygen coordination, as found by ESR data. The ESR
spectra of Fe-containing silicalites exhibit three signals with
g factors of 4.23, 2.3 and 2.0.The first one is attributed (ref.

5,6) to Fe^{3+} ions in the tetrahedral coordination and the other two to Fe^{3+} in cation positions in the octahedral coordination.

It has been found for silicalites containing Mo^{6+}, Eu^{3+}, Mn^{2+}, Cu^{2+}, V^{4+}, Cr^{3+} cations that methanol conversion is very low and is nearly the same as that on So-silicalite prepared from non--modified silica gel. Final products contain dimethyl ether, water, CO and CH_4.

On silicalites containing Ga^{3+}, Al^{3+}, Fe^{3+}, B^{3+} and Be^{2+} cat-ions conversion of methanol results in the formation of hydrocar-bons (see Table 2). On silicalites with Ga^{3+} and Al^{3+} high frac-tions of aromatic hydrocarbons are obtained.

DISCUSSION

A number of ideas concerning the nature of the catalytically active crystalline alumosilicates, e.g. zeolites, are based on the isomorphous substitution of silicon atoms in the silicon-oxygen framework by aluminium atoms (ref. 7,8,9).

If this assumption is valid, one might expect that any hetero-valent isomorphous substitution in a silicate framework will re-sult in the formation of a catalytically active center, and that the protonation degree of structural hydroxyl groups will depend on the strength of electron-acceptor interactions in the field of the Me^{n+} cation-substitute, i.e. in a rough approximation on its value of e/r. Therefore the concentration of active centers and, consequently, the activity of silicalite will depend on the nature of the cation introduced in its framework and will increase with increasing its e/r.

Based on the general ideas reported it could be expected that in the series In^{3+}, Fe^{3+}, Ga^{3+}, Al^{3+}, Be^{2+}, B^{3+}, boron silicalite will be the most active, and indium silicalite the least active. Crystalline SiO_2 - silicalite (So) should be catalytically inert. It is seen from the results presented in this communication that hydrocarbons are not formed on silicalites without modifying ele-ments (So). Silicalites containing V^{4+}, Cr^{3+}, Mn^{2+} and Cu^{2+}, fixed mainly in octahedral environment, are of the same selectivity of reaction with respect to aliphatic products as the sample So.

Table 2

Catalytic properties of silicalites synthesized with I-VIII group cations

No	Modifying element	$\frac{SiO_2}{Me_2O_3}$	Reaction temperature, °C	Content of methanol in reaction mixture, wt %	Time of contact s	Degree of methanol conversion	Composition of hydrocarbon products of reaction, wt %.					
							CO,CH_4	DME^{xx}	Olefins C_2-C_4	i-paraffins C_4-C_8	$\sum C_2 - C_{10}$ n-paraffins C_2-C_7	Aromatic hydrocarbons
1	-	1400	380	14.6	10.7	2.5	0.4	99.6	-	-	-	-
2	B^{3+}	46	410	15.2	7.2	66.7	-	99.30	0.73	-	-	-
3	Be^{2+}	36	410	15.1	6.9	82.4	1.41	95.59	2.73	0.09	-	0.19
4	Be^{2+}	60	410	14.1	7.4	87.2	1.14	50.33	41.14	6.11	0.90	0.35
5	Al^{3+}	59	450	13.2	7.6	99.6	1.16	-	68.52	10.14	6.85	13.3
6	Ga^{3+}	67	410	11.7	1.0	97.0	0.9	1.0	65.3	19.8	2.5	10.6
7	Fe^{3+}	57	410	15.6	2.5	91.0	3.5	6.6	72.0	9.2	2.71	6.0
8	Fe^{3+}	108	410	11.6	1.0	85.0	1.2	37.4	49.4	7.0	-	4.9
9	V^{4+}	62	410	11.0	12.3	70.0	0.63	58.51	31.56	3.98	1.07	4.25
10	Cr^{3+}	112	450	14.0	8.9	60.9	22.89	75.18	1.69	-	0.24	-
11	Mn^{2+}	77	380^x	100	6.5	2-5	mainly CO and H_2 are formed					
12	Cu^{2+}	91	380^x	100	6.5	5		100	-		-	-
13	Eu^{3+}	104	450	12.2	7.5	80.5	0.85	99.15	-	-	-	-

^{x}Samples were tested using a flow reactor; feed – methanol, space velocity 1-2 h^{-1}.

DME^{xx} – dimethyl ether

The formation of aromatic hydrocarbons, paraffins and olefins from methanol has been stated for silicalites containing at least a part of Fe^{3+}, Ga^{3+}, Al^{3+}, Be^{2+}, B^{3+} cations in tetrahedral oxygen environment.

The dependence of the degree and selectivity of conversion on the e/r value of these cations is, however, not of linear, but of an extreme character. This is probably due to a lower amount of cations fixed in the active center than corresponding to the total content of the element introduced, because of a part of cations releasing the framework during the thermal activation of the samples. This process should be favoured by distortions taking place in the silicon-oxygen framework when the length of Me-O bond differs from the length of the Si-O bond.

According to the sequence of changes in bond lengths B-O Be-O Si-O Al-O Ga-O Fe-O In-O the greatest distortions in the framework should be expected in the case of silicalites containing B^{+3} In^{+3}, Fe^{+3}. Besides, the release of cations from the framework may be caused by a lower thermodynamic stability of cations in tetrahedral environment compared to another, a higher or a lower, coordination. As has been seen above, V^{4+}, Cr^{3+}, Mn^{2+}, Cu^{2+} ions entering into the silicalite framework were not revealed at all because of their fixation in octahedral environment. Fixation in trigonal environment is highly probable for B^{3+} cations.

Thus, the character of the catalytic data presented is determined by a number of factors. In any case, these data permit to suggest that a catalytically inert oxide system may by highly active in reactions following the acid-base mechanism, provided isomorphous substitution of its cations by atoms of another nature with a different charge value takes place.

REFERENCES

1 Kh.M. Minachev and Ya.I. Isakov, in Zeolite Chemistry and catalysis, (J.A.Rabo,Ed.) ACS Monograph 171, 1976, pp.552.
2 K.G. Ione, L.A. Vostrikova, E.A. Paukshtis, E.N. Yurchenko and V.G. Stepanov, Dokl. AN SSSR, 5 (1981) 1160-1162.
3 L.A. Vostrikova, N.G. Maksimov and K.G. Ione, React. Kinet. Catal. Lett., 17 (1981) 397-400.
4 K.G. Ione, L.A. Vostrikova and V.G. Stepanov, U.S.S.R. Patent No 3339026.
5 B. Wichterlová, P. Jírů, React. Kinet. Catal. Lett., 13 (1980) 197-200.
6 E.G. Derouane, M. Mestdagh and L. Vielvoye, J. Catal., 33 (1974) 169-175.
7 C.D.Chang, Catal. Rev. Sci. Eng., 25 (1983) 1-118.

158

8 P.A. Jacobs, W.J. Mortier, J.B. Uytterhoeven, J. Inorg. Nucl.
  Chem.; 40, No 11 (1978) 1919-1923.
9 P.A. Jacobs. Carboniogenic Activity of Zeolites, Elsevier,
  Amsterdam 1977.

P.A. Jacobs et al. (Editors), *Structure and Reactivity of Modified Zeolites*
1984 Elsevier Science Publishers B.V., Amsterdam — Printed in The Netherlands

A NEW ROUTE TO ZSM-5 ZEOLITE: SYNTHESIS AND CHARACTERIZATION

E. MORETTI, G. LEOFANTI, M. PADOVAN, M. SOLARI, G. DE ALBERTI and
F. GATTI

Montedipe Research Center - 20021 Bollate Mi (Italy)

ABSTRACT

A new synthesis method of ZSM-5 zeolites, developed by us, by starting from microspheroidal silica impregnated with an aqueous solution of tetrapropylammonium bromide, $NaAlO_2$ and NaOH, has been compared with a traditional one, starting from colloidal silica.

In both cases a reaction mechanism consisting of an initial crystallization of an Al-poor product, followed by a slow rearrangement of the Al into the crystalline lattice, has been suggested. In the case of our new technique the total time of synthesis, to obtain the best catalyst for the m-xylene isomerization (nucleation + crystallization + Al rearrangement), is much shorter than in the case of the traditional method.

INTRODUCTION

In previous works [1-2] the possibility of synthesizing Silicalites or ZSM-5 zeolites, by using a new method which consists of a dry impregnation of a microspheroidal silica with an aqueous solution of tetrapropylammonium bromide (TPA-Br), $NaAlO_2$, and NaOH, has been described. This new technique, compared with the conventional ones, offers some advantages, because it reduces considerably the volume of the reagents and effluents, the quantity of the expensive TPA-Br and the synthesis temperature. Generally one can obtain advantages in terms of energy consumation and ecology.

By using this particular method, studies on the kinetics of formation of Silicalite and ZSM-5 zeolites at different Si/Al ratios have been performed [2-4]. At any starting ratio, an initial crystallization of ZSM-5 zeolite which after H^+-exchange exhibits low acidity has been emphasized. During the following synthesis time the acidity

and correspondingly the TPA decomposition temperature increase
until a maximum.

We interpreted this transformation as the Al entering the lattice progressively, after an initial crystallization of a product
with a low content of Al (Silicalite-like product); this transformation seems to occur through processes of solubilization and recrystallization.

In the present work we have intended to verify above all whether
this type of mechanism is characteristic only of our particular
silica source and method of preparation of the starting mixture or
whether, on the contrary, it can be valid also for conventional
syntheses.

A further aim ha been to compare the effectiveness of the two
different techniques in terms of total time of synthesis and of
catalytic activity.

EXPERIMENTAL PART

Synthesis. The samples prepared following our new technique (A)
were synthetized starting from microspheroidal silica (Akzo F-7)
with a pore volume of 1.7 cm^3/g. It was impregnated with an aqueous
solution of $NaAlO_2$, NaOH and TPA-Br. Its volume was equal to that
of the silica pores.

The samples obtained following the conventional way (B) were
prepared by dissolving the TPA-Br in the silica-sol (Ketjensol
AS 40) and then adding, slowly and under vigorous stirring, a solution of NaOH and $NaAlO_2$;during this last step a heavy gel precipitates.

In both cases the composition, expressed as atomic ratios, was:
0.3 TPA : 0.2 Na : 1/R Al : 1 Si : 7.4 H_2O
where R was the Si/Al ratio of the starting mixture and was made
change from 20 to $>$ 1000 (Silicalite samples).

Each initial sample was divided into some parts and kept at 393K
in sealed tubes for various times. Samples were then filtered,
washed free of extraneous salts with water and dried for 15 h at

393K. Their acid form was obtained by calcinating at 813K for 15 h
and exchanging three times with a 0.5 N HCl solution, then by washing and drying again for 15 h at 393K.

Characterization. The samples were characterized with XRD,
thermal and chemical analysis. By the XRD, apart from the identification of the formed structure, it was possible to calculate the
degree of crystallinity through a method recently developed in our
laboratories [5] and the size of the crystallites by using the
classical equation of Scherrer [6]. By thermal analysis the temperature of decomposition of TPA was determined.

Catalytic activity. The catalysts were prepared by extrudating
the zeolite in H-form with 35%w of silica as a binder and by act-
ivating for 2 h at 813K. The apparatus, to test the catalytic activity in the m-xylene isomerization, consisted of a fixed-bed downflow glass reactor (10 mm diameter). The reaction conditions were:
feed = m-xylene + N_2 (1:4 molar ratio); temperature = 548K and
598K; pressure = 1 Ate; space velocity = 10 WHSV.

RESULTS AND DISCUSSION

Degree of crystallinity. The crystallization of samples at various R (R = 20, 40 and >1000), prepared according to the two tecni-
ques, was followed by means of XRD. The typical curves of the crystallinity degree of the ZSM-5 zeolite (the only one crystalline
phase present) vs. the synthesis time, are shown in figure 1a.

From the comparison between the two techniques it emerges that
our one (A) requires crystallization times shorter than the conventional one (B).

In both cases, particularly in the B method, the nucleation time
of the R=20 samples is much higher than that of the other ones, in
agreement with the previously reported inhibiting effect of the Al
content [2, 7-9].

TPA decomposition peak temperature (T_p). As in the previous
work (2), we employed this techniques to follow the transformations,
which continue also after the complete crystallization and which

162

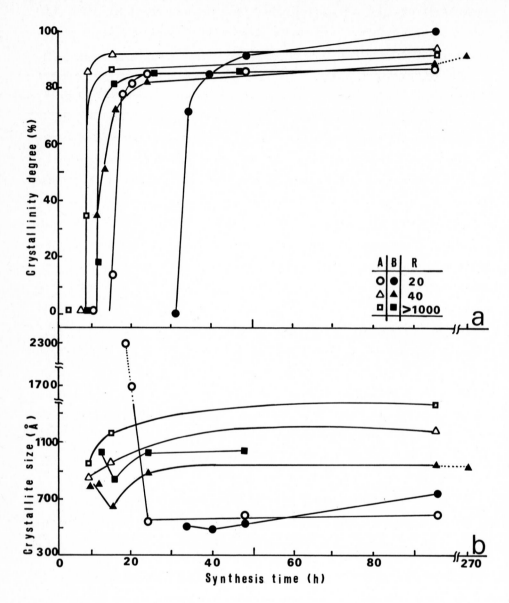

Fig. 1. Effect of the synthesis time on (a) Crystallinity degree.
(b) Crystallite size

are due to the different degree of interaction of the Al with the structure. Except for the Silicalite-like products where the T_p is constant (648K), in all the other cases the T_p increases with the synthesis time (see fig. 2a). Just when the crystallization is quite complete, the A samples show very low values, which then quickly increase towards a maximum. At the beginning of the crystallization of each series the B samples start from values of T_p higher, but then they remain slightly lower than the corresponding A ones.

This means that in both cases a progressive entrance of the Al into the framework occurs during the synthesis, but with the A method at the beginning a Al-poor product crystallizes with an Al content much lower than in the B case.

The higher times of crystallization of the B method above described in comparison with the A method could be explained by the higher quantity of the Al initially involved in the crystallization.

Chemical analysis. A comparison between the compositions of the initial solids shows that the Si/Al ratios of the A and B samples

TABLE 1

Chemical analysis of ZSM-5 samples (amorphous + crystalline phase)

R	Method A		Method B	
	Synthesis time	Si/Al	Synthesis time	Si/Al
at/at	h	at/at	h	at/at
40	4	36.3	10	40.2
	96	40.1	270	47.0
20	4	19.3	20	21.6
	96	23.3	96	25.3

are respectively lower and higher than R (see Tab. 1). This means that the solutions in the first case are poorer in Al than those in the second one; our analyses of the mother-liquors of some initial A samples confirm that only traces of Al are present. This result can explain the initial formation in the B case of zeolites with higher content of structural Al than in the A case, as suggested by the previous comparisons of the T_p and crystallization time data. The high content of Al in the starting solids does not contradict our hypothesis of formation of products with low content of structural Al, but, on the contrary, could help to understand it. It is known [10] that, when one adds the NaAlO$_2$ solution to the silica, a very insoluble amorphous aluminosilicate compound precipitates, which removes most of the Al from the solution.

In both cases the Si/Al ratios of the solids tend to increase with the synthesis time, exceeding anyway the R values. In the A case and probably also in the B case, these trends mean that the Al transfers from the solid to the solution, making itself available to the entrance into the structure. The different rates of this last transformation with the two techniques can be explained by different rates of solubilization of the Al, as the crystallization proceeds, depending on the characteristics of the starting silica.

Crystallite sizes. The values of the crystallite sizes of the samples B are generally slightly lower than the corresponding A ones (see Fig. 1b). When we happened to isolate some samples in early hours of the crystallization, it seems possible to observe an initial decrease of their sizes; this is evident in the R=20A sample, where a high variation of structural Al content occurs. This decrease can be explained only by a solubilization of the first formed crystals and new crystallization, just as previously emphasized for other sample [2-4]. When the crystallization is slowly proceeding (crystallinity from 80 to 90%), the crystallite sizes of all the samples tend to remain constant or to increase very slightly.

Catalytic activity. The catalytic activity has been tested in the m-xylene isomerization which is a typical carbocationic reaction. The activities of the B samples are generally lower than the corresponding A ones (see Fig. 2b).

While the R $>$ 1000 samples are inactive and the activity of the R=40 B ones rises slowly with the synthesis time, the m-xylene conversions of the other samples increase quickly to a maximum and then decrease. These trends confirm that the transformation after the crystallization, evidenced by the T$_p$ measurements, is much slower in the B case. The decrease of activity in some samples after the

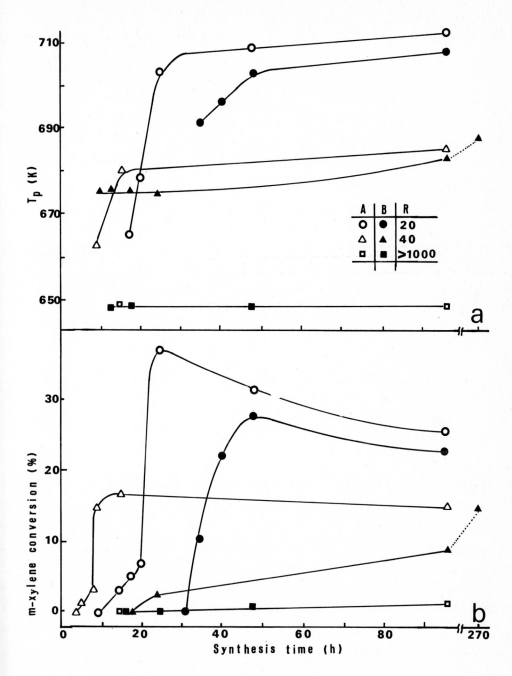

Fig. 2. Effect of the synthesis time on (a) TPA decomposition peak temperature (T_p). (b) M-xylene conversion.

initial quick increase can be attributed to the increase of the sizes of the particles as emphasized in previous studies [2, 11].

CONCLUSIONS

The mechanism of crystallization of ZSM-5 zeolites is similar in both ways of synthesis: after the initial quick crystallization, a second slow transformation occurs with an increase of acidity, interpreted as an increase of the structural Al. In this sense, as demonstrated by the trends of T_p and the catalytic activity, the A technique leads to initial crystallization of a product with a much lower Al content, which however, after a short time, increases dramatically exceeding that of the B samples. The kinetics of this mechanism is much quicker following the A method than the B one, with reference to the crystallization times and to the next entrance of Al into the lattice.

The method A appears more effective than the B one because it allows us to get more active catalysts in a much shorter time.

AKNOWLEDGEMENTS

The authors thank Mr.s W.Grazioli, A.Pelorosso, A.Ponzoni, A.Targa and G. Volpi for their contributions in the various techniques.

REFERENCES

1 E.Moretti, M.Padovan, V.Zamboni, G.Paparatto and M.Solari (Montedison-Monte-dipe), Ital.Pat.Appl. 21,050 A/82 (1983).
2 E.Moretti, G.Leofanti, M. Padovan, M.Solari, G.De Alberti and F.Gatti, submitted for publication in the Proceedings of the 8th International Congress on Catalysis, Berlin, July 2-6, 1984.
3 M.Padovan, G.Leofanti, M.Solari and E.Moretti, Proceedings of 4th Italian-Czechoslovach Symposium of Catalysis, Turin, September 20-23, 1983, University, Turin, 1983, p. 125.
4 M.Padovan, G. Leofanti, M.Solari and E.Moretti, in press on "Zeolites".
5 G.Carazzolo, F.Gatti and A.Ponzoni, in preparation.
6 H.P.Klug and L.E.Alexander, X Ray Diffraction Procedures, J.Wiley, London, 1954, p. 491.
7 L.D.Rollmann and E.W.Valyocsik (Mobil Oil), Eur.Pat. 21,674 (1981).
8 K.J.Chao, T.C.Tasi, M.S.Chen and I.Wang, J.Chem. Soc. Farad. Trans. 1, 77 (1981) p. 547.
9 R.Mostowicz and L.B.Sand, Zeolites, 2 (1982), p. 143.
10 R.K.Iler, The Colloidal Chemistry of Silica and Silicates, Cornell University Press, Ithaca (N.Y.), 1955, p. 31.
11 G.Leofanti, F.Colombo, G.DeAlberti and M. Padovan, in preparation.

P.A. Jacobs et al. (Editors), *Structure and Reactivity of Modified Zeolites*
1984 Elsevier Science Publishers B.V., Amsterdam — Printed in The Netherlands

SYNTHESIS OF GALLOSILICATE AND ALUMOGERMANATE ZEOLITES AND INVESTIGATION OF THEIR ACTIVITY IN THE REACTION OF ALCOHOL DEHYDRATION

Z.G. ZULFUGAROV, A.S. SULEIMANOV, CH.R. SAMEDOV

Institute of Inorganic and Physical Chemistry, Academy of Sciences of the Azerbaijan SSR, 370143, Narimanov Prospect 29, Baku (USSR)

ABSTRACT

The gallosilicate and alumogermanate analogues of X and ZBK zeolites were obtained by direct synthesis. In the synthesized zeolites aluminium and silicon are completely substituted by gallium and germanium, respectively. The influence of aluminium and silicon substitution by other atoms (Ga, Ge) on the catalytic activity, selectivity and stability of zeolite catalysts in the conversion of alcohols was investigated. It was established that over gallium zeolites the yield of aromatic hydrocarbons is increased in comparison with the alumosilicate zeolites.

The alumogarmanate zeolite exhibits a high selectivity to the propylene yield.

Some possible reasons for the changes in the activity and selectivity of the zeolites on substitution of aluminium by gallium and silicon by germanium are discussed.

INTRODUCTION

In (ref.1,2,3,4), the catalytic properties of zeolites in the alcohol conversion are connected with the state of aluminium in zeolites. The authors of (ref.2) established the dependence of the negative charge on the oxygen atom on the bond length of atoms in the tetrahedra T-O-T.

In (ref.3), the determined heterogeneity of acid centres is connected with that of the states of aluminium atoms.

In (ref.4), the dependence of process selectivity on the strength of acid centres has been established. The authors of (ref.5,6) conclude that aluminium atoms in the zeolites with octahedral coordination and deficiency in coordination must act correspondingly as strong Bronsted and Lewis acids.

The author of (ref.1) suggested that secondary conversions of olefins take place only in the presence of catalysts containing aluminium atoms in cation positions.

From the literature review it follows that at present many special catalytic properties of zeolites are explained by the electron and coordination state of aluminium.

In some articles (ref.7,8,9) methods of the synthesis of gallosilicate and alumogermanate analogues of zeolites are described.

The research devoted to the study of the influence of silicon and aluminium substitution by germanium and gallium, respectively, on the zeolite properties is still incomplete (ref.10). In this connection the synthesis of zeolites, where aluminium is isomorphously substituted by other atoms, is of great interest for the elucidation of the action mechanism of zeolite catalysts.

In the course of our work gallosilicate and alumosilicate analogues of X and ZBK zeolites (zeolites with a high content of silica,i.e.ZSM type) were synthesized and their catalytic activity, selectivity and stability in the conversion of alcohols were studied.

THE METHODS OF RESEARCH AND ANALYSIS

The zeolites were synthesized by the methods described in (ref.7,8,9). In order to obtain zeolites of higher crystallinity, the initial gels were kept under the conditions described in (ref.11).

The obtained samples were subjected to chemical, X-ray, IR-spectroscopic and thermal analysis.

To identify the synthesis products containing Ga and Ge the X-ray method was used.

Dehydration of alcohols of 99.5% chemical purity was studied in a flow reactor at $360-380^{\circ}C$, the space velocity of the alcohol supply ranging from 1 to 5 h^{-1}. The reactor was charged with 5 cm^3 of zeolite granules of 2 mm diameter. Before the experiment the zeolites were treated with dried air at $550^{\circ}C$ for 4 hours.

Calcium ion-exchanged and hydrogen forms of X and ZBK zeolites were used as catalysts.

RESULTS AND DISCUSSION

In Table 1 the data on the composition of the initial gel and the synthesized zeolites are presented. From them it follows that the gallosilicate zeolite crystallizes at a lower Na_2O/SiO_2 ratio than the alumosilicate. As it follows from Table 1, for the synthesis of a pure alumogermanate analogue of X zeolite one should take germanium oxide 1.5 times less than is the equivalent quantity of SiO_2 (3.2 moles) necessary for the synthesis of the alumosilicate X zeolite.

On substitution of the silicon by germanium the parameter of a unit cell increases more than on substitution of aluminium by gallium (Table 1). As it is seen from Table 1 in the synthesis of gallium analogues of zeolites with a high content of silica the initial compositions of gels and the synthesis conditions of aluminium- and gallium-containing zeolites are the same.

Gallium analogues of zeolites are crystallized from reaction mixtures with a surplus content in silicon. The synthesis of gallium analogues of zeolites, where in the initial gels SiO_2/Al_2O_3 <3, is connected with great experimental difficulties and sometimes fails.

In Fig.1 IR-spectra of adsorption of initial alumosilicate zeolites and their gallium and germanium analogues are presented.

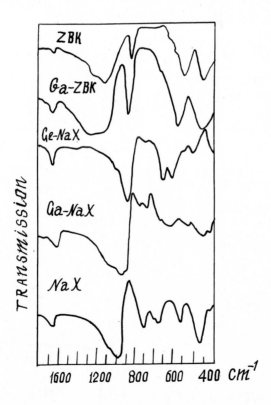

Fig. 1. IR-spectra of adsorption of initial alumosilicate zeolites and their gallic and germanic zeolites.

From the comparison of the spectra it follows that in the spectrum of gallosilicate zeolite (Ga-NaX) the band, corresponding to the bond vibration Si(Al)-O$_4$ and located at 1000 cm^{-1}, undergoes a noticeable change,i.e.it expands considerably. The observed changes reflect a mixed character of the tetrahedra. The absorption bands of silicon tetrahedra interfere with the bands of gallium tetrahedra.

As it is seen from Fig.1,for the substitution of silicon by germanium (Ge-NaX) the absorption bands, when compared to an alumosilicate zeolite spectrum, shift towards low-frequency regions of the spectrum. The peaks of absorption bands shift approximately by 100cm^{-1}. Such experimental results are in a good agreement with the theory of IR-spectra. From the theory of IR-spectra (ref.12,13) it follows that a variation of atomic mass, in general, affects the frequency of atomic vibrations. In Fig.1, IR-spectra of absorption of the initial and gallium zeolites of ZBK type are also presented. IR-spectra of these zeolites are in a satisfactory mutual agreement.

Fig.2 illustrates the curves of differential thermal analysis of alumosilicates and their gallium- and germanium-substitu-

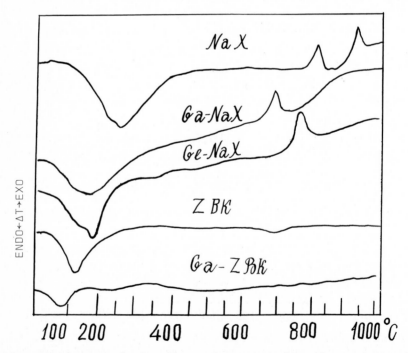

Fig.2. DTA curves of the alumosilicate, gallosilicate and alumogermanate zeolites of X and ZBK types.

ted forms. Endoeffects corresponding to the dehydration of sodium forms of the alumosilicate (NaX), alumogermanate (Ge-NaX) and gallosilicate (Ga-NaX) zeolites appear at 246°C, 190°C and 160°C, respectively. Mass losses when heating the alumosilicate and its alumogermanate and gallosilicate analogues are correspondingly equal to 25.3%, 22.5% and 18%, respectively. The decrease of the adsorption volume of a zeolite seems to be determined by the difference in the ionic radii of the isomorphously substituted atoms (see Table 1). Gallium and germanium atoms occupy a relatively larger volume in the zeolite lattice and thus decrease the volumes of free cavities. After heating at 760°C the zeolite NaX becomes amorphous and recrystallizes into carnegeite at 800°C, and into nepheline at 1000°C (ref.14).

In contrast to NaX, the alumogermanate and gallosilicate analogues of X zeolite recrystallize only once into germanate and sodium gallate and at 800°C their crystal structure is destroyed (see Fig.2). Thus, it is established that on substitution of aluminium and silicon in X zeolites by gallium and germanium, respectively, the thermostability of zeolites decreases in the order: NaX (760°C) > Ge-NaX (740°C) > Ga-NaX (650°C).

In Fig.2 the curves of differential thermal analysis of the initial and gallium zeolites of ZBK type are presented too. Fig.2 shows that both gallosilicate ZBK and alumosilicate form are characterized by a high thermostability.

Catalytic properties of the gallosilicate and alumogermanate analogues of X and ZBK zeolites were investigated in the dehydration reactions of isopropyl and methyl alcohols.

The catalytic results are given in Table 2. Comparing the activities (Table 2) of the sodium form of the gallium and germanium X zeolites with the initial zeolite we may see that the sodium form of gallium- and germanium-containing X zeolites has not a very high activity. It is shown in Table 2 that the activity of calcium forms of all three zeolites (NaX, Ga-NaX and Ge-NaX) rises with increasing degree of exchange by exchange for calcium. X zeolite and gallium zeolite exhibit a high activity when 70% of sodium has been exchanged by calcium. On the contrary, the alumogermanate zeolite exhibit a high activity at lower degree of sodium exchange by calcium (50%). All these three zeolites (CaNaX, Ga-CaNaX and Ge-CaNaX) show almost the same high activity in isopropanol conversion. In the case of alumosilicate and gallosilicate zeolites, the activity was sta-

TABLE 1: The composition of the reaction mixture and of the zeolite X and analogues

The composition of the reaction mixture, mole/Al_2O_3	Temperature of the synthesis, °C	Phase composition	The composition of zeolite, mole/Al_2O_3	Ion type	Radius of the ion Å	Parameter of u.c. a Å
7,6Na_2O Al_2O_3 5,2SiO_2 500H_2O	78	NaX	$Na_2O Al_2O_3$,6SiO_2,6H_2O	Al^{3+}	0,57	24,86
6,5Na_2O Ga_2O_3 4,5SiO_2 500 H_2O	80	Ga-NaX	Ga-NaX Na_2O Ga_2O_3 24 SiO_2 5H_2O Ga^{3+}	Ga^{3+}	0,62	25,04
5,21Na_2O Al_2O_3 2,2GeO_2 347 H_2O	95	Ge-NaX	Ge-NaX Na_2O Al_2O_3 2GeO_2 5H_2O	Ge^{4+}	0,44	25,51
10Na_2O 9,3TBA Ga_2O_3	120	Ga-ZBK	Ga-ZBK 0,9Na_2O Ga_2O_3 62SiO_2	Ga^{3+}	0,62	a=b=20,2 c=13,5

100 SiO_2 2700 H_2O

TABLE 2: Isopropanol dehydration over zeolite X and analogues[a]

Substitution of sodium by calcium %	CaNaX				Ga-CaNaX				Ge-CaNaX			
	Conversion % initial	Selectivity to propylene	[b]Conversion %	Selectivity to propylene	Conversion % initial	Selectivity to propylene	[b]Conversion %	Selectivity to propylene	Conversion % initial	Selectivity to propylene %	[b]Conversion %	Selectivity to propylene
0	37	74	20	72	30	56	20	53	27	86	18	84
20	79	77	75,2	74	85,7	58	83,6	54	94,6	88	87	82
50	86,6	78	83,4	77	57	57	88,5	56	98	96	81	92
70	92	78	92	75	93	57	90,3	55	93	95	72	91
100	93	77	92	73	92,4	56	90,1	55	94	94	69	91

[a] Temperature of the reaction is 350°C, space velocity is 1 hour^{-1}.
[b] Conversion was taken after a cycle of periodic regeneration, total time of which was 50-60 hours.

ble during the investigation, while the activity of the germanate zeolite decreased with time. As germanium zeolites are less thermostable, they undergo a partial destruction on the heat treatment and activation.

In Table 2 the selectivity to propylene in dehydration of isopropyl alcohol is also presented. From the table it follows that among the investigated zeolites the alumogermanate zeolite exhibits the highest selectivity. Other conditions being the same, on complete substitution of aluminium by gallium the selectivity to propylene yield decreses when compared with alumosilicate zeolite (see Table 2). Different selectivity in the propylene yield can be explained perhaps by the different strength of acidity of the investigated zeolites and by their ability to give rise to secondary reactions. A similar picture is observed in the case of the gallium ZBK. In Table 3, the data on the catalytic properties of alumosilicate and gallium zeolites of the ZBK type are presented. Comparing the activities of the hydrogen forms of gallosilicate zeolites BK and alumosilicate ZBK we may see that the yield of liquid hydrocarbons decreases by 3%. Comparing the compositions of the hydrocarbon fraction of the reaction products we may see that in the case of gallium zeolite more aromatic and isoparaffine hydrocarbons are formed on Ga-ZBK.

Thus, it is established that activity, selectivity and stability of zeolite catalysts depend noticeably on the cationic composition of the zeolite frame.

TABLE 3

The selectivity of methanol conversion over alumosilicate and gallosilicate analogues of BK zeolite

Reaction conditions	ZBK*	Ga-ZBK	The composition of hydro-carbons, %		
				ZBK	Ga-ZBK
T, $^{\circ}$C	380	380	methane	1	–
space velocity, hr^{-1}	2	2	ethane-ethylene	1	–
conversion, %	100	100	propane	15.2	12.1
Reaction products, %			propylene	2	–
Gas	14.5	16.4	i-butane	16.7	19.6
Hydrocarbons > C_5	25	22	n-butane	6.6	3.5
Methanol	0	0	butenes	2.3	1
Water	56	56	hydrocarbons C_5-C_9	14.6	12.6
Coke	2.7	3.6	aromatics	39.8	48

*SiO_2 : Al_2O_3 = 62   (hydrogen forms)

174

REFERENCES

1 К.Г.Ионе Полифункциональный катализ на цеолитах. Изд."Наука" Новосибирск, 1982 , 68-84.
2 G.V. Gibbs, E.P. Meagher, J.V. Smith, J.J. Pluth, in Molecular Sieves. II ACS Symp.Ser., 40 (1977), 19-20.
3 R. Beaumont, D. Barthomeuf, J. Catal., 26 (1972),218-225
4 P. Jacobs, M.Tielen, J.B. Uytterhoeven, J. Catal., 50(1977) 98-108.
5 E. Dempsey, J.Catal., 39 (1975) 155-157.
6 D.W. Breck, G.W. Skeels, Preprints of 6th Intern.Congr.on Catalysis, 2 (1976), 1-155.
7 R.M. Barrer, J.W.Baunham, F.W. Bultitude and W.M. Meier, J. Chem.Soc., 1 (1959) 195-208.
8 L. Lerot, G.Poncelet, J.J. Fripiat, Mater.Res.Bull., 9 (1974) N7, 979-987.
9 J. Selbin and R.B. Mason, J.Inorg.Nucl.Chem., 20 (1961) 222-228.
10 G. Poncelet, M.L. Dubru, G.Somme, L.Lerot, Acta Phys. et Chem. (Szeged), 24 (1978) 273-280
11 Х.Р.Самедов, А.С.Сулейманов, Кинетика и катализ 20 (1979) 1188-1193.
12 Л.С.Маянц, Теория расчета колебаний молекул. Изд.АНСССР, Москва 1960
13 К.Макамато Инфракрасные спектры неорганических и координационных соединений. Изд.Мир. Москва, 1960.
14 D.W. Breck, Zeolite Molecular Sieves, Wiley, New York 1974, p. 493-498.

P.A. Jacobs et al. (Editors), *Structure and Reactivity of Modified Zeolites*
© 1984 Elsevier Science Publishers B.V., Amsterdam — Printed in The Netherlands

STEREOCHEMICAL EFFECTS IN SHAPE-SELECTIVE CONVERSION OF HYDROCARBONS ON ZEOLITES

P.A. JACOBS, M. TIELEN and R.C. SOSA

Laboratorium voor Oppervlaktescheikunde, Katholieke Universiteit Leuven, Kardinaal Mercierlaan 92, B-3030 Leuven (Belgium)

ABSTRACT

A literature overview is made on the conversion of mono-, bi- and tricyclonaphtenes on zeolites and new data are shown on the effect of pore size and geometry on the selectivity in the bifunctional conversion of cyclooctane over faujasite, ZSM-5, and ZSM-11 zeolites.

The strcuture of the pores definitely induces remarkable changes in selectivity although interpretation of them in terms of spatio-stereo shape-selectivity or product diffusion is not always straightforward.

INTRODUCTION

Molecular shape selectivity in zeolites has been used to characterize medium pore zeolites (ref.1) and is also at the basis of several industrial processes as Mobil methanol to gasoline (ref.2), Mobil dewaxing (ref.3), Mobil xylene isomerization, toluene disproportionation (ref. 4), and Mobil-Hoechst polyparamethylstyrene (ref. 5). The molecular selectivity in these processes reflects either effects of spatioselectivity in the formation of the transition state or is the result of mass transport selectivity of products (ref.6-8). In highly active zeolites however, diffusion and desorption may disguise true kinetic effects in the formation of the transition state, even at low conversions (ref. 9). It results that care must be made in elucidating reaction mechanisms from rate data. An obvious method to discriminate between transition state shape selectivity and product diffusion selectivity is to vary the diffusional path length of the reacting molecules or products by changing the crystallite particle size of the zeolite (ref.9).

The molecules studied on this type of zeolites mechnistically undergo the following type of rearrangements : methyl shifts, carbon-carbon bond shifts, cationic oligomerization and cyclisation. For most of these reactions product diffusion is possibly the selectivity determining event. It results that the adaptation of the reaction transition state to the geometry of the pores and pore intersections is not fully exploited this way. Therefore, the aim of this paper is twofold :

- to review the literature which on zeolites treats the stero-conversion of

molecules, as naphtenes, mono-, bi- or tricyclic in nature.
- to report on the chemistry of cyclooctane molecules in particular on shape selective zeolites.

The scope of the work was to investigate whether the adaptation of transition state to pore sizes and geometry of pore intersections is able to give unusual selectivities and to learn the rules governing these conversions. The reactions are not free from industrial interest since the reactants are available as side-products in Reppe-type of synthesis (ref.10).

STATE OF THE ART

Conversion of bicyclo- and tricyclo-naphtenics over zeolites

Faujasite-type zeolites catalyze the gasphase conversion of norbornene (NB) (bicyclo-{2.2.1}hept-2-ene) to nortricyclene (NTC) (tricyclo-{2.2.1.0^{2,6}} heptane) with high selectivity compared to amorphous silica-alumina. The polymer formation and subsequent catalyst deactivation is entirely suppressed when the reaction occurs in the faujasite supercages :

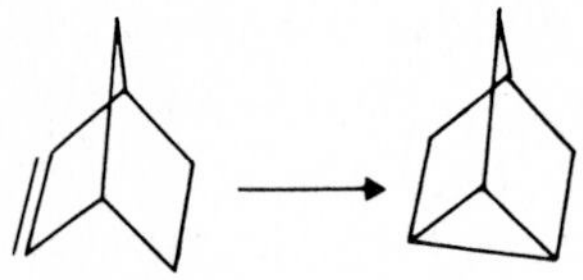

(NB)   **(NTC)**

Some pertinent reaction data (ref.11) are shown in Table 1. It results that the acid-catalysed conversion of the ethylenic bond in NB into a cyclopropane structure constitutes a selective way of synthesizing NTC.

The faujasite supercage seems also particularly well suited for the formation of the adamantane (Ad) structure from tricyclo-{5.2.1.0^{2,6}} decane (TCD) :

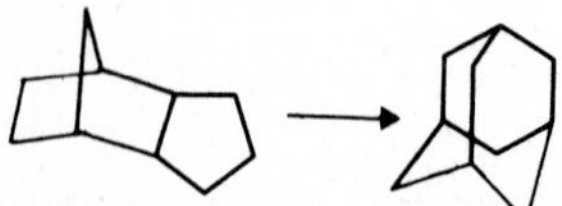

(TCD)   **(Ad)**

TABLE 1

Norbornene (NB) to nortricyclene (NTC) isomerization[1]

Catalyst	GHSV/h	react. temp./K.	NB conversion/%	NTC selectivity/%
silica-alumina	0.25	560	57.7	78.5
η-alumina	0.30	580	64.3	56.6
CaNaX	0.30	630	58.1	96.2
0.5 Pt/MgY	0.30	570	54.3	99.0

[1]after ref. 11.

Via carbocation intermediates, this rearrangement requires four successive alkylshifts, in the right order and at least one occurring over a bridgehead. This is a highly demanding conversion with respect to the geometry of the environment in which the transition state can be formed (ref.12). The advantage of the faujasite-type zeolite is such that again much less polymeric by-products are formed and Ad selectivites up to 50 % are reached on LaY. The formation of carbonaceous deposits is eliminated when a nobel metal is added to this catalyst.

1,3-dimethyladamantane and 1-ethyladamantane can be formed from cyclododecatrienes in a two-step reaction over faujasite-type zeolites (Pt-Re-Ni-LaY) (ref.12).

(AN)

In a first step, at almost 100 % conversion, 80 % yields of acenaphtenes (AN) are obtained (see scheme above). The remaining products however are still polymeric in nature and cause catalyst deactivation. A better tuning of the site geometry to that of the transition state might reduce this. In a second step AN can then be converted with similar selectivities to alkyl- and dimethyladamantanes.

The same alkyladamantanes and in particular 1,3-dimethyladamantane can be synthesized through hydro-isomerization of tetracyclo-$\{6.2.1.1.^{3,6}0^{2,7}\}$ -dodecane (TCDD) over the same catalyst (ref.12) :

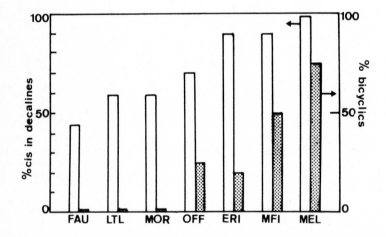

The TCDD feedstock is easily prepared via a Diels-Alder condensation of ethylene and cyclopentadiene and hydrogenation of the intermediate formed this way.

Conversion of monocyclo-naphtenes over zeolites

The conversion of cyclodecane over different acid zeolite structures loaded with a Pt metal phase has been reported (ref.13) as well as a study on the selectivity differences when the same probe is converted via a metal catalyzed or a bifunctional reaction (ref.14).

Fig. 1. Yield of bicyclics from cyclodecane at 80 % conversion (shaded area) and relative content of cis-decaline in decalines at 250°C.

Cyclodecane over bifunctional zeolites is converted into 2 different sets of products :

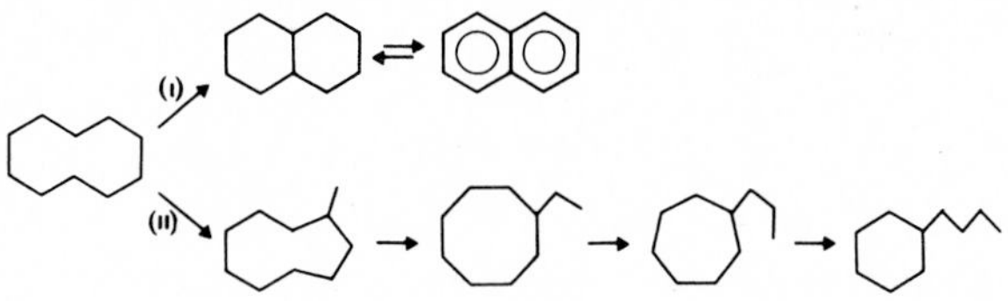

Via an internal ring alkylation of a dienic-type intermediate, cyclodecane is converted to mainly cis-decaline, in hydrogenation/dehydrogenation equilibrium with naphtalene :

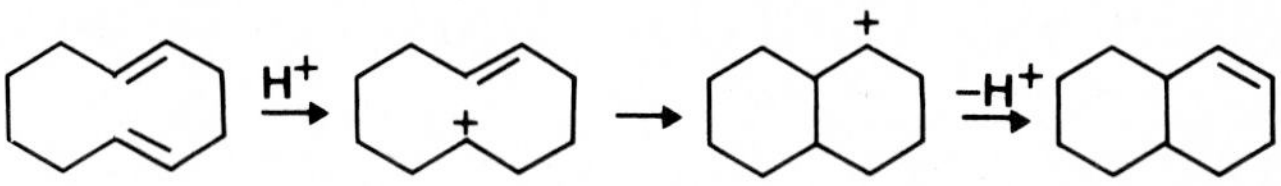

The absolute yield of bicyclics is shown in Fig. 1 for several zeolite structures together with the relative amount of cis-decaline in the bicyclics. Both figures are strongly zeolite-dependent. It seems that internal ring alkylation preferentially occurs in medium pore size zeolites. In such structures geometrical isomerization of the primarily formed cis-isomer for sterical reasons seems to be hindered.

The relative abundance of the alkylcycloalkane isomers formed is also seen to be structure dependent. It is impossible to decide whether this is the result of transition state selectivity more or less disguised by preferential diffusion of some components. Anyway, the data are undoubtedly of diagnostic value for zeolite characterization in catalytic conditions. Mechanistically, the successive ring contraction is initiated via a protonated cyclopropane structure in the naphtenic ring and may then continue via internal alkyl shifts and fast hydride shifts :

TABLE 2

Distribution of monocyclic naphtenes from cyclodecane at 25 % conversion[1]

	FAU	LTL	MOR	OFF	ERI	MFI	MEL
methylcyclononane	32	24	24	41	11	15	23
ethylcyclooctane	50	76	76	46	77	74	77
propylcycloheptane	15	0	0	10	9	8	0
butylcyclohexane	3	0	0	3	3	3	0

[1]after ref. 13.

On acid-free Pt zeolites (ref.14) of the ZSM-5/silicalite-type bis{5.3.0}cyclodecane (azulene)(AZ) is the primary product. It is then converted into decaline and at higher reaction temperatures finally to naphtalene. Mechanistically this can be viewed as a classical metal-catalyzed 1,5-transannular dehydrocyclisation followed by an internal isomerization. The latter may be metal-catalyzed as well as proton-catalyzed by residual Brønsted acid sites :

In dealuminated Y zeolite, the major products are multiply-branched cyclohexanes. Given the high number of this kind of isomers, no mechanistic information can be withdrawn from the product distribution.

BIFUNCTIONAL CONVERSION OF CYCLOOCTANE OVER ZEOLITES

Catalysts and methods used

The acid forms of ZSM-5 and ZSM-11 (denoted as MFI and MEL, respectively) with a Si/Al ratio of 60 and a crystallite size of 4 x 6 μm were used. The OH-stretching spectrum of the sample silanol groups were at 3720 and 3600 cm^{-1}, indicating that extra-lattice material was present (ref.15). A HY* zeolite (further denoted as FAU*) with the same chemical composition was prepared from an ultrastable-Y zeolite using a SiCl$_4$ treatment at 823 K. The samples were impregnated with an aqueous solution of Pt(NH$_3$)$_4$Cl$_2$ so as to obtain a metal loading of 1 % by weight. Catalyst activation was done in situ in the reactor using a calcination in oxygen at 673 K followed by a reduction at the same temperature. Identical samples have been used in the conversion of n-decane (ref.16).

The cyclooctane conversion was done in a continuous flow reactor at a WHSV between 0.1 and 1.5 h^{-1}. Analysis of the reaction products was done using high-resolution gaschromatography. Product identification occurred via GC/MS and was confirmed with the use of reference compounds.

Primary products from cyclooctane over bifunctional zeolites

The Pt/H zeolites used here, in the isomerization and hydrocracking of n-decane behave as ideal bifunctional catalysts (ref.16). Therefore, the results from hydroconversion of cyclooctane will also be interpreted in terms of a classical bifunctional mechanism, in which the rate of carbocation conversion is rate determining. Moreover, the criteria for a bifunctional mechanism are also obeyed here (ref.17) :
- isomerization and hydrocracking are consecutive phenomena
- the isomerization selectivity increases to a maximum value for increasing platinum contents on the zeolites
- on a pure acidic zeolite or on a pure metal on neutral zeolite, the products are distinctly different.

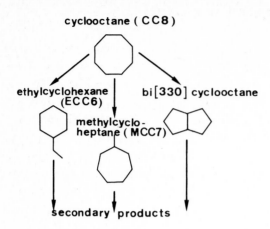

Using the extrapolation method to zero conversion, the eventual primary products from cyclooctane are methylcycloheptane, ethylcyclohexane and bicyclo{3.3.0}octane as shown in the above scheme. Typical is the behavior of bicyclo{3.3.0}octane formation (see Fig. 2). This product is preferentially formed in MEL, compered to its content found in MFI and FAU*. It is hardly probable that this is a case of disguised kinetic behavior, since otherwise a much higher yield of this product should be observed in faujasite. It seems rather a case of stereoselectivity : the geometry of the pore intersections of MEL is needed to stabilize the formation of the product and inhibits its consecutive degradation.

According to the data of Figure 3, ethylcyclohexane is another primary product from cyclooctane as well as methylcycloheptane (see Fig. 4).

In terms of a bifunctional reaction mechanism the formation of bicyclo{3.3.0}octane from cyclooctane can only be explained via an internal alkylation (IA) in a cyclooctene-carbocation. It seems to be a preferential reaction in ZSM-11. The appearance of methylcycloheptane from cyclooctane is easily accounted for via opening of protonated cyclopropane elements in the naphtenic ring, just as explained earlier for the cyclodecane case. By a subsequent internal alkyl shift (IAS) ethylcyclohexyl cations are formed, avoiding in this way the formation of primary carbocations. In FAU* methylcycloheptane apparently is an unstable intermediate. If this would be a case of disguised kinetic behavior (the diffusion of methylcycloheptane being slower than that of ethylcyclohexane), it is expected to be more pronounced in pentasil zeolites which is not the case. It rather seems that the selectivity of the Figures 2, 3 and 4 is in some way related to stereoselectivity. Therefore, it is likely that in MFI, ethylcyclohexane is a primary product from cyclooctane.

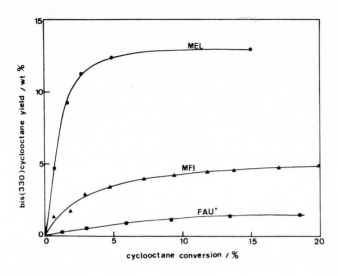

Fig. 2. Yield of the bicyclo{3.3.0}octane formation at increasing cyclooctane conversion

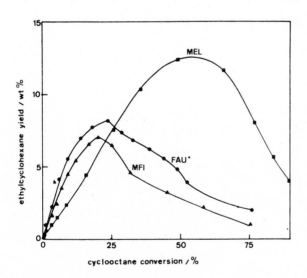

Fig. 3. Yield of ethylcyclohexane from cyclooctane.

184

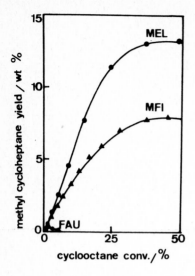

Fig. 4. Methylcycloheptane from cyclooctane

This can only be visualized when protonated cyclobutane intermediates (PCB) exist. The existence of cyclobutyl carbonium ions has already been proposed (ref. 18). All this is schematically represented in the next scheme :

The relative rates of formation of the primary products from cyclooctane, as determined at low conversions are shown in Table 3. These data illustrate the preferential formation of bicyclo{3.3.0}octane over the Pentasil zeolites, more in particular over H-ZSM-11. For the other two products the primary nature could still be considered to be questionable on the basis of the data of Table 3.

TABLE 3

Relative initial rates of formation of primary isomerization products from cyclooctane at 450 K.

Catalyst	Ethylcyclohexane	Methylcycloheptane	Bicyclo{3.3.0}octane
FAU*	7.8	1.0	2.3
MFI	5.0	4.5	15.9
MEL	3.0	6.2	75.0

In order to shed more light on this problem, the conversion rates of both cyclooctane and methylcycloheptane are compared on the three catalysts (see Fig. 5). In faujasite, the disappearance of methylcyclohexane occurs at much lower temperatures than that of cyclooctane, which indicates that ethylcyclohexane is definitely not a primary decomposition product from cyclooctane, but its decomposition occurs via an unstable intermediate, namely methylcycloheptane. On the two pentasil-zeolites (MFI and MEL) both reactants show a very similar temperature dependency. This in our opinion is strong evidence for ethylcyclohexane being a primary reaction product on this zeolite and indicates that possibly PCB-structures are formed in these medium pore zeolites.

186

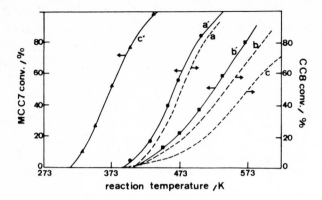

Fig. 5. Conversion of methylcycloheptane (MCC7) and cyclooctane (CC8) at different reaction temperatures on a,a' : MEL; b,b' : MFI and c,c' : FAU*.

CONCLUSIONS

This overview of a few literature data indicate that all zeolites because of their tri-dimensional pores or pore intersections are particularly suitable for selective conversion of mono-, bi-, and even tricyclonaphtenic molecules. There are indeed stereo-effects which direct a conversion into a particular pathway or allow new unusual products to be formed.

On the other hand, the systems are not easily approached mechanistically since stereo-effects may be obscured by so-called disguised kinetics : secondary products which diffuse faster than primary, but more bulky ones may always have a primary appearance.

Nevertheless, it seems that zeolites are much more selective than more open classical catalysts in synthesis of tridimensional chemicals and merit a more systematic study.

ACKNOWLEDGMENTS

The authors acknowledge financial support from the Belgian Government in the frame of a "Concerted Action on Selective Catalysis". P.A.J. is grateful to N.F.W.O. (Belgium) for a position as "Onderzoeksleider". R.C.S. acknowledges a grant from the K.U.Leuven.

REFERENCES

1 P.B. Weisz, in T. Seiyama, K. Tanabe (Eds.), Catalysis A, Kodasha-Elsevier Scientific, 1980, p.3.
2 C.D. Chang, Catal. Rev. Sc. Eng., 25(1983) 1.
3 N.Y. Chen, R.L. Gorring, H.R. Ireland and T.R. Stein, Oil Gas J., (June 6, 1977) 75.
4 F.G. Dwyer, P.J. Lewis and F.M. Schneider, Chem. Eng., 83(1976) 90.
5 W.W. Kaeding, L.B. Young and A.G. Prapas, Prepr. A.C.S. Div. Fuel Chem., 28(2)(1983) 158.
6 E.G. Derouane, in M.S. Whittingham and A.J. Jacobson (Eds.), Intercalation Chemistry, Academic Press, 1982, p.101.
7 S.G. Scissery, Prepr. A.C.S. Div. Fuel Chem., 28(2)(1983) 116.
8 H. Heinemann, Catal. Rev. Sc. Eng., 23(1981) 315.
9 W.O. Haag, R.M. Lago and P.G. Rodewald, J. Molec. Catal, 17(1982) 161.
10 K. Weissermel and H.J. Arpe, in "Industrial Organic Chemistry", Verlag Chemie, 1978, p. 214.
11 K.L. Moll, K. Gloede, C.E. Ruerup and M. Weber, D.D.R. Patent 158, 391(VEB Leuna-Werke) 1983.
12 K. Honna, H. Ichikawa and M.Sugimoto, Prepr. A.C.S. Div. Petrol. Chem., Honolulu Meeting, 1979, p.304.
13 P.A. Jacobs and M. Tielen, Proceed. 8th Int. Conf. Catal., Berlin, 1984, submitted for publication.
14 P.A. Jacobs, M. Tielen and H.K. Beyer, Proceed. Western-Europe-U.S. symposium on the Chemistry in Intracrystalline Environments, Bruges, 1983, J. Molec. Catal., (1984) submitted for publication.
15 P.A. Jacobs and R. Von Ballmoos, J. Phys. Chem., 86(1982) 3050.
16 P.A. Jacobs, J.A. Martens, J. Weitkamp and H.K. Beyer, Disc. Faraday Soc., 72(1982) 353.
17 P.A. Jacobs, R.S. Sosa and M. Tielen, unpublished results.
18 A. Maccoll, "The Transition State", Chem. Soc. Spec. Publ., 16(1962) 166.

P.A. Jacobs et al. (Editors), *Structure and Reactivity of Modified Zeolites*
© 1984 Elsevier Science Publishers B.V., Amsterdam — Printed in The Netherlands

SHAPE-SELECTIVE EFFECTS ON THE METAL CATALYZED REACTIONS OF PREHNITENE OVER ZEOLITES

J.A. MARTENS and P.A. JACOBS

Laboratorium voor Oppervlaktescheikunde, Katholieke Universiteit Leuven, Kardinaal Mercierlaan 92, B-3030 Leuven Heverlee (Belgium)

ABSTRACT

The temperature dependent conversion of prehnitene (1,2,3,4-tetramethylben-zene) was followed over alumina with (2.5 nm) pores, NaY and silicalite loaded with 3.3 and 1 % by weight of Pt, respectively. The hydroconversion was carried out in a continuous tubular flow reactor with on-line high resolution GC fa-cilities. From the product distribution details on the reaction network and its dependence on the pore dimensions of the support are derived.

At low temperatures a a-specific stereoselective hydrogenation is followed by a support dependent geometric cis-trans isomerization. At higher temperatures prehnitene is isomerized via subsequent methyl shifts. On silicalite effects of product shape selectivity are apparent. At the same time a parallel dealkyla-tion reaction occurs which is a-selective.

All these products are obtained mainly via metal catalysis, although on NaY residual acidity adds products from acid-catalyzed isomerizations of naphtenic and aromatic intermediates.

INTRODUCTION

Zeolites exhibit some control of reactants and products through their catalytic cavities which are of molecular dimensions. These shape selective effects have been recognized and described in the case of acid catalyzed reactions. Distinction has been made between reactant, product and restricted transition state shape selectivity (ref.1,ref.2). Reactant shape selectivity was first described on a Linde 5A zeolite : it was shown that n-butanol could be dehydrated in the presence of isobutanol (ref.3).Enhanced selectivities for p-xylene are a result of product shape selectivity in reactions such as xylene isomerization and toluene disproportionation on H-ZSM-5 (ref.4). Restrictions in the nature of the transition state prevent the further conversion of the p-xylene to polyaromatics and coke.

In the methanol-to-gasoline process, the synthesis of hydrocarbons over H-ZSM-5 is restricted to the gasoline range, with a sharp cut-off in the products at the level of tetramethylbenzenes. In this fraction of products, the selectivity for durene is far in excess of its equilibrium value, while the yield of the more bulky isodurene and prehnitene is far below this value. On 12-ring zeolites the equilibrium between the tetramethylbenzenes is observed (ref. 5). The pore dimensions of pentasil zeolites exclude bulky molecules such as 1,2,3,4-tetramethylbenzene. Its hydrogenation products, however, are expected to be more flexible and may have access to the intracrystalline catalytic sites. It has been observed that trimethylcyclopentanes formed through a cyclization reaction on Pt particles, external to the zeolite crystals, enter easily the H-ZSM-5 pores, while the feed molecule doesnot (ref.6). Shape selective effects which occur during the dehydrocyclyzation of n-decane and aromatization of cyclodecane over Pt metal, finely disperced in silicalite and non-acidic faujasite have been reported recently (ref.7).

In this work, the hydrogenation, isomerization and degradation of 1,2,3,4-tetramethylbenzene is investigated on Pt metal, disperced in the pores of silicalite. In order to certify that shape selective effects are observed, the reaction products are compared with those obtained over Pt dispersed on a large cage zeolite, NaY and on a non-acidic γ-alumina.

EXPERIMENTAL

Materials

The following materials were used as supports for the Pt metal :
- Linde NaY zeolite after treatment with 0.1 M NaCl aqueous solution.
- Silicalite, synthetized at 150°C for 96 h using a diluted gel with the following molar composition : (tetrapropylammoniumbromide)$_{0.83}$ (SiO$_2$)$_{0.33}$ (NaOH)$_{0.13}$ (NH$_3$)$_{0.13}$ (H$_2$O)$_{48}$ (Glycerol)$_{6.6}$.
- Spheralite, a γ-alumina from Rhône-Poulenc with a surface area of 350 m^{3}g^{-1} and mean pore diameter of 2.5 nm.

For these three supports the following abbreviations are used throughout the text : NaY, SIL for silicalite and SPHE for the γ-alumina.

Pt(NH$_3$)$_4$Cl$_2$ from Degussa was impregnated on these supports as to obtain respectively 1 % by weight of metal on silicalite and 3 % on NaY and spheralite. The catalyst powders were compressed, crushed and sieved. The 0.3-0.5 mm fraction of the pellets was retained for use in the reactor. Prehnitene (1,2,3,4-tetramethylbenzene) was purchased from Aldrich and had a purity of 97 %. The main impurities were durene (1,2,4,5-tetramethylbenzene) and isodurene (1,2,3,5-tetramethylbenzene).

Methods

The catalytic tests were done in a fixed bed continuous flow reactor in H_2 at atmospheric pressure. The catalysts were pretreated in flowing O_2 at 400°C and reduced with H_2 at the same temperature. The W/F_o ratio for prehnitene in all reactions was 1040 kg s mol^{-1}, W being the amount of catalyst in kg and F_o the flow rate of the feed in mol s^{-1}. The reactor outlet was analysed using a Hewlett Packard 5830A GC equipped with a fused silica capillary column of 50 m length, 0.3 mm internal diameter and coated with cPtSil5 from Chrompack. The product distributions were corrected for the durene-isodurene impurities in the feed by substraction.

RESULTS AND DISCUSSION

Reaction Selectivity

Fig. 1 shows for the three catalysts the product yield from prehnitene at different reaction temperatures. Up to 200°C, prebnitene is completely hydrogenated into naphtenes over the Pt on alumina and NaY catalysts. On the Pt on silicalite, only 35 % of the prebnitene is hydrogenated under these conditions. This is possibly the result of either the lower Pt loading on this sample or the reduced accessibility of the feed to the Pt particles located in the zeolite pores. The direct hydrogenation products of prehnitene are the 1,2,3,4-tetramethylcyclohexanes which constitute the main fraction of the naphtenes.

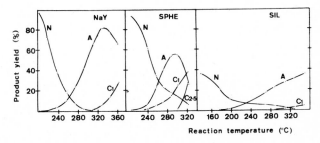

Fig. 1. Overall product yield from prehnitene over 3 % Pt on NaY, 3 % Pt on spheralite-γ-alumina and 1 % Pt on silicalite : N, C_{10}-cyclohexanes, A, aromatics, C_1, methane and C_{2-5}, light paraffins from ethane to pentane.

Apart from these 1,2,3,4-tetramethylcyclohexanes, other C_{10}-naphtenic products are formed at 200°C which amount to 8, 14 and 23 % for the Pt on alumina, silicalite and faujasite-type supports, respectively. The 8 % of C_{10}-naphtenes on alumina contain mainly tetramethycyclohexanes which can be derived from the primary hydrogenation products via a metal catalyzed methyl-shift. This yield of C_{10}-naphtenes other than 1,2,3,4-tetramethylcyclohexanes is enhanced on silicalite and becomes even significant on NaY. It is not only composed of tetramethylcyclohexanes but contains also ethyl- and propyl-branched C_{10}-cyclohexanes. These differences in our opinion can be accounted for if it is assumed that an acid-catalyzed ring contraction-expansion mechanism is operative next to a metal-catalysed methyl-shift. This bifunctional isomerization of 1,2,3,4-tetramethylcyclohexanes occurs relatively fast on NaY. It results that all the precautions taken to get rid of residual acidity in NaY (NaCl treatment and $Pt(NH_3)_4Cl_2$ impregnation) have not been sufficient. The residual acidity of the silicalite material is caused by the incorporation of aluminum present at impurity levels in the reactants used to synthetize the material.

From 260°C on, aromatics are the main reaction products in all cases. At still higher reaction temperatures more and more methane is formed on alumina and faujasite resulting from a hydrodealkylation reaction. Fig. 2 shows the composition of this aromatic fraction more in detail. Pt on alumina is a very active hydrodealkylation catalyst, producing trimethylbenzenes, xylenes, toluene and benzene successively. In contrast, Pt on NaY and on silicalite shows much less hydrodealkylation activity, since durene and isodurene are then the main aromatics. The succession of events indicates that for Pt on NaY and Pt on silicalite a dealkylation reaction is also the main degradation route. It remains to be proven whether this route is also metal-catalyzed or occurs partially via carbocation intermediates.

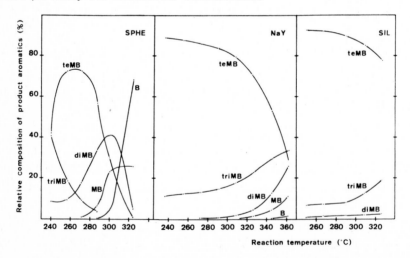

Fig. 2. Relative composition of the aromatics from prehnitene on 3Pt/NaY, 3Pt/spheralite and 1Pt/silicalite : teMB, tetramethylbenzenes; triMB, trimetehylbenzenes; diMB, dimethylbenzenes; MB, methylbenzene; B, benzene.

Cis-trans isomerization of 1,2,3,4-tetramethylcyclohexanes

The chromatogram of the hydrogenation products of prehnitene shows 5 peaks. Based on GC/MS data, boiling points and equilibrium compositions (ref.8), an assignment of these peaks to the six 1,2,3,4-tetramethylcyclohexane stereoisomers was possible.

In Fig. 3, the composition of these isomers is plotted against the reaction temperature for the three catalysts. It can be deduced that prehnitene is hydrogenated firstly into 1,2,3,4-tetramethyl-cis,cis,cis-cyclohexane. The latter molecules at higher temperatures on Pt on alumina and Pt on NaY are then completely transformed into the other isomers. Sterochemical effects during hydrogenation are known (ref.9). It seems that whenever different reaction products can be expected from the cis or trans addition of 2 H atoms, the cis addition is always favored since adsorption of the reactant on the substrate occurs either via α,β-diadsorbed species or via π-adsorbed complexes. Given also the data of Fig. 3, it is logical to assume that 1,2,3,4,-tetramethyl-cis,cis,cis-cyclohexane is the primary hydrogenation product in all cases. It follows that this reaction sequence is not influenced by the pore size nor the structure of the supports.

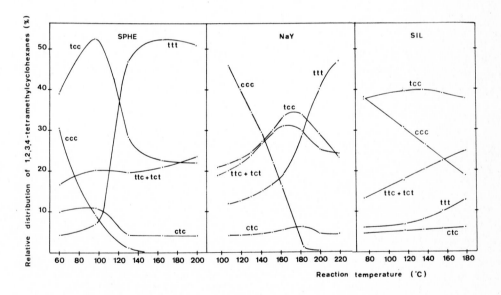

Fig. 3. Relative distribution of the geometric isomers of 1,2,3,4-tetramethyl-cyclohexane after hydrogenation of prehnitene : c, cis; t, trans.

Above 200°C the composition of the 1,2,3,4-tetramethylcyclohexanes attains equilibrium over Pt on alumina and Pt on NaY. Typical data are shown in Table 1 to illustrate this. It is known that the thermodynamic stability of an isomer decreases with an increase of the number of axially oriented substituents on the chair cyclohexane ring (ref.8).

TABLE 1

Relative distribution at 220°C of the 1,2,3,4-tetramethylcyclohexane isomers from 1,2,3,4-tetramethyl-cis,cis,cis-cyclohexane by geometric cis-trans isomerization

catalyst	isomer			
	ctc[a]	tcc	ttc+tct	ttt
SPHE	4.2	22.0	23.3	50.5
NaY	4.5	23.6	24.4	47.5
Sil	7.0	34.9	34.9	23.2

a : t, trans; c, cis

On silicalite, the equilibrium value is only reached for the ctc isomer. For the others strongly deviating values are obtained, which are probably the result of both restricted transition state and product shape selectivity.

Using a minimum number of steps, the cis-trans isomerization of 1,2,3,4-tetramethyl-cis,cis,cis-cyclohexane can be rationalized with the following scheme :

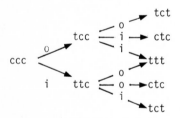

On the three supports, cis-trans isomerization from the ccc-isomer gives the tcc-isomer in excess. This indicates that isomerization in which an "outer" methylgroup (in position 1 or 4) is involved is clearly a more favored route than isomerization of an "inner" methylgroup (position 2 or 3) to produce ttc.

Over Pt on NaY, the overall cis-trans isomerization rate is slowed down compared to Pt on alumina as shown by the higher reaction temperatures needed to isomerize the ccc-isomer. Given its smaller pore size and lower Pt content, it could have been anticipated that Pt on silicalite would still be less active in this cis-trans isomerization reaction, as is shown in Fig. 3.

Isomerization of prehnitene

At higher reaction temperatures (Fig.1), the 1,2,3,4-tetramethylcyclohexane isomers are dehydrogenated again and isomerization of 1,2,3,4-tetramethylbenzene via methyl-shifts as well as dealkylation become important (Fig.2). Successive methyl-shifts are expected to give the following product sequence :

    prehnitene → isodurene → durene

    (1,2,3,4-)   (1,2,3,5-)   (1,2,4,5-tetramethylbenzene)

In Fig. 4, the durene:isodurene ratio is plotted as a function of the reaction temperature for the three catalysts. Over Pt on silicalite, this ratio remains allways above thermodynamic equilibrium, although the production of durene from prehnitene has to proceed via isodurene. Therefore, this two step methyl-shift has to occur at the level of the pore intersections of this zeolite, but desorption of the most bulky isomer is slowest. Indeed, the kinetic diameter for durene is the same as for o-xylene, and is much smaller than for isodurene and prehnitene. The preferential formation of durene over Pt on silicalite is therefore the result of product shape selectivity and consequently the catalysis occurs mainly inside the silicalite pores.

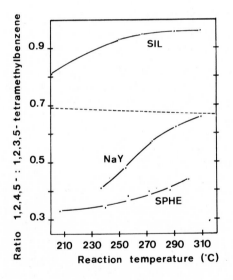

Fig. 4. Ratio of the positional isomers from 1,2,3,4-tetramethylbenzene. The dashed line represents this ratio at thermodynamic equilibrium

Over Pt on NaY, durene and isodurene are equilibrated at higher reaction temperatures. The continuously increasing ratio of the 2 isomers (Fig.4) clearly indicates that the mechanism is a step-by-step methyl-shift.

However, the fact that on alumina this equilibrium seems to be approached much slower indicates that between NaY and alumina mechanistic differences are operative : indeed, as already suspected from the overall distribution of the C_{10}-cyclohexanes, it might well be that on NaY a fast acid-catalyzed methyl-shift is added to the metal-catalyzed shift.

Dealkylation of tetramethylbenzenes

In Fig. 5 the relative amount of 1,2,3- and 1,2,4-trimethylbenzene are plotted as a function of the reaction temperature. For the three supports, the yield of 1,2,3-trimethylbenzene is found to be largely in excess of the thermodynamic equilibrium value. 1,2,4-trimethylbenzene is below its equilibrium value as well as 1,3,5-trimethylbenzene, as can be deduced from Fig. 5.

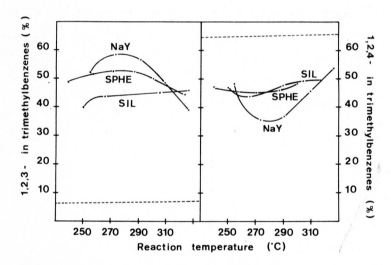

Fig. 5. Relative distribution of the trimethylbenzenes after dealkylation of the tetramethylbenzenes. The dashed line shows the equilibrium composition.

Assuming equal probability for each methylgroup to be removed from the aromatic ring, the following trimethylbenzenes can be obtained :

1,2,3,4-tetramethylbenzene
→ 1,2,3-trimethylbenzene
→ 1,2,4-trimethylbenzene

1,2,3,5-tetramethylbenzene
→ 1,2,4-trimethylbenzene
→ 1,2,3-trimethylbenzene
→ 1,3,5-trimethylbenzene

1,2,4,5-tetramethylbenzene ⟶ 1,2,4-trimethylbenzene

Dealkylation of 1,2,3,4-tetramethylbenzene should then give equal amounts of 1,2,3- and 1,2,4-trimethylbenzene. This is actually observed for the three supports, despite the higher amounts of isodurene and durene produced over Pt on NaY and Pt on silicalite (Fig.2). Clearly dealkylation and isomerization are parallel reactions and equilibration within the group of trimethylbenzenes via methyl-shifts is very slow. This is in line with recent work (ref.10), where it is shown that trimethylbenzenes do not isomerize extensively over Pt on non-acidic alumina.

Under the present reaction conditions, acid catalyzed equilibration is expected to be much faster and as a result it doesnot seem probable that this reaction is catalyzed on the residual acidity of NaY and silicalite.

Dealkylation of trimethylbenzenes

It was reported recently that the demethylation products from trimethylbenzenes reflect demethylation without isomerization and without any positional selectivity (ref.10).

Assuming equal probability for the dealkylation of each methylgroup of the trimethylbenzenes, the following xylenes should be obtained :

```
                        o-xylene
1,2,3-trimethylbenzene
                        m-xylene


                        o-xylene
1,2,4-trimethylbenzene
                        p-xylene
```

1,3,5-trimethylbenzene ⟶ m-xylene

Given the actual concentrations of these trimethylbenzenes on the three catalysts, o-xylene yields of at least 50 % should be observed. In Fig. 6, the percentage o-xylene in the total xylene fraction is given as a function of the reaction temperature. In all cases, more than 50 % o-xylene is produced. Moreover, these yields can be predicted using the relative yields of trimethylbenzenes and assuming equal probability for dealkylation.

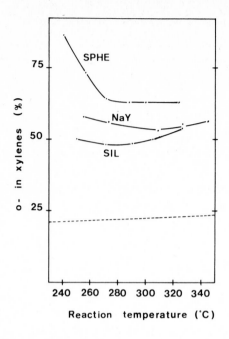

Fig. 6. The relative abundance of o-xylene in the xylenes after dealkylation of the tetramethylbenzenes. The dotted line gives the thermodynamic amounts.

CONCLUSIONS

The overall reaction network of 1,2,3,4-trimethylbenzene on alumina, NaY and silicalite impregnated with Pt is very similar. At low reaction temperatures a stereoselective hydrogenation consisting of a cis-addition is observed. Subsequently, the 1,2,3,4-tetramethyl-cis,cis,cis-cyclohexane is isomerized via a selective cis-trans isomerization involving the methyl-groups in position 1 and 4. The relative distribution of these isomers is dependent on the support porosity, indicating that restricted transition state or product shape selectivity come into play. It also indicates that the catalysis is mainly going on in the pores of the support, which at least for silicalite is not a priori predictable. On NaY, residual acidity causes secondary isomerization of these primarily formed tetramethylcyclohexanes.

At higher temperatures, the tetramethylbenzenes are equilibrated step-by-step via metal catalyzed methyl-shifts. On NaY this equilibration is faster as a result of the low residual acidity present. On silicalite, an excess of durene is formed via product shape selectivity. At the same time, a parallel gradual dealkylation reaction occurs, which is non-selective since it attacks each of the methyl-group positions with equal probability.

ACKNOWLEDGMENTS

J.A.M. and P.A.J. acknowledge the Belgian National Fund of Scientific Research for a grant as Research Assistant and Senior Research Associate. A grant for the study of "shape selective catalysis" from the same institution and from the Belgian Government (Service of Scientific Affairs) is highly appreciated.

REFERENCES

1 S.M. Ciscsery, in J.A. Rabo (Ed.), A.C.S. Monograph N°171, "Zeolite Chemistry and Catalysis", A.C.S., Washington D.C., 1976, 680.
2 P.B. Weisz, Pure & Applied Chemistry, 52 (1980) 2091.
3 P.B. Weisz, V.J. Frilette, R.W. Maatman, E.B. Mower, J. Catal., 1 (1962) 307.
4 E.G. Derouane, in B. Imelik et al. (Eds.), Catalysis by Zeolites, Elsevier Scientific Publ. Co., Amsterdam, 1980, 5.
5 C.D. Chang, Hydrocarbons from Methanol, Chemical Industries vol. 10, Marcel Dekker Inc., New York, 1983, 16.
6 P.A. Jacobs, J.A. Martens, J. Weitkamp and H.K. Beyer, Faraday Disc. Chem. Soc., 72 (1982) 373.
7 P.A. Jacobs, M. Tielen and J. Martens, J. Mol. Catal., to be published.
8 S.S. Berman, V.A. Zakharenko, Al.A. Petrov, Neftekhimiya, 7(6) (1957) 850.
9 J.E. Germain, Catalytic Conversion of Hydrocarbons, Academic Press, New York, 1969, p. 68.
10 B.H. Davis, J.K. Shou, Applied Catalysis, 1 (1981) 277.

P.A. Jacobs et al. (Editors), *Structure and Reactivity of Modified Zeolites*
© 1984 Elsevier Science Publishers B.V., Amsterdam — Printed in The Netherlands

RELATION BETWEEN ACIDIC PROPERTIES OF ZSM-5 AND CATALYST PERFORMANCE OF METHANOL CONVERSION TO GASOLINE

T. INUI, T. SUZUKI, M. INOUE, Y. MURAKAMI, and Y. TAKEGAMI

Department of Hydrocarbon Chemistry, Faculty of Engineering, Kyoto University, Sakyo-ku, Kyoto 606 (Japan)

ABSTRACT

Various H-ZSM-5's having different acidic properties were prepared by changing the Si/Al ratio, the proton-exchange method, and the degree of proton exchange. Strong acid sites increased in number with increases of the Al content and the degree of proton exchange. The ZSM-5 with higher concentration of strong acid sites showed longer life time and gave larger amount of gasoline fraction. However, the ratio of total carbon atoms found in the product until catalyst deactivation to the carbon atoms in the coke was constant irrespective of acidic properties of the zeolite.

INTRODUCTION

Since the excellent catalytic performance of ZSM-5 for methanol conversion

to gasoline was found (ref.1), many studies have been carried out on this novel

type of zeolite. The Al content in zeolite is usually assumed to be closely

associated with acidic properties, and it has been reported that the hexane-

cracking activity of ZSM-5 is proportional to the Al content (refs.2,3), and

that the product selectivity in methanol conversion depends on the Al content

(refs. 4,5). However, relation between acidic properties and catalyst life

(another catalyst performance) is left still unresolved. In order to obtain

detail insight into the relation between acidic properties of ZSM-5 and its

catalytic performance, various ZSM-5 catalysts having different acidic proper

ties were synthesized by changing (a) Si/Al ratio, (b) proton-exchange reagent,

and (c) degree of proton-exchange, and effects of acidic properties on the

catalyst performance were explored.

EXPERIMENTAL

The ZSM-5 zeolites were prepared from solution containing tetrapropyl ammo

nium bromide according to the procedure described elsewhere (ref.6). Synthesized

zeolite was calcined at $540^{\circ}C$ for 5 h. The (Na,H) form of the zeolite was then

ion-exchanged as follows (unless otherwise mentioned); the calcined zeolite was

soaked in 1 M solution of ammonium nitrate, washed with water, dried at $90^{\circ}C$,

and calcined at $500^{\circ}C$ for 5 h.   Concentrations of Na and Si were determined by an atomic adsorption spectrometer (Shimazu AA-640-01) after dissolving the zeolite using hydrofluoric acid. The Si/Al ratio in zeolite was calculated taking 96(Si+Al) atoms per unit cell, and assuming that there is no significant contribution from other amorphous or crystalline impurity phases. Number of zeolitic water molecules per unit cell was arbitrarily assumed to be ten. Temperature-programed desorption (TPD) of ammonia was recorded on a thermo-gravimetric analyzer (Rigaku TG-8002). A weighed amount (150 mg) of sample was placed on a Pt cell and dried by heating up to 350°C. After cooling down to 80°C in a N_2 stream, the zeolite was allowed to saturate with NH_3 at that tem-perature using 5.1% NH_3 diluted with N_2. Following elution with N_2 at 80°C, the temperature was raised at a constant heating rate of 5°C/min up to 630°C. The NH_3 TPD profile was obtained by numerical differential of the weight loss curve. Methanol conversion reaction was carried out in an ordinary flow reactor at atmospheric pressure. A 0.2 g portion of the catalyst (tableted and then crashed into 15-20 mesh) was placed in a Pyrex tubular reactor. Two experimental condi-tions were used: (1) To determine the catalyst selectivity, a mixture of metha-nol (20 mol%) and N_2 was fed at 360°C and GHSV = 1000 h^{-1}; (2) to determine the catalyst life, methanol without diluent was fed at 375°C and WHSV = 7 h^{-1}. Products were analyzed by gas-chromatography.   The amounts of coke deposited in the deactivated catalysts were determined by the weight loss of the sample in a flow of air at 300°C after complete removal of volatile materials in a N_2 stream at 150°C.

RESULTS AND DISCUSSION
Effects of Si/Al ratio on acidic properties

Figure 1 illustrates the TPD pro-files of NH_3 from H-ZSM-5's having various Si/Al ratios. Two peaks at 120 and 460°C with a dividing mini-mum at 330°C were observed corre-sponding to adsorption on weak and strong acid sites, respectively. Three peaks in the NH_3 TPD profile were reported (refs.8-10); however, a peak corresponding to medium acid sites was not detected by our exper-iments. This may be attributed to relatively high Si/Al ratio of our samples. Absolute concentrations of weak and strong acid sites were

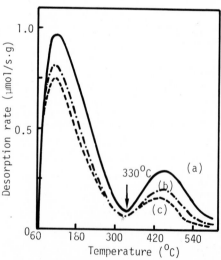

Fig. 1. Typical profiles of TPD of NH_3 from H-ZSM-5. (a) Si/Al=35.9, (b) Si/Al =55.5, (c) Si/Al=67.6 (see TABLE 1).

TABLE 1

Effect of Si/Al ratio on acidic property of H-ZSM-5

Si/Al charged ratio	Chemical analysis		Si/Al found	Amount of desorbed ammonia		
	Na (atom/uc)	Al (atom/uc)		below 330°C (mmol/g)	above 330°C (mmol/g)	total (mmol/g)
50	0.1	2.6	35.9	1.81	0.59	2.46
100	0.0	1.7	55.5	1.45	0.39	1.84
200	0.0	1.4	67.6	1.33	0.26	1.59
400	0.1	0.4	239.0	0.75	0.04	0.78

calculated from the weight loss of the sample below and above 330°C, respectively. Results are summarized in Table 1. Both concentrations of weak and strong acid sites increased with the increase of Al content (with the decrease of Si/Al ratio). The concentration of strong acid sites was proportional to the Al content. However, the number of the strong acid sites per unit cell was 40% larger than the theoretical number of acid sites calculated from the Si/Al ratio. It was reported that weak acid sites do not depend on the Si/Al ratio (ref.4); however our results show that the concentra-

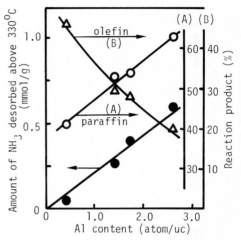

Fig. 2. Effects of Al content on strong-acid site concentration and on product selectivity.

tion of weak acid sites also increased with the increase of Al content. There-fore, it seems to be reasonable to assign it to interaction of NH_3 molecules with frame oxygen atoms by hydrogen bonding, because an increase of Al content in the zeolite network results in an increase of negative charge density of frame oxygen atoms.

Figure 2 shows the relation between the Si/Al ratio and the product distri-bution for methanol-to-gasoline (MTG) reaction, together with the effect of the Si/Al ratio on the concentration of strong acid sites. As can be seen in Fig.2, strong acid sites seems to promote the formation of paraffins from olefins pre-sumably by a process of oligomerization of olefin followed by hydrocracking of the oligomers (ref.4). As the concentration of weak acid sites also increased with increasing Al content, effects of weak acid sites on the product selectiv-ity may be considered; however, this could be ruled out by the experiment of changing the degree of proton exchange (vide infra).

TABLE 2

Effect of proton-exchange reagent on acidic property of H-ZSM-5

Catalyst	Reagent for proton exchange	Chemical analysis		Amount of desorbed ammonia		
		Na (atom/uc)	Al (atom/uc)	below 330°C (mmol/g)	above 330°C (mmol/g)	total (mmol/g)
Na-ZSM-5		2.5	2.5	19.2	0.28	2.20
H-ZSM-5(AN)	1.0N NH$_4$NO$_3$	0.0	2.5	1.69	0.83	2.52
H-ZSM-5(N)	1.0N HNO$_3$	0.0	2.4	1.94	0.54	2.48
H-ZSM-5(ACl)	2.0N NH$_4$Cl	0.1	2.4	1.95	0.64	2.59
H-ZSM-5(Cl)	2.0N HCl	0.0	2.4	1.90	0.61	2.51
H-ZSM-5(AS)	2.0N (NH$_4$)$_2$SO$_4$	0.0	2.5	1.93	0.63	2.56
H-ZSM-5(S)	2.0N H$_2$SO$_4$	0.0	2.3	1.79	0.74	2.53

Effects of the reagent for ion-exchange

Six reagents were examined for ion-exchange of sodium form of ZSM-5 zeolite. All the reagents gave more than 95% proton-exchanged zeolites, and slight de-alumination was observed (less than 10%), especially when free acids were used. Acidic properties of the proton-exchanged zeolite varied significantly by the reagent used (Table 2). The use of NH$_4$NO$_3$ gave the zeolite with highest concentration of the strong acid sites. The use of HCl or NH$_4$Cl afforded relatively small amounts of strong acid sites, which may be due to the chloride ion remaining in the zeolite network.

The product distribution in MTG reaction is given in Table 3, and carbon number distributions of the unsaturated products are shown in Fig. 3. In the figure, relative yield to ethylene is given since the absolute yield of

TABLE 3

Effect of proton-exchange reagent on MTG reaction

Catalyst	Strong-acid site conc. (number/uc)	Product distribution			
		aromatics (wt%)	olefin (wt%)	paraffin (wt%)	gasoline (wt%)
Na	1.68	16.7	35.0	48.4	49.1
H-ZSM-5(AN)	4.94	27.9	20.5	51.6	51.5
H-ZSM-5(N)	3.21	20.6	22.1	57.3	46.5
H-ZSM-5(ACl)	3.81	25.2	19.5	55.3	50.9
H-ZSM-5(Cl)	3.62	21.4	22.5	56.1	48.6
H-ZSM-5(AS)	3.75	22.8	21.2	56.0	47.3
H-ZSM-5(S)	4.40	27.1	20.7	52.2	51.6

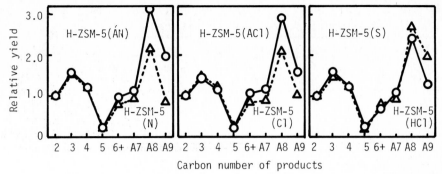

Fig. 3. Effects of the reagent for proton exchange on carbon number distribution of unsaturated product.

ethylene scarcely depends on the ion-exchange reagent (ca. 4.4%). As can be seen in Fig. 3, the product distribution in C_2-C_5 olefins was essentially constant, while the yield of aromatic compounds varied by the proton-exchange reagents. The yield of aromatic products increased with an increase of strong-acid site concentration (see Fig. 4). On the other hand, the yield of paraffinic products decreased as the concentration of strong acid site increased, and the yield of

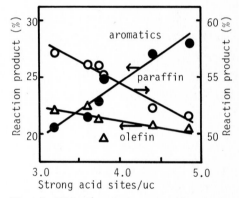

Fig. 4. Effects of strong-acid concentration on MTG reaction (see Table 3).

olefins was little affected by the ion-exchange reagents. These results show a sharp contrast against the results shown in Figs. 2 and 5, where an increase of the strong-acid site concentration resulted in an increase of the yield of paraffins, a decrease of the olefin yield, and a slight increase of the yield of aromatic products. The cause of this descrepancy is not clear and remains for further studies.

Effects of proton-exchange degree on acidic properties

Six zeolites having different degrees of proton exchange were prepared from the same lot of sample by changing the concentration of NH_4NO_3 (Table 3). At the high proton-exchange degree, slight dealumination was observed. The concentration of strong acid sites increased linearly with an increase of proton-exchange degree, while the total acid concentration remained constant irrespective of the proton-exchange degree, which indicates that a part of weak acid sites in the sodium form was converted into strong acid sites by proton exchange.

TABLE 4

Effect of proton-exchange degree on acidic property of ZSM-5

Reagent for ion-exchange	Chemical analysis		proton exchange (%)	amount of desorbed ammonia		
	Na (atom/uc)	Al (atom/uc)		below 330°C (mmol/g)	above 330°C (mmol/g)	total (mmol/g)
1.0M NaNO$_3$	2.9	2.9	0	1.59	0.34	1.93
as calcined	2.4	2.9	17	1.58	0.43	2.01
0.01M NH$_4$NO$_3$	0.9	2.9	69	1.43	0.57	2.00
0.01M NH$_4$NO$_3$	0.8	2.8	71	—	—	—
1.0M NH$_4$NO$_3$	0.3	2.7	89	1.26	0.79	2.05
5.0M NH$_4$NO$_3$	0.2	2.7	93	—	0.74	—

Some strong acid sites were found in the Na-form of the zeolite, the concentration of which seems to be proportional to the Al content. Therefore, whole sets of data (Table 1,2,4) for the zeolites ion-exchanged with NaNO$_3$ and NH$_4$NO$_3$ were analyzed by the following equation: $Z=(a+bY)X$; where X is Al content (atom/uc), Y is proton exchange degree (%), and Z is strong-acid site concentration (site/uc). The least square analysis gave a = 0.683 and b = 0.00935, which means an Al atom in unit cell creates 0.68 strong acid site in sodium form and additional 0.94 strong acid site is generated by complete proton exchange. This analysis suggests that there exist two types of strong acid sites; one depends on only Al content (type A) and the other depends on both Al content and proton exchange degree (type B). Because (Al content)x(proton exchange degree) corresponds to the number of formal acidic protons in a unit cell, the concentration of strong acid sites of type B is proportional to the number of formal acidic protons. As the calculated value 0.94 (site uc^{-1} atom^{-1}) is close to unity, formal acidic protons seem to behave actually as strong (presumably Brønsted type) acid sites.

Effects of ion-exchange on the catalyst life

Catalyst lives were determined for a series of zeolites mentioned in Table 4. The reaction was continued until the conversion of methanol to hydrocarbon became less than 20%, and the catalyst life was expressed in terms of total amounts of product formed until deactivation of the catalyst. The Na-from of the zeolite exhibited no catalytic activity for MTG reaction and only a small amount of dimethyl ether was detected. The other catalysts examined here showed the complete conversion of methanol at the initial stage of the reaction. Therefore, among this series of the catalysts, the activity seems to depend on the strong acid sites of type B. The selectivities for gasoline fraction were higher than 60%.

The catalyst life was proportional to the proton-exchange degree at the region of lower proton-exchange, and gradually dropped off from the proportional line. On the other hand, the olefin selectivity decreased with an increase of proton-exchange degree (or strong acid site concentration), while the paraffin selectivity increased. As a result the cummulative yield of paraffins was apparently proportional to the proton-exchange degree, and the cummulative yield of olefin attained a plateau level at near 50% degree of proton exchange (Fig. 5).

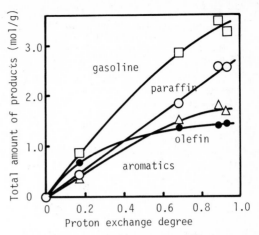

Fig. 5. Influence of proton exchange degree on the catalyst life.

As can be seen in Fig. 6, the amount of coke-deposits found in the deactivated catalyst was proportional to the total amount of the products formed until deactivation of the catalyst (i.e. catalyst life). This fact suggests that the rate ratio of product formation to coke formation is constant during the course of the reaction (ref.11), irrespective of the proton exchange degree. At the region of lower proton exchange degree, the catalyst life was proportional to the proton exchange degree, which in turn is proportional to the content of formal acidic protons in a unit cell. As the amount of coke deposits was proportional to the catalyst life, then the amount of the coke deposits is proportional to the content of formal acidic protons. Therefore, at this region, catalyst deactivation seems to be caused by coke deposits covering the strong acid sites of type B, and the catalyst life is determined by the concentration of strong acid sites of type B. If a large enough amount of strong acid sites were to be present in a catalyst, the catalyst life would not be determined by the amounts of the strong acid sites any more, but would be determined by the pore capacity.

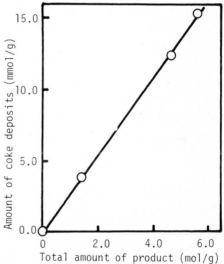

Fig. 6. Relationship between catalyst life and amount of coke deposits.

This seems to be the reason for the dropping off from the proportinal line at the region of higher proton exchange degree.

REFERENCES

1 S. L. Meisel, J. P. McCullough, C. H. Lechthaler and P. B. Weisz, CHEMTECH, 6 (1976) 86-89.
2 I. Wang, T. Chen, K. Chao and T. Tsai, J. Catal., 60 (1979) 140-147.
3 D. H. Olson, W. O. Haag and R. M. Lago, J. Catal., 61 (1980) 390-396.
4 P. Dejaifve, J. C. Védrine, V. Bolis and E. G. Derouane, J. Catal., 63 (1980) 331-345.
5 H.-J. Doelle, J. Heering and L. Riekert, J. Catal., 71 (1981) 27-40.
6 T. Inui and Y. Takegami, Pan-Pacific Synfuel Conference, Vol 1 (1982) 145-151
7 G. T. Kokotailo, S. L. Lawton, D. H. Olson and W. M. Meier, J. Phys. Chem., 85 (1981) 2238-2243.
8 J. R. Anderson, K. Foger, T. Mole, R. A. Rajadhyaksha and J. V. Sanders, J. Catal., 58 (1979) 114-130.
9 N. Topsøe, K. Pedersen and E. G. Derouane, J. Catal., 70 (1981) 41-52.
10 G. P. Babu, S. G. Hegde, S. B. Kulkarni and P. Ratnasamy, J. Catal., 81, (1983) 471-477.
11 P. Dejaifve, A. Auroux, P. C. Gravelle and J. C. Védrine, J. Catal., 70 (1981) 123-136.

P.A. Jacobs et al. (Editors), *Structure and Reactivity of Modified Zeolites*
© 1984 Elsevier Science Publishers B.V., Amsterdam — Printed in The Netherlands

INFLUENCE OF Ni, Mg AND P ON SELECTIVITY OF ZSM-5 CLASS ZEOLITE CATALYSTS IN TOLUENE-METHANOL ALKYLATION AND METHANOL CONVERSION

M.DEREWIŃSKI, J.HABER[2], J.PTASZYŃSKI, V.P.SHIRALKAR[1], S.DŹWIGAJ
Institute of Catalysis and Surface Chemistry, Polish Academy of Sciences, ul.Niezapominajek, 30-239 Kraków, Poland.
[1]Permanent address: National Chemical Laboratory, Pune 411 008, India.

[2]To whom all correspondence should be addressed.

ABSTRACT

Physicochemical and catalytic properties of ZSM-5 zeolites modified with nickel, magnesium and phosphorus have been studied. Conversion of methanol and alkylation of toluene with methanol were used as test reactions. Measurements of sorption capacity of argon, water, cyclohexane, para- and ortho-xylene revealed plugging of channels and modication of hydrophilicity due to impregnation. Introduction of phosphorus blocked all strong acid centers, whereas Ni and Mg had only little effect on their distribution. Unmodified HZSM-5 and Ni/HZSM-5 gave near-equilibrium composition of xylene isomers, whereas with P/HZSM-5 95-100% of para-isomer was formed. Introduction of nickel resulted in considerable increase of the content of aromatics in methanol conversion.

INTRODUCTION

One of the interesting properties of ZSM-5 zeolite catalysts discussed recently is their ability to enhance the formation of para-substituted benzene derivatives in such reactions as xylene isomerization, toluene-methanol alkylation or toluene disproportionation. By modifying the catalyst with boron, magnesium or phosphorus and altering reaction parameters, very high concentrations of para isomer in the xylene product have been obtained (1,2). It was argued (3) that different reaction paths followed by the para-selective catalysts are consistent with diffusional control of product selectivites. On the basis of ir spectroscopic, ESCA and microcalorimetric data it was suggested that phosphorus neutralizes acidic sites primarily at the entrance of the channels of the zeolite particles, the strongest acid sites remaining unmodified (4).

A fundamental question remains however unanswered as to whether the role of modifying elements consist only in the physical phenomenon of blocking the channel openings and creating thus the diffusional hindrances or also in the chemical phenomenon of altering the concentration, strength and localization of acid sites. It is also not fully clear which is the distribution pattern of the modifying agents in teh crystallites of the zeolite.

We wish to report some new data on ir spectra, thermodesorption of ammonia, adsorption and catalytic activity in alkylation of toluene and conversion of methanol bearing upon these questions.

EXPERIMENTAL

The ZSM-5 catalyst was synthetized by the method discussed elsewhere (5,6). Si/Al molar ratio was 70, the particle size distribution 2-5 μm. Modified ZSM-5 catalysts designed; Ni/HZSM-5, Mg/HZSM-5 and P/HZSM-5 were prepared from diammonium hydrogen phosphate, magnesium acetate and nickel nitrate. The slurry was dried in air with continuous stirring at 60°C, then calcinated at 550°C in air for 8 h. Pellets, 10-20 mesh were activated in flow of nitrogen at 550°C for 10 h. The catalysts contained 2% wt. of Ni, 5% wt. of Mg and 3% wt. of P. Amorphous aluminosilicates (13% Al_2O_3-87%SiO_2, 87%Al_2O_3-13%SiO_2) were obtained by coprecipitation of respective hydroxides from isopropanol solution of aluminium isopropoxide and tetraethoxy silicon.

The surface areas were determined using BET method with argon as adsorbate. The sorption measurements were carried out using Mc Bain ballance. The catalysts were degassed at 400°C at about 10^{-6} torr for 4 h. In TPD of ammonia experiments the 100 mg samples were evacuated to 5.10^{-3} torr at 550°C for 3 h, then cooled to room temperature and NH$_3$ was introduced. The sample was then evacuated to 5.10^{-3} torr for 1 h and TPD was carried out at the rate of 10°C min^{-1}. Ir spectra were registered by means of UR-20 Carl Zeiss Jena spectrometer in the 4000-700 cm^{-1} region. Selfsupported discs (6-10 mg cm^{-2}) of zeolite were activate in vacuum of 10^{-5} torr for 3 h at 360°C. The ir spectra were registered at room temperature and at 250°C (7). The catalytic reactions were carried out in a fix-bed continuous-flow microreactor with 0.4 ml (0.28 g) of catalyst. Experiments were carried out at atmospheric pressure. For activation before the catalytic run, the catalysts were heated under dry air at 550°C during 3 h.

Liquid reactants were introduce with a syringe pump into a prehe-
ater, where they were vaporized and passed through the catalytic
bed. The reaciton products were analyzed by on-line gas chromatograph using flame-ionization detector and DC-550 on Chromosorb PAW
column. Benton 34 and didodecylphtalide (10%) on Chromosorb W was
used to separate isomers of xylene, ethylbenzene and trimethylbenzene.

RESULTS AND DISCUSSION

Results of the sorption experiments are summarized in Table 1.
Several interesting features may be noted. The impregnation with
nickel, magnesium and phospohorus results in the more and more
pronouced decrease of the equivalent surface area as determined
from the adsorption of argon. As the argon atoms have the diameter considerably smaller than the diameter of the ZSM-5 channnels
and the measurements were carried out in the static conditions,
it may be concluded that the impregnation with Ni, Mg and P
results in the increasing plugging of channels. This is confirmed
by the observation that a considerable decrease of the equivalent
surface area takes place in the course of the reaction, but the
final values are much less influenced by the presence of the impregnating agent than the initial ones. Apparently, the deposition
of coke has a similar effect as impregnation, but the degree of
plugging of the zeolite channels has a limiting value. It is
interesting that neither of this effects is observed in the ma-
croporous amorphous aluminosilicate.

At variance with the behaviour of argon the sorption of water
increases in the order HZSM-5 < Ni/HZSM-5 < Mg/HZSM-5. This may be
explained by invoking the increase of hydrophilicity of silicates
due to the presence of lower valent ions (8,9). Taking into account that the zeolites contained 2% Ni and 5% Mg it may be noted
that the increase of hydrophilicity resulting from the introduction of one divalent cation is the same for Ni/HZSM-5 and Mg/HZSM-5.
Thus, this effect is of electrostatic character, independent of
the chemical nature of cation. Similar effect is observed in case
of aluminosilicates, whose hydrophilic properties increase proportionally to the content of aluminium.

The assumption that hydrophilicity is increased by introduc-
tionof divalent cations is confirmed by the observation that adsorption of cyclohexane is decreased in the same proportion.

Table 1

Equivalent BET surface area and sorption capacities

Catalyst	Equivalent BET area /m^2g^{-1}		Sorption capacaties[c] moles/100 g of catalyst			
			Water	Cyclo-hexane	para-Xylene	ortho-Xylene
	a	b				
HZSM-5	428.0	241.5	0.319	0.051	0.138	0.030
Ni/HZSM-5	411.0	175.4	0.357	0.046	0.123	0.019
Mg/HZSM-5	368.0	239.2	0.417	0.031	0.125	0.024
P/HZSM-5	310.0	190.8	0.362	0.049	0.100	0.009
P/13Al_2O_3-87SiO_2	132.3	119.5	0.240	0.137	0.087	0.126
P/87Al_2O_3-13SiO_2	191.9	190.3	0.903	0.222	0.180	0.292

a=fresh catalyst
b=catalyst after reaction: Methanol conversion, Atmospheric pressure; Temperature = 400°C; LHSV = 2.5; Time on stream = 2 h.
c/ Temperature = 25°C; p/p_o = 0.8

The influence of phosphorus is more complex, because it increases the adsorption of water, but has very little affect on the adsorption of cyclohexane. Interesting conclusion may be drawn from the comparison of the adsorption of para-xylene and ortho-xylene. Sorption capacity of para-xylene is several time larger than that of ortho-xylene and decreases on impregnation to the same degree as the equivalent specific area what indicates that in both cases it is the decrease of the void volume which may be responsible for the change of adsorption. Different behaviour is observed on adsorption of ortho-xylene, which decreases dramatically in P/HZSM -5 as compared to the initial HZSM-5. Apparently in this case the changes of pore dimensions render the majority of the channel volume inaccessible for the branched ortho isomer. It may be noted that adsorption of para-xylene is threefold greater than that of cyclohexane although the sizes of these molecules are comparable. It seems that para-xylene being a base with polarizable π-electrons interacts much more strongly with the zeolite, its specific adsorption being thus greater than that of cyclohexane.
Fig.1 shows the TPD curves of ammonia. Two peaks are observed in the case of HZSM-5 and Ni/HZSM-5: the lower energy one with maximum at about 160°C and the higher energy one with maximum at about

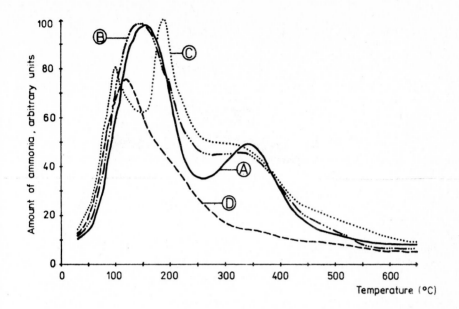

FIGURE 1   Temperature-programmed desorption of ammonia from ZSM-5 catalysts; sample weight 0.1 g, heating rate 10°C min⁻¹. (A) HZSM-5 (B) Ni/HZSM-5, (C) Mg/HZSM-5, (D) P/HZSM-5

340°C. The latter corresponds to the γ-state assigned by Topsoe et al. (10) to strong acid centers, whereas the former may be a super position of the α- and β-states, corresponding to weaker acid centers. Indeed, when magnesium was introduced into the zeolite, the higher temperature peak remained practically unchanges, whereas that at lower temperature became split into two peaks: less intensive one at about 100°C and a more intensive one with maximum at about 180°C in good agreement with the positions α- and β- states of Topsoe et al. (10). Introduction of phosphorus results in a complete change of the acidity of the zeolite. The two types of stronger acid centers (β- and γ-states) become almost completely neutralized and only the weak centers (α-state) remain their concentration being comparable to that in Mg/HZSM-5 and much lower than in HZSM-5.

This picture is confirmed by the ir spectra shown in Fig.2. The ir spectrum of HZSM-5 agrees well with those quoted by Vedrine et al. (11) and Topsoe et al. (10). Three bands are observed in the OH stretching frequencies range: 3620 cm⁻¹, 3680 cm⁻¹ and 3740 cm⁻¹ as well as a band at 1660 cm⁻¹ of adsorbed water and a very weak band at 1490 cm⁻¹, probably due to residual ammonium groups. No much change is observed with Ni/HZSM-5 and Mg/HZSM-5 zeolites,

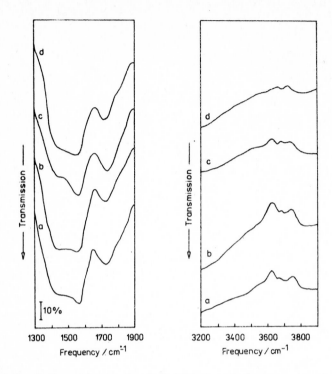

FIGURE 2 Infrared spectra of ZSM-5 catalysts; samples evacuated for 3 h at 360°C in vacuum of 10^{-5} torr. (a) HZSM-5, (b) Ni/HZSM-5 (c) Mg/HZSM-5, (d) P/HZSM-5.

except that in the case of the latter the band at 3740 cm^{-1} has much smaller intensity, what may account for the separation of the two TPD peaks of ammonia at 100°C and 180°C. A drastic change is noticed on passing to the spectrum of P/HZSM-5, in which the band at 3620 cm^{-1}, assigned to the strong acid centers (γ-state) disappears completely and the bands at 3680 cm^{-1} and 3740 cm^{-1} have much smaller intensity.

It may be thus concluded that three distinctly different types of acid centers exist in the HZSM-5 zeolite, in good agreement with earlier reports (12,13) . The introduction of nickel or magnesium has no influence on the strong acid centers, only the concentration of weakest centers is reduced in the case of magnesium. If we assume the assignment of the α-state to the sites located on the external surface or to interaction of NH$_3$ molecules with surface oxide or hydroxyl groups (10), we may explain the decrease of their concentration by heavier loadings of the zeolite with divalent ions, as in the case of magnesium. On introduction of phosphorus all strong acid centers become neutralized. It must

TABLE 2

Methanol conversion on modified ZSM-5 zeolites

Product distribution / wt %	Catalyst			
	HZSM-5	Ni/HZSM-5	Mg/HZSM-5	P/HZSM-5
Methanol	–	–	–	47.3
Dimethyl ether	–	–	–	23.6
Methane, ethane, ethene	4.9	9.2	4.7	2.5
Propane, propene	20.8	8.8	12.6	5.0
Isobutane	21.7	8.0	9.4	0.4
Butane, 1-butene	7.5	5.2	6.3	2.2
trans, cis-2-Butene	2.4	2.8	4.7	2.1
Isopentane	8.3	5.1	4.7	–
C_5	2.4	3.2	5.3	2.0
C_6^+, C_6	9.3	13.6	12.6	13.6
Benzene	0.6	trace	2.5	–
Toluene	4.7	5.3	6.3	0.4
Ethylbenzene	trace	trace	0.7	–
p-Xylene	3.0	5.0	9.0	0.75
m-Xylene	5.6	11.2	6.1	0.20
o-Xylene	2.2	6.3	3.0	0.15
Ethyltoluenes	1.5	3.8	6.9	–
Trimethylbenzenes	4.7	11.7	5.0	–

Temperature = 400°C; LHSV = 2.5 h^{-1}; Time on stream = 2 h.

TABLE 3

Reaction of toluene with methanol on modified ZSM-5 catalysts

Catalyst	Methanol conversion / %	Toluene conversion / %	Xylenes in products / %	Xylene isomer composition / %		
				para	meta	ortho
HZSM-5	100	16.3	82.9	29.1	51.4	19.5
Ni/HZSM-5	100	19.3	84.1	27.0	54.1	18.9
Mg/HZSM-5	100	13.9	85.9	38.3	46.9	14.8
P/HZSM-5	92	8.3	84.3	89.3	8.0	2.7

Atmospheric pressure; Temperature = 500°C; LHSV = 2.5 h^{-1};
Toluene/Methanol, molar ration = 2. Traces of gas were observed.

be thus concluded that phosphorus is located at the channel intersections, where the strongest acid centers are belived to be situated. This is contrary to the conclusion reached by Vedrine et
al. (4), who locate phosphorus mainly in the outer layers of zeolite crystalites.

Results of the studies of the catalytic activity of zeolites
in the conversion of methanol and alkylation of toluene with methanol are summarized in Table 2 and 3, respectively. Several interesting features should be emphasized:

introduction of nickel or magnesium drastically increases the
amount of aromatics formed in the conversion of methanol; the composition of xylenes is near to equilibrium in HZSM-5 and Ni/HZSM-5
whereas in the presence of magnesium the amount of para-isomer
increases both in the case of conversion of methanol and alkylation of toluene;

with P/HZSM-5 mainly the para-xylene is formed in the alkylation
of toluene, this zeolite showing only a very low activity in the
conversion of methanol, in which mainly C_3 and C_6 olefins are
formed. This is in line with the conclusion that strong acid centers were to a large extent neutralized by the introduction of
phosphorus, because such centers are required for the first C-C
bond to be formed in the methanol conversion, and then for the
transformation of oligomers into cycloolefins and then aromatics
(14). However, in the small amount of xylenes formed the para-
isomer predominates. In the case of amorphous aluminosilicates
impregnation with phosphorus has no influence on the xylene composition, which in both cases is determined by the inductive effect
of the methyl groups.

REFERENCES

1   N.Y.Chen,W.W.Kaeding,F.G.Dwyer,J.Amer.Chem.Soc.,101 (1979) 6783.
2   W.W.Kaeding,C.Chu,L.B.Young,B.Weinstein,S.A.Butter,J.Catal.,
    67 (1981) 159.
3   L.B.Young,S.A.Butter,W.W.Kaeding, J.Catal.,76 (1982) 418.
4   J.C.Vedrine,A.Auroux,P.Dejaifve,V.Ducarme,H.Hoser,S.Zhou,
    J.Catal.,73 (1982) 147.
5   R.J.Argauer,G.R.Landolt, U.S.Patent 3702886 (1972) Mobil Oil
    Corp.
6   S.B.Kulkarni,V.P.Shiralkar,A.N.Kotasthane,R.B.Borade,P.Ratna-
    samy, Zeolites,2 (1983) 313.
7   J.Haber,H.Piekarska-Sadowska,T.Romotowski,Bull.Acad.Polon.Sci.,
    Ser.Sci.Chim.,26 (1978) 967.
8   D.H.Olson,W.O.Haag,F.M.Lago, J.Catal.,61 (1980) 390.
9   R.J.Mc Intosh,D.Seddon, Appl.Catal.,6 (1983) 307.
10 N.Y.Topshoe,K.Pedersen,E.G.Derouane, J.Catal., 70 (1981) 41.
11 J.C.Vedrine,A.Auroux,V.Bolis,P.Dejaifve,C.Naccache,P.Wierzcho-
    wski,E.G.Derounae,J.B.Nagy,J.P.Gilson,J.H.C.van Hooff,
    J.P.van den Berg,J.Wolthuizen, J.Catal.,59 (1979) 248.
12 J.R.Anderson,K.Foger,T.Mole,R.A.Rajadhyaksha,J.V.Sanders,
    J.Catal.,58 (1979) 114.
13 P.A.Jacobs,J.B.Uytterhoeven,M.Stevns,G.Froment,J.Weitkamp, in
    Proc.5 th Int.Conf.on Zeolites,Naples,Heyden,London 1980 p.607.
14 S.Dźwigaj,J.Haber,T.Romotowski,Zeolites (in print).

P.A. Jacobs et al. (Editors), *Structure and Reactivity of Modified Zeolites*
© 1984 Elsevier Science Publishers B.V., Amsterdam — Printed in The Netherlands

REACTION OF SMALL AMOUNTS OF METHANOL ON HZSM-5, HY AND MODIFIED Y ZEOLITES

LUDMILA KUBELKOVÁ, JANA NOVÁKOVÁ AND PAVEL JÍRŮ

The J. Heyrovský Institute of Physical Chemistry and Electrochemistry, Czechoslovak Academy of Sciences, Máchova 7, CS - 121 38 Prague 2

ABSTRACT

The interaction of methanol with HY, AlHY, dealuminated Y and HZSM-5 zeolites
containing various amounts of extralattice Al and acidic OH groups was studied.
Surface species with increasing thermal stability in the sequence MetOH ads,
Al-OMet and O(zeol)-Met were identified from IR spectra. Under low-pressure flow
conditions (1-3 Pa) at 670 K the products of methanol transformation consisted
mainly of lower olefins and aromatics over Y zeolites; HZSM-5 yielded predominantly formaldehyde and methane, however, pressure increase to 100 Pa resulted
in the appearance of hydrocarbons. Possible reaction mechanism involving mentioned surface species is discussed.

INTRODUCTION

The conversion of methanol over zeolites has become a promising process in
the preparation of hydrocarbons as well as high-octane-number gasoline. The reaction mechanism of this transformation, especially its primary steps, has not
been clarified, although many theories have appeared in last few years (ref. 1).
The main difficulty in obtaining direct experimental evidence on the reaction
pathway emerges from the fast kinetics accompanied by a large number of parallel
and consecutive reactions. In our previous paper (ref. 2), we tried to avoid
this problem studying the methanol transformation on modified Y and HZSM-5 zeolites by temperature programmed desorption (TPD) of very small preadsorbed
amounts and by low-pressure flow (LPF) measurements. The product composition was
found to be different over HZSM-5 and HY zeolites, but the same at the same temperature in both TPD and LPF experiments. In a continuation of this study, we
have more closely considered the characterization and behaviour of adsorbed methanol species during desorption at various temperatures and the effect of a
pressure increase on the composition of products obtained under flow conditions.
Methanol-d_3 was also used in this work to help elucidate the reaction steps.

EXPERIMENTAL PART

Only the most important factors will be described here as the majority of the
data on zeolite preparation, characterization and methods used were detailed
elsewhere (refs. 2,3). AlNH$_4$Y was obtained by ion exchange of NH$_4$Y, zeolites denoted Y-C and Y-B by dealumination of NaY with SiCl$_4$. HZSM-5 was prepared in

TABLE 1

Characterization of the zeolites

Zeolite	Si/Al[a] (total)	Si/Al (lattice)	H/g[c] $\times 10^{-20}$	L/g[d] $\times 10^{-20}$	T_{max}^e (NH$_3$) (K)	capacity (mmol Ar/g)
H(70)Na(30)-Y	2.5	2.5[b]	7.9	0.16	470	10.9
Al(25)H(25)-Y	2.0	2.9[a]	3.3	2.6	440	9.0
Y-C	5.7	10.8[b]	3.8	2.0	550	9.2
Y-B	19.9	–	0.9	0.75	–	9.4
HZSM-5	17.5	–	2.4	0.6	570	5.0

[a]From chemical analysis.
[b]From mid-infrared data.
[c,d]H-the number of strong proton-donor sites, L-the number of strong electron-
-acceptor sites determined from the IR spectra of adsorbed pyridine.
[e]Temperature of the maximum peak of NH$_3$ desorption, for Y-C and HZSM-5 of the
second maximum.

the usual way. Before the interaction with methanol, the zeolites were treated
in vacuo at 10^{-4} Pa and 690 K. Table 1 demonstrates that HY zeolite contains almost exclusively strongly acidic hydroxyls in the greatest amount while the AlHY,
dealuminated Y and HZSM-5 zeolites contain extralattice Al exhibiting electron-
-acceptor properties, along with proton-donor sites. The Lewis acidity has been
found to be most pronounced for AlHY. According to the TPD of ammonia, the HZSM-
-5 zeolite exhibits the highest acid strength of the zeolites studied (refs.2,3).

In the IR experiments, methanol or methanol-d$_3$ was dosed at ambient temperature up to an amount of 3.5-4 mmol/g onto the zeolite plates 8-10 mg/cm^2 in
thickness. Then the zeolite was heated at a rate of 6°/min in vacuo, after reaching 380, 460, 580, 690 and 790 K, respectively, it was rapidly cooled and IR
spectra were recorded on Fourier transform infrared spectrometer Nicolet MX-1E.
In some cases, the desorbed products were analyzed mass spectrometrically.

The conversion of methanol over zeolites was studied under low-pressure flow
(LPF) conditions (1-3 Pa, 3x 10^{-4} l hr^{-1}) in a reactor placed in front of a modified MI 1305 mass spectrometer. A „cap" with a leak and a by-pumping system pul-
led over the reactor allowed to increase the methanol pressure to 100 Pa (flow
rate : 3x 10^{-3} l hr^{-1}).

RESULTS

Interaction with methanol - IR experiments

Adsorption of methanol at ambient temperature followed by desorption of surface species up to 790 K leads to appreciable changes in the spectrum of OH
groups as is demonstrated in Fig. 1 for HY, AlHY and HZSM-5 zeolites. Simultaneously, typical bands of CH$_3$ groups appear with position, shape and intensity
depending on the number of acid sites and on the desorption temperature. This

can be seen from Fig. 2, where the CH_3 stretching vibration bands are compared for HY, AlHY and HZSM-5 zeolites.

The formation of methoxyl groups by the reaction of methanol with structural hydroxyls of the HY zeolite has already been proved in the literature (ref.4,5). The sharp bands at 2977 and 2870 cm^{-1} found in our spectrum of HY zeolite after evacuation at 390 K, together with the appreciable intensity decrease of the 3645 cm^{-1} band of structural hydroxyls are in good agreement with (ref. 4) emphasizing high reactivity of the zeolitic OH groups and thus ready substitution of their hydrogen by a methyl group from methanol: $(OH)_z + CH_3OH \rightarrow O_zCH_3 + H_2O$. However, some of the physisorbed methanol interacting with the structural hydroxyls remains in the HY zeolite at this temperature. It is evidenced by weaker CH_3 bands at 2960 and 2850 cm^{-1} and by the appearance of broad bands in the region 3200 – 2800 cm^{-1} and at 2430 cm^{-1} originating from perturbed structural hydroxyls and methanol OH groups. These species are gradually removed by outgassing at temperatures up to 590 K while the O_z-CH_3 species are considerably more stable. They are decomposed mainly above 670 K and this process is accompanied by the re-formation of structural hydroxyls (Fig. 1,2).

The surface species found on the AlHY zeolite differ from those on HY, which is again apparent from the spectra in CH_3 stretching vibration region. The most easily removable species are characterized by the band at 2960 and 2850 cm^{-1} reflecting the presence of physisorbed methanol at lower temperatures (380, 460 K). The new bands at 2968 and 2857 cm^{-1} appear even at 380 K but they are strongly overlapped with the bands of physisorbed methanol. They become most pronounced at 460 K and then decrease at higher outgassing temperatures but they are still present at 690 K. In an early work (ref. 6) R.G. Greenler identified surface methoxides on alumina. The appearance of the new bands at 2968 and 2857 cm^{-1} with appreciable intensity on the AlHY zeolite, which contains a large amount of cationic Al with electron-acceptor ability, strongly suggests that $AlOCH_3$ species are formed here. The adsorption of methanol as $Al \leftarrow \overset{H}{O}CH_3$ is most probably the preceeding step. Along with $AlOCH_3$, O_zCH_3 species are also formed, but in a low amount consistent with the low number of hydroxyls available for the reaction. The corresponding CH_3 bands (2977 and 2870 cm^{-1}) are best resolved as weak bands at high temperatures. The participation of the OH groups in the interaction with methanol is again reflected by changes in the OH bands (Fig. 1,2).

The same type of species with similar behaviour as described for AlHY is found in dealuminated zeolites and in HZSM-5. However, the amount is lower, especially with highly dealuminated Y-B and HZSM-5 zeolites, in accordance with the decrease in the number of acid sites. Figs. 1 and 2 show the spectral changes for the HZSM-5 zeolite. The difference spectra presented here indicate that the zeolite hydroxyls reappear during the decomposition of methoxyl groups above 590 K. The typical reaction product released into the gas phase in the tempera-

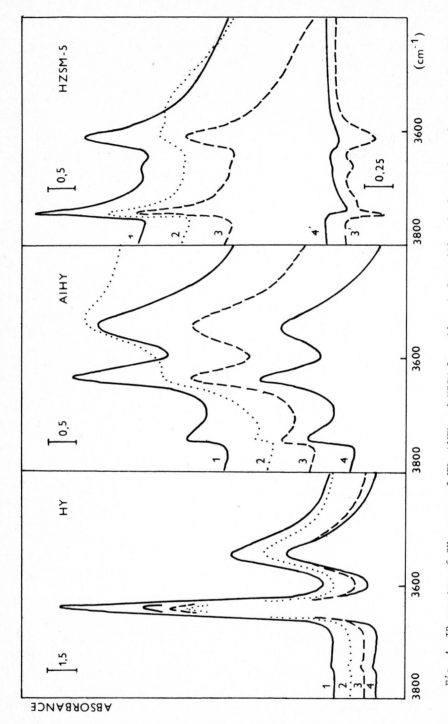

Fig. 1. IR spectra of OH groups of HY, AlHY and HZSM-5 zeolites before (1) and after adsorption of methanol at ambient temperature followed by desorption at 390 K (2), 590 K (3) and 790 K (4). Spectra (3´),(4´) were obtained after substraction of original spectrum (1) from (3) and (4), resp.

ture interval 390 – 590 K is dimethyl ether (DME). Olefins and aromatics started to appear above 590 K, not only with Y zeolites but also with HZSM-5. IR study of surface species formed by interaction of HZSM-5 with CD_3OH shows that decomposition of $AlOCD_3$ and O_zCD_3 groups above this temperature is accompanied by isotopic exchange of hydroxyls, predominantly the structural ones (3610 cm^{-1} – OH band, 2663 cm^{-1} – OD band).

Low-pressure flow experiments

The LPF conversion of methanol at 670 K and 2 Pa over HZSM-5 and Y-C is depicted in Figs. 3a,b. While predominantly formaldehyde and methane are formed over HZSM-5, Y-C yields C_{3-7} olefins and aromatics up to C_9. Using CD_3OH over

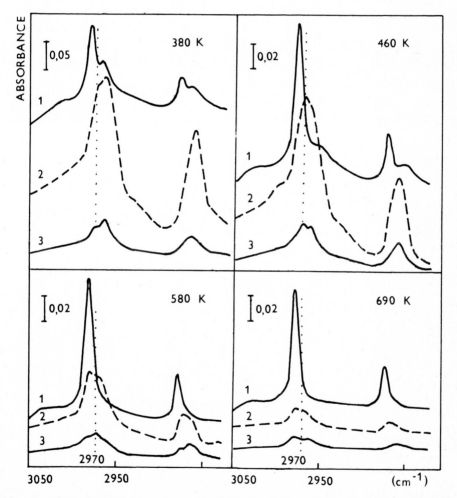

Fig. 2. IR spectra of CH_3 groups remaining on the surface after desorption of preadsorbed methanol at 380 K, 460 K, 580 K and 690 K. (1) HY, (2) AlHY and (3) HZSM-5.

HZSM-5 fully deuterated products are obtained. Therefore, hydrogen from the methanol OH group is not incorporated into formaldehyde, DME and methane during
the catalytic reaction. Nevertheless, the formation of some CD_2H_2 cannot be fully excluded because of overlapping of the molecular peaks of CD_2H_2 and water.
The amount of deuterated water is small or none. Closing the CD_3OH inlet leads to
a decrease in the deuterium content in the products, which are now only desorbed from the zeolite (Fig. 4). The highest amount of hydrogen is observed in
methane, where CDH_3 predominantes. In contrast to the latter experiment in which
deuterium exchange was observed only in the desorbed products from the zeolite
after the reaction, an immediate random exchange of deuterium is found using a
mixture of CD_3OH and CH_3OH over HZSM-5 at 670 K.

The influence of increased methanol pressure on the composition of the products over HZSM-5 at 670 K is shown in Fig. 3c. At 100 Pa, the products are
composed of unsaturated and saturated hydrocarbons and aromatics which disappear
when lowering the pressure. Formaldehyde and methane are formed instead. The
product composition can be continuously changed on the same zeolite sample by
lowering and increasing the methanol pressure; the conversion is high in all
cases.

DISCUSSION

DME formation and its role in the methanol-to-hydrocarbon conversion. It is
known that acidic catalysts exhibit activity in the formation of DME from metha-

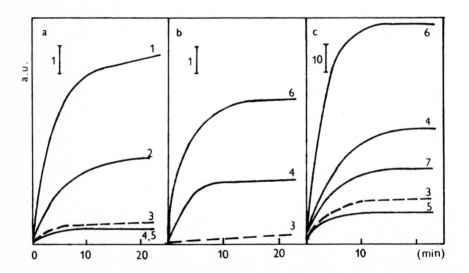

Fig. 3. Low-pressure flow measurements of methanol (2 Pa) at 670 K over HZSM-5
(a) and Y-C (b) and effect of increased pressure (100 Pa) over HZSM-5 (c).
(1) Formaldehyde, (2) methane, (3) methanol, (4) aromatics, (5) DME, (6) olefins,
(7) saturated hydrocarbons.

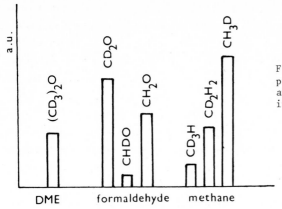

Fig. 4. Isotopic composition of products released from HZSM-5 at 670 K after closing the CD_3OH inlet at the same temperature.

nol. With zeolites, this step has often been assumed to be the first one in the transformation of methanol to hydrocarbons (ref. 1). The data obtained previously using TPD and LPF measurements (ref. 2) did not support this assumption as the activity of various zeolites in the conversion of methanol to DME could not be correlated with the activity in hydrocarbon formation. In addition, the temperature regions of DME and hydrocarbons release were well separated.

With the HY zeolite, the physisorbed methanol along with O_zCH_3 species were found by IR experiments in the temperature region typical for DME formation. In accordance with ref. 4, it can thus be assumed that the reaction on this zeolite proceeds with the participation of physisorbed methanol either between CH_3OH_{ads} and O_zCH_3 or between two adsorbed molecules. In AlHY, dealuminated Y and HZSM-5 zeolite relatively large amounts of $AlOCH_3$ species were identified, which decreased with increasing desorption temperature, even though this species was of higher stability than adsorbed methanol also present in these zeolites. This suggests that $AlOCH_3$ groups also participate in the formation of DME. Consequently, it may be concluded that the catalytic activity of the zeolites is influenced by the presence of extralattice Al exhibiting electron-acceptor properties.

Under normal catalytic conditions, methanol feed is continuous; therefore, some adsorbed methanol is always present and the formation of DME is facilitated. DME can be then converted into hydrocarbons and aromatics in a side-reaction of methanol-to-hydrocarbon transformation.

Hydrocarbon formation. Two types of methoxyls were found to be formed directly from methanol on the zeolites under study: O_zCH_3 and $AlOCH_3$ by the reaction with zeolite hydroxyls and extralattice Al, respectively. The release of hydrocarbons into the gas phase is accompanied by the disappearance of both these methoxyls and with the re-formation of surface hydroxyls. This process occurs at higher temperatures when the C-H bond in CH_3 is weakened, as was documented by deuterium exchange between CD_3OH and zeolitic hydroxyls and between CD_3OH and CH_3OH. It is tempting to assume that the reaction takes place among several methoxy groups with the primary formation of lower olefins ($C \geq 3$) whose further reaction via carbonium ions to higher hydrocarbons can proceed readily. The $AlOCH_3$ methoxyls are decomposed at lower temperatures than O_zCH_3. The high activity of AlHY and dealuminated zeolites (ref. 2) can thus be attributed to the presence of extralattice Al.

Special behaviour of HZSM-5. A large difference was observed between HZSM-5 and Y zeolites studied under LPF conditions at 1-2 Pa as well as TPD conditions with a very small amount of preadsorbed methanol (ref. 2). This is connected with the ability of HZSM-5 to yield formaldehyde and methane at about 670 K. In TPD, this reaction probably occurs between two methoxy groups without the rupture of one C-O bond; formaldehyde might thus be formed from $Al-OCH_3$ as these methoxyls were also found on HZSM-5. At higher temperatures, the interaction of more methoxyls takes place, accompanied by C-O bond rupture and higher hydrocarbons are formed. In LPF measurements, the absence of deuterium exchange in CD_3OH suggests a pathway via gaseous or adsorbed methanol and methoxy group. The appearance of the normal gas phase composition (aliphatics, aromatics) with an increased methanol pressure might easily be explained by an increased number of methoxyls and thus by enhanced probability of their inter-reaction. This is also confirmed by the release of aliphatics and aromatics in IR desorption experiments using relatively high amounts of preadsorbed methanol. The LPF experiments with B-Y, exhibiting a similar small number centers for the formation of methoxyls did not indicate the formation of formaldehyde and methane. This thus suggests that the unique properties of HZSM-5 related to its special structure and strong acid sites are responsible for the formation of the latter products from methanol.

REFERENCES

1　C.D. Chang, Catal. Rev.-Sci. Eng., 25 (1983) 1-118, and refs. herein.
2　J. Nováková, L. Kubelková, K. Habersberger and Z. Dolejšek, J. Chem. Soc. Faraday 1, (1984) in press.
3　L. Kubelková, V. Seidl, J. Nováková, S. Bednářová and P. Jírů, J. Chem. Soc. Faraday 1, (1984) in press.
4　P. Salvador and W. Kladnig, J. Chem. Soc. Faraday 1, (1977) 1153-1168.
5　P. Salvador and J.J. Fripiat, J. Phys. Chem., 79 (1975) 1842-1849.
6　R.G. Greenler, J. Chem. Phys., 37 (1962) 2094-2100.

P.A. Jacobs et al. (Editors), *Structure and Reactivity of Modified Zeolites*
1984 Elsevier Science Publishers B.V., Amsterdam — Printed in The Netherlands

ON THE FORMATION OF HYDROCARBON CHAINS IN THE AROMATIZATION OF
ALIPHATIC OLEFINS AND DIENES OVER HIGH-SILICA ZEOLITES

G.V. ISAGULIANTS, K.M. GITIS, D.A. KONDRATJEV and Kh.M. MINACHEV
N.D. Zelinski Institute of Organic Chemistry, USSR Academy of
Science, Moscow, USSR

ABSTRACT
    The transformations of C_4-C_8 olefins and dienes over ZSM-5
type high-silica zeolites were investigated. The aromatization
of these hydrocarbons proceeds via preliminary fragmentation followed by the assemblage of fragments to aromatic molecules rather
than via intermediate formation of oligomers. The proposed mechanism is supported by the isotopic data and satisfactorily explains
the product distribution.

INTRODUCTION, RESULTS AND DISCUSSION
    The aromatization of aliphatic hydrocarbons carried out over
oxide or metal catalysts continues to arouse interest, and in
recent studies (ref. 1) this reaction was shown to proceed according to the consecutive dehydrogenation mechanism via the intermediate formation of olefins, dienes and trienes. Among the new
aromatization catalysts ZSM-type high-silica (HS) zeolites are of
particular interest since they allow to obtain aromatics from
lower olefins. In contrast to the conventional process of aromatization of C_6 and higher hydrocarbons, the process catalyzed by
HS-zeolites involves chain lengthening of lower alkenes providing
precursors for the subsequent benzene ring formation. According
to (ref.2,3) initial steps of this process are olefin oligomerization followed by the hydrogen loss due to the intermolecular
transfer. Hence the mechanism of lower alkene aromatization is
reduced to the sequence of dehydrogenation steps as discussed in
(ref.4). However, the number of C-atoms in hydrocarbons predominantly formed from lower alkenes over HS-zeolites does not correspond to that suggested by the simple oligomerization mechanism.
On the basis of general consideration and indirect evidence this
phenomenon was ascribed (ref.5-7) to the processes of cracking
and disproportionation. The question of whether the feed hydrocarbon, intermediate oligomer, the diene formed therefrom or C_9-C_{10}

alkylaromatics undergo cracking, remains, however, to be elucidated. It was therefore of interest to investigate the conversion
of C_4-C_8 olefins and dienes, to establish the role and participation of cracking (fragmentation) in the overall aromatization process and to check the reaction mechanism using ^{14}C-labelled compounds.

Studies of propylene and isobutylene conversion over HS-zeolites showed (ref.8) that oligomerization is the prevailing reaction up to 300°C. At higher temperatures aromatization products
appear and then become predominant. An increase in arene formation is accompanied by the enhanced cracking of previously formed oligomers. Both reactions can be assumed to be parallel and
independent processes simultaneously accelerated by temperature
rise, but in fact both processes were found to be interrelated.
It can be seen in Fig.1 that the yields of arenes from isobutylene are higher than those from octene. Although the yields of
arenes from both hydrocarbons at high contact time (τ) attain
the same level, the mass distribution of arenes is essentially
different. That indicates certain differences in the reaction
pathways of arene formation at high τ. In octene aromatization
the bulk of arenes beginning with τ = 10 are formed not from C_6-
-C_9 aliphatic hydrocarbons but from low-molecular fragments
which are detected in the gas phase as hydrocarbons C_1-C_4. Two
factors corroborate this contention, viz.; 1) the increase in
arene yield (15%) after τ = 10 can not stem from C_6-C_9 aliphatics
since their content to this moment is ca. 6% only, 2) the yield
of octene cracking products vs. τ has a maximum which is particularly pronounced in the case of unsaturated hydrocarbons as
exemplified by Σ C_4H_8. As a decrease in the yields of C_4H_8 is
accompanied by a half as great increase in the yield of Σ C_4H_{10},
it is evident that a considerable portion of C_4H_8 is involved in
aromatization. Hence at higher τ values octene breaks into fragments which interact   mutually and form the precursors of
various aromatic hydrocarbons.

The participation of lower-molecular fragments in aromatization becomes even more evident in transformation of dienes
which undergo aromatization more readily than olefins (Table 1)
and yield a set of aromatic hydrocarbons, where the number of
carbon atoms fails to match that of the starting diene or its
dimer (Fig.2). The data of Table 1 show that the yield of aromatics passes through a maximum with increasing molecular weight

of the diene, with the yield of aromatics being higher from non-conjugated dienes than from conjugated ones. On the other hand, piperylene which is capable of izomerizing into non-conjugated 1,4-pentadiene, yields more arenes than butadienes does. Therefore the ability of a diene to form aromatics is determined by its sensitivity to fragmentation rather than by its susceptibility to oligomerization. Direct experiments using both linear

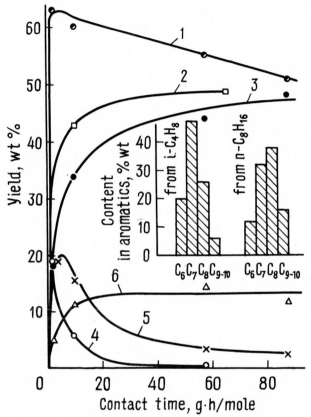

Fig. 1. Yields of octene and isobutene conversion products as a function of contact time.
Octene products: total yield of aromatics (3), C_6-C_9 aliphatics (4); C_1-C_4 hydrocarbons (1); ΣC_4H_8 (5); ΣC_4H_{10} (6). Isobutene products: total yield of arenes (2). In the inset: distribution of arenes from isobutene ($\tau = 65$) and octene ($\tau = 58$) aromatization.

and cyclic dimers of piperylene support this suggestion, since with decatriene the yield of aromatics is below that with piperylene, while alkenylcyclohexene conversions to aromatics display an essentially different mass distribution (Fig.2).

In the same way the transformations of 1,3-pentadiene at

228

different temperatures and τ indicate that the ability of dienes to undergo fragmentation is a necessary step for diene aromatization. The equivalent amounts of the cracking products (C_2-C_4 hydrocarbons; no methane) were found to appear prior to aromatization. Indeed, the ratio of the number of moles of C_3 and C_6 hy-

TABLE 1

Conversions of aliphatic olefins and dienes over HS zeolite
Dilution: 1:5 mole; hydrocarbon feed rate 0.15 mole/h

Hydrocarbon	Temperature, °C	Conversion, %wt	Liquid % wt	Yield of arenes, % wt
Propylene	400	61.9	8.8	2.1
1,4-Butadiene	400	81.9	20.0	1.6
1-Pentene	400	94.5	30.4	1.5
1-Hexene	400	91.1	30.6	1.6
1,3-Pentadiene	400	100.0	61.3	47.3
1,4-Butadiene	300	48.3	19.4	5.5
1,3-Pentadiene	300	62.5	69.1	14.2
1,4-Pentadiene[x]	300	99.5	63.2	30.3
2,4-Hexadiene	300	61.9	72.2	7.0
1,5-Hexadiene[xx]	300	96.0	78.1	14.7

[x] The products contain maximum 0.5% of 1,3-pentadiene.
[xx] The products contain 6.1% of hexadiene isomers.

drocarbons formed from C_3 fragments to that of moles of C_2 and C_4 hydrocarbons formed from C_2 fragments, i.e. the ratio
$\varphi = (n_{C_3} + 2n_{C_6})/(n_{C_2} + 2n_{C_4})$ is close to unity at the reaction temperature of 300° and at the lowest values of τ (Fig.3). As the aromatization proceeds, the gaseous products become enriched in C_2 hydrocarbons (the ratio φ and also n_{C_4}/n_{C_2} decrease), while among the arenes the toluene portion increases and the content of C_9 compounds diminishes (ref.9). These findings can be readily interpreted in terms of piperylene breakdown into C_2 and C_3 fragments, which subsequently participate in aromatization. It is conceivable that these fragments combine with another piperylene molecule or some other diene from fragment recombination[*] and yield a precursor, e.g. a dienyl radical, which then undergoes cyclization and dehydrogenation to form an aromatic hydro-

[*] Conjugated C_7 dienes and, to a lesser extent, C_6 dienes were identified in the product of pentadiene conversion by the disappearance of corresponding chromatographic peaks on product treatment with maleic anhydride.

carbon. In the framework of this concept, φ diminution can be explained by the fact that C_3 fragments are more readily involved in aromatization than are C_2 fragments. This is in line with the observation that C_3H_6 aromatization is facile as compared with C_2H_4 (ref.5,7).

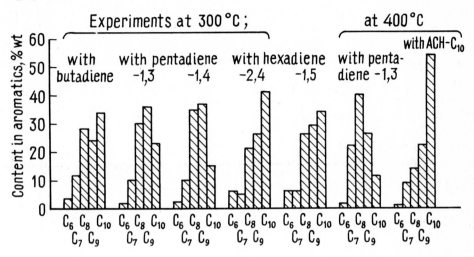

Fig.2. Mass distribution of aromatic hydrocarbons from aromatization of various dienes.

The formation of aromatics via the fragmentation pathway was supported by experiments with piperylene-3-[14]C. The choice of label position is dictated by the necessity of retaining the location of ^{14}C in the C_3 fragment irrespective of bond migration in the starting molecule:

$$CH_3{-}CH{=}CH{-} + CH_2{=}CH{-} \qquad CH_2{=}CH{-} + {-}CH_2{-}CH{=}CH_2 \qquad CH_2{=}CH{-} + CH{=}CH{-}CH_3$$

In fact, ^{14}C atom was found to be incorporated into ethylene as well, presumably due to isotopic isomerization (ref.10). It is, however, essential that the sum of molar radioactivities (MRA) of ethylene and propylene equals the MRA of piperylene taken as unity, the gases in question being, therefore, formed from piperylene fragments. On the other hand, the value of MRA for C_4 hydrocarbons is approximately twice as great as that of ethylene and, hence, points to the formation of butylenes by the dimerization of C_2 fragments. As can be seen in Table 2, the value of MRA found for p-xylene is nearly coincident with that expected from the formation of p-xylene from two C_3 and one C_2 fragments. A similar result can be attained by assuming the addition of a C_3 fragment to a piperylene molecule. In other assemblage versions, e.g. if four C_2 fragments are involved, the observed and calculated values of MRA are essentially different. In the case of C_9 arenes, it is likewise feasible to select an assemblage version, for which the MRA values of the arenes would correspond to the sum of MRA for the fragments. Here the maximum discrepancy between assemblage versions are observed at a lower reaction temperature, when isotopic isomerization occurs to a lesser extent.

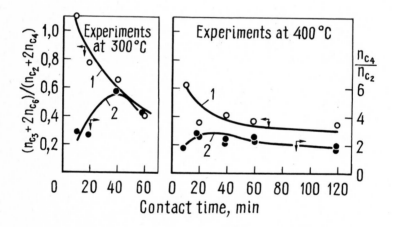

Fig.3. Yields of light hydrocarbons as a function of contact time in piperylene aromatization over HS-zeolite.

1 : $\varphi = (n_{C_3} + 2n_{C_6})/(n_{C_2} + 2n_{C_4})$ ratio; 2 : n_{C_4}/n_{C_2} ratio.

The suggested aromatization mechanism makes it possible to elucidate the effect of contact time and diene nature on the composition of the arenes formed. As emphasized above, the C_3 are more readily involved in aromatization than C_2 fragments. This fact is presumably responsible for the predominance (at low τ values) of xylenes and pseudocumene, viz., the products formed, according to the data of Table 2, with the predominant or exclusive participation of C_3 fragments. In the case of hexadienes, apart from C_3 fragments, the formation of C_2 and C_4 fragments is also possible. In addition to the starting diene, the C_4 fragments are expected to give a mixture of aromatics, in which C_{10}

arenes are predominant, and that was found to be actually the case (Fig.2). Owing to conjugation and the absence of a non-conjugated analogue, butadiene fragmentation should be hindered and
a more likely course would be butadiene dimer destruction and
the formation of C_2-C_4 fragments, which on combination with butadiene or its dimer produce the precursors of respective arenes. This seems to provide an explanation of the increased content of C_{10} arenes in the butadiene reaction products.

TABLE 2

Possible versions of aromatics formation from hydrocarbon
fragments

C_8-C_9 arenes			
Possible assemblage types			
MRA found, 300°, 1h^{-1} MRA calcd.	1.77 1.78　　1.44	2.26 2.13　　1.84	2.16 2.13　　1.84
MRA found, 400°, 2h^{-1} MRA calcd.	1.48 1.47　　1.32	1.51 1.71　　1.55	1.61 1.71　　1.55

Hence, the precursors of aromatics are not diene oligomers,
not even partly cracked, but the products of combination of fragments with a diene molecule. This conclusion is substantiated by
the pattern of changes in the composition of pentadiene conversion products as a function of contact time, by a higher reactivity of non-conjugated dienes, and by the data of tracer studies.
Moreover, the latter technique permits to find the reaction pathways, in which the fragments interact and yield various aromatic
hydrocarbons. The results of C_6-C_8 olefin aromatization experiments are likewise indicative, at least at longer contact times,
of the role played by fragments. In diene aromatization the cracking (fragmentation) processes are, therefore, not side reactions but they constitute active stages in the overall aromatization process. Oligomers may be regarded as intermediates only
as long as they serve as the sources of fragments when the starting molecule fails to yield fragments as in case of ethylene
and butadiene.

EXPERIMENTAL

Hydrocarbon conversions were investigated on a flow-type unit with HS-zeolite (made in the USSR) at 300-500° and in a wide range of contact times. For dienes, the diene: inert diluent molar ratio was 1:5. Liquid and gaseous products were analysed by capillary GLC. Piperylene-3-^{14}C conversion products were analysed using a chromatograph with the katharometer outlet connected to a flow-type radiation counter. Use was made of commercial propylene, 1-butene, isobutene and butadiene (purity, 97.5% or better) and of rectified 1,3-pentadiene and 1,5-hexadiene (purity, 98.5-99%). 2,4-Hexadiene was prepared by the dehydration of unsaturated alcohol with oxalic acid, the alcohol being obtained by the Grignard synthesis from crotonaldehyde and EtBr; 1,4-pentadiene was synthesized according to (ref.11). To prepare piperylene-3--^{14}C, the unsaturated alcohol synthesized from propanal-3-^{14}C and CH_2=CHBr by the Grignard reaction was dehydrated over $MgSO_4$.

REFERENCES

1 G.V. Isaguliants, M.I. Rosengart, and Yu.G. Dubinsky, Catalytic Aromatization of Aliphatic Hydrocarbons, Nauka P.H., Moscow, 1983, 160 pp.
2 W.W. Kaeding and S.A. Butter, J. Catal., 61 (1980) 155-164.
3 Kh.M.Minachev, D.A. Kondratjev, A.A. Dergachev, T.V. Borovinskaya, T.N. Bondarenko, B.K. Nefedov and T.V. Alekseyeva, Izv. AN SSSR, Ser. Khim., 8 (1981) 1833-1838.
4 F. Fajula and F.G. Gault, J. Catal., 68 (1981) 312-328.
5 J.R. Anderson, K. Foger, T. Mole, R.A. Rajadhyaksha and J.V. Sanders,J. Catal., 58 (1979) 114-130.
6 O.V. Bragin, B.K. Nefedov, T.V. Vasina, V.N. Lutovinova, T.V. Alekseyeva, A.V. Preobrazhensky and Kh.M. Minachev, Dokl. AN SSSR, 255 (1980) 103-106.
7 P.Dejaifve, J.C. Vedrine, V. Bolis and E.G. Derouane, J. Catal., 63 (1980) 331-345.
8 Kh.M. Minachev, D.A. Kondratjev, B.K. Nefedov, A.A. Dergachev, T.N. Bondarenko, T.V. Alekseyeva and T.B. Borovinskaya, Izv. AN SSSR, Ser. Khim., 11 (1980) 2509-2513.
9 G.V. Isaguliants, K.M. Gitis, V.N. Kornyshev, B.K. Nefedov and T.V. Alekseyeva, Izv.AN SSSR, Ser. Khim., 2 (1984), in press.
10 F. Chevalier, M. Guisnet and R. Maurel, The Sixth International Congress on Catalysis, London, 1976, Paper A-38.
11 B.N. Mikhailov and Yu.N. Bubnov, Zh. Obshch. Kim., 41 (1971) 2309.

P.A. Jacobs et al. (Editors), *Structure and Reactivity of Modified Zeolites*
© 1984 Elsevier Science Publishers B.V., Amsterdam — Printed in The Netherlands

CONVERSION OF LINEAR BUTENES TO PROPYLENE ON H–ZSM–5 ZEOLITES: EFFECT OF
REACTION PARAMETERS AND ZEOLITE MORPHOLOGY ON CATALYTIC ACTIVITY

Fausto COLOMBO and Giordano DE ALBERTI

Montedipe S.p.A., Research Center of Bollate, Via San Pietro 50, 20021 Bollate
(Milano) Italy

ABSTRACT

The conversion of n-butene on H–ZSM–5 zeolites was studied at high temperatures and high space velocities. At 823 K and WHSV greater than 50 h^{-1} selectivities towards propylene of 40% at 80% n-butene conversion were obtained. Zeolites characterized by very small crystallites and spherulitic morphology showed a better resistence to decay.

INTRODUCTION

Propylene is industrially produced by thermal or catalytic cracking of naphtha.   A shortage of this product is expected at the end of the 80's because of a progressive shift of steam cracking feeds from naphtha to LPG |1-3|. The only alternative process available for propylene production is propane dehydrogenation |4|.   In the present work we describe a new route to propylene, consisting in the high temperature-high space velocity conversion of n-butenes on H–ZSM–5 zeolites |5|.

The high temperature conversion of butenes on these catalysts in continuous flow reactors has been studied sporadically so far.   A few authors e.g./6–8/ made experiments under particular reaction conditions in order to elucidate the mechanism of methanol conversion to light olefins or gasoline, butenes being intermediates and products. Other experimental data appeared in the patent literature concerning respective aromatization reactions e.g./9,10/.

Olefins are extremely reactive on acidic zeolites, about two orders of magnitude more reactive than the corresponding paraffinic hydrocarbons |11|.

Bednarova et al. |12| compared the performance of different acidic zeolites for the conversion of the C$_4$ olefinic cut, after isobutene extraction, and showed the remarkable stability of H-ZSM-5 zeolites.

In the present work we have investigated the influence of temperature and contact time on the conversion of n-butenes on H-ZSM-5 zeolites; we have also studied the effect of the crystallite dimensions on catalyst decay.

EXPERIMENTAL

Five ZSM-5 samples of SiO_2/Al_2O_3 molar ratio of 28 were synthesized following two different general procedures, reported in the patent literature: i.e. using tetra-propyl-ammonium bromide (TPA-Br) |13| and tri-ethanol-amine (TEA) |14| as a templating agent, respectively. The synthesis parameters (temperature, time, stirring rate etc....) were changed in order to obtain samples with different morphologies and crystallite dimensions.

All the zeolites were calcined at 813 K for 10 hours in air, then exchanged with NH_4Cl (1.0 N) at 343 K five times and deammoniated at 813 K for 10 hrs. Before testing for catalytic activity all catalysts were extruded using 20% SiO_2 (Ketjensol AS 40) as a binder and activated for 2 hrs at 813 K.

XRD measurements of the calcined and activated samples were recorded using CuKα radiation by means of a Philips 1140 diffractometer. The crystallite sizes were estimated by the classical Scherrer equation |15|.

The morphological characteristics of the five catalysts employed for experiments are reported in table 1.

TABLE 1

Crystallite size and morphology of ZSM-5 samples

Samples	Templating agent	Crystallite size (nm)	Morphology
A	TEA	78	spherulites
B	TEA	170	euhedral intergrowths
C	TPA-Br	30	spherulites
D	TPA-Br	34	spherulites
E	TPA-Br	40	spherulites

The apparatus used for testing the catalytic activities consisted in a fixed--bed stainless steel downflow reactor (10 mm i.d.).

Philips pure grade 1:1 mixtures of 2-butenes were used as reagents. The products emerging from the reactor were cooled by means of a water condenser. Liquid and gas products were analyzed separately using a Carlo Erba Model C and a Carlo Erba Fractovap gaschromatograph, respectively.

RESULTS AND DISCUSSION

The conversion of a 1:1 mixture of cis and trans 2-butene on H-ZSM-5 zeolites was studied in a continuous flow reactor at atmospheric pressure (no inert diluent was used) at various temperatures and weight hourly space velocities (WHSV). Tests carried out with 1-butene as reagent in the same reaction conditions have shown that, at temperatures greater than 573 K, the product distribution is independent of the particular n-butene employed.

Because the n-butenes in the product mixture were always in thermodynamic equilibrium concentration among themselves, the conversions have been calculated considering all the n-butenes flowing out of the reactor as unreacted reagents.

Variation of the product distribution with time on stream

In all the experimental conditions tested a quasi-stationary state with respect to the reaction selectivity was reached. The time needed varied, depending on temperature and space velocity. At a given temperature, by increasing the WHSV, the time needed to reach steady-state became progressively shorter.

In fig. 1 the variation of reaction conversion and selectivity with time on stream is shown in a test carried out at 623 K and WHSV = 4.0 $g \cdot g_{cat}^{-1} \cdot h^{-1}$ using catalyst B.

Conversion of n-butenes was above 90% and decreased a few per cent only within 12 hrs. The selectivity towards $\leqslant C_4$ paraffinic hydrocarbons and aromatics (mainly toluene, benzene and xylenes) decreased sharply in the first three hours; correspondingly the selectivity towards aliphatic liquid products, mainly linear and branched C_5-C_8 olefins, increased.

Aromatics and paraffinic light gases are produced together through olefin oligomerization-dehydrocyclization reactions, which require the strongest acid sites of the zeolite. Such catalytic sites become deactivated first because aromatics are the precursors for the formation of coke |16|.

The catalyst aging was comparatively slower in the temperature range between

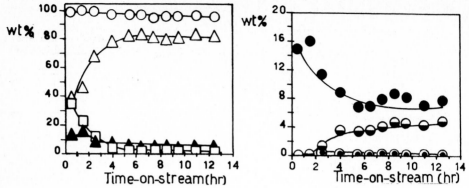

Fig. 1 Variation of conversion and selectivity with time on stream; T = 623 K, WHSV = 4.0 hr^{-1}. ($\bigcirc$ n-C$_4^-$ Conv.; $\triangle$ Sel. to aliph. liquids; $\blacktriangle$ Sel. to aromatics; $\square$ Sel. to $\leqslant$ C$_4$ sat. gases; $\bullet$ Sel. to C$_3^-$; $\ominus$ Sel. to iso-C$_4^-$; $\odot$ Sel. to C$_2^-$).

573 K and 823 K.   At lower temperatures the loss of activity was probably due to olefin polymerization and diffusional limitations in the zeolite channels, which produced a progressive pore blockage.   At temperatures higher than 873 K carbon deposits became the predominant agents of catalyst decay |16|.

When the zeolites were reactivated in air flow for 2 hrs at 823 K the initial catalytic activity was restored.   After regeneration the reaction conversion and selectivity showed the same trends with time on stream of the fresh catalysts.

All these general observations about the influence of time on stream on product distribution are in general agreement with those reported by Fajula and Gault |17,18| relative to the conversion of methylpropene-2-^{13}C on H-Mordenite.

Effect of temperature on reaction conversion and selectivity

Catalyst B was used to investigate the effect of temperature on conversion and selectivity, when operating at high space velocity (see figure 2).   The n-butene conversion measured at the 6th hour on stream, reached maximum values between 623 K and 723 K;   at higher temperatures, the conversion lowered progressively because the catalyst aging became faster.

As temperature increased the amount of liquid products decreased but their content in aromatics was progressively higher;   correspondingly the selectivity for propylene formation rose to maximum values between 773 K and 823 K.   Iso-butene was in equilibrium with the n-butenes at temperatures higher than 723 K. Selectivity towards $\leqslant$ C$_4$ paraffinic hydrocarbons was always rather low.

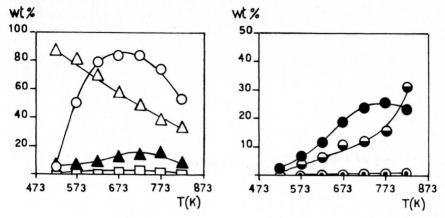

Fig. 2 Effect of temperature on conversion and selectivity; WHSV = 50 hr^{-1}.
(Symbols are defined in Fig. 1).

Effect of crystallite size on catalyst decay at high space velocities and temperatures

The remarkable selectivities for propylene formation at high space velocities
and temperatures greater than 673 K were joined with rapid catalyst decay.   The
effect of crystallite size on catalyst aging was therefore investigated at 823 K
(see figure 3).

There was a remarkable difference among the catalysts;   the most stable were
the ones with smaller crystallite size which had spherulitic morphology.

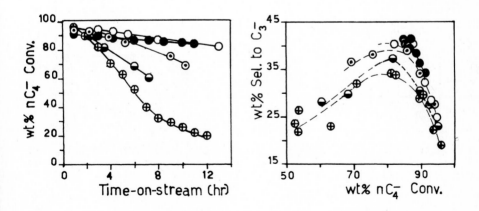

Fig. 3 Effect of zeolite crystallite size on catalyst decay and on selectivity
towards propylene;   T = 823 K, WHSV = 50 hr^{-1} (⊖ cat. A; ⊕ cat. B; ○ cat. C;
● cat. D; ⊙ cat. E).

Furthermore the smaller were the crystallites the better was the propylene yield.
The trend of selectivity towards propylene with time on stream was strongly
dependent on n-butene conversion. With catalyst C the conversion was fairly
stable around 90% and correspondingly the selectivity for propylene formation
was constantly above 40%. It is interesting to observe that the maximum concentrations of propylene in the products were found at 80% n-butene conversion.

Effect of space velocity on conversion, selectivity and catalyst decay at 773 K

Various experimental runs were made at 773 K varying the WHSV from 23.5 to
329 $g.g_{cat}^{-1}.h^{-1}$ using sample C as catalyst. The experimental results shown in
figure 4 are a clear evidence of the very high reactivity of olefins on H-ZSM-5
zeolites and also of the better stability of samples with very small crystallites. In fact the influence of space velocity on n-butene conversion was
surprisingly small; catalyst decay became only slightly faster at higher
space velocities.

As previously said, aromatics are the precursors of heavy polycyclic unsatu-
rates and of coke, which cause catalyst aging at high temperatures |16|. As
WHSV was increased the concentration of aromatics in the products lowered
progressively, compensating in part the faster catalyst decay due to the higher
amount of reagents passing on the zeolite per unit time.

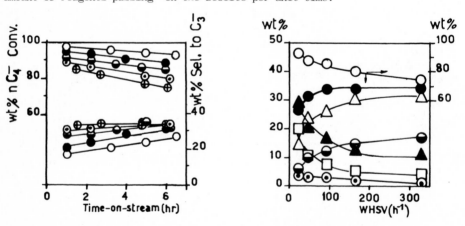

Fig. 4 A.: effect of WHSV on catalyst aging at 773 K (WHSV: ○ 23 hr^{-1}, ● 50 hr^{-1},
◐ 93 hr^{-1}, ◉ 165 hr^{-1}, ⊕ 330 hr^{-1}). B.: effect WHSV on conversion and selectivity at the 6th hr on stream. (Symbols are defined in Fig. 1).

The selectivity towards propylene and isobutene increased with WHSV; these olefins were in equilibrium with the n-butenes in the product flowing out of the reactor at the higher space velocities.

CONCLUSIONS

The experimental results of the present work have shown that good propylene yields may be achieved by reacting n-butenes at high temperatures and space velocities on H-ZSM-5 zeolites. In these operative conditions the formation of aromatics and paraffinic light gases is greatly reduced; apart from propylene the other main reaction products are isobutene and C_5-C_8 olefins, ethylene is produced only in minor amounts.

At conversions higher than 80% propylene and isobutene are in equilibrium with the n-butenes, therefore the propylene yield is limited by the thermodynamic equilibrium among the $\geqslant C_3$ olefins. In fact, when pure isobutene [19] or any other linear or branched C_5-C_{10} olefin [20] is converted on H-ZSM-5 zeolites at high temperature and space velocity, the same remarkable conversions to propylene are obtained. Therefore, if all the $\geqslant C_4$ olefins are recycled to the reactor and mixed with the fresh feed of n-butenes, propylene may be produced in total yields greater than 80% [21].

This catalytic transformation could be usefully coupled with a conventional thermal cracking plant in order to increase the propylene/ethylene ratio according to the market demands.

ACKNOWLEDGEMENTS

The authors wish to thank Dr. E. Moretti and Mr. M. Padovan for the synthesis of catalysts used in this study, Prof. G. Carazzolo and Dr. F. Gatti for XRD measurements and Dr. M. Solari for SEM photographs. The contributions of Mr.s A. Pelorosso and E. Pogliani in carrying out the experimental work are also gratefully acknowledged.

REFERENCES

1   Chem. Eng., July 26 (1982) pp. 12D-12E
2   Eur. Chem. News, Febr. 28 (1983) p. 11
3   A. Mol, Oil G.J., Mar. 28 (1983) p. 78
4   P.R. Pujado, B.V. Vora and R.C. Berg, Oil G.J., Mar. 28 (1983) p. 71

5   F. Colombo, G. De Alberti and E. Moretti, (Montedipe), Ital. Pat. Appl.
    24,152 A/83 (1982)
6   P. Dejaifve, J.C. Vedrine, V. Bolis and E.G. Derouane, J. Catal. 62 (1980)
    p. 331
7   Y. Ono and T. Mori, J. Chem. Soc., Faraday Trans. 1, 77 (1981) p. 2209
8   J.C. Vedrine, P. Dejaifve, E.D. Garbowski and E.G. Derouane, St. Surf. Sci.
    Catal. 5, Catalysis by Zeolites, Elsevier, Amsterdam (1980) p. 29
9   N.Y. Chen and W.O. Haag, (Mobil Oil), U.S. Pat. 4,100,218 (1978)
10 E.E. Davies and A.J. Kolombos, (B.P.) U.S. Pat. 4,180,689 (1979)
11 W.O. Haag, R.M. Lago and P.B. Weisz, Faraday Disc. Chem. Soc. 72 (1981)
    p. 317
12 S. Bednarova, P. Jiru, L. Kubelkova and B. Wichterlova, Proc. 4th Italo-
    Czechoslovak Symp. of Catal., Torino, Sept. 1983, ed. Turin University,
    Torino (1983) p. 125
13 R.J. Argauer and G.R. Landolt, (Mobil Oil), U.S. Pat. 3,702,886 (1972)
14 E. Moretti, M. Padovan, M. Solari, C. Marano and R. Covini, (Montedipe),
    Ital. Pat. Appl. 19,238 A/82 (1982)
15 H.P. Klug and L.E. Alexander, X Ray Diffraction Procedures, J. Wiley,
    London (1954) p. 491
16 E.G. Derouane, St. Surf. Sci. Catal. 5, Catalysis by Zeolites, Elsevier,
    Amsterdam (1980) p. 5
17 F. Fajula and F.G. Gault, J. Catal. 68 (1981) p. 291
18 F. Fajula and F.G. Gault, J. Catal. 68 (1981) p. 312
19 F. Colombo, S. Contessa and G. De Alberti, (Montedipe), Ital. Pat. Appl.
    19,292 A/83 (1983)
20 F. Colombo, G. Paparatto and E. Moretti, (Montedipe), Ital. Pat. Appl.
    24,550 A/82 (1982)
21 F. Colombo and G. De Alberti, (Montedipe), Ital. Pat. Appl. 20,896 A/83
    (1983)

P.A. Jacobs et al. (Editors), *Structure and Reactivity of Modified Zeolites*
© 1984 Elsevier Science Publishers B.V., Amsterdam — Printed in The Netherlands

STUDY OF ETHYLENE OLIGOMERIZATION ON BRONSTED AND LEWIS ACIDIC
SITES OF ZEOLITES USING DIFFUSE REFLECTANCE IR SPECTROSCOPY

L.M. KUSTOV, V.YU. BOROVKOV and V.B. KAZANSKY
N.D. Zelinsky Institute of Organic Chemistry, Academy of Sciences
of the USSR, Moscow (USSR)

ABSTRACT

Different types of Lewis acidic sites in zeolites (three-coordinated aluminium and silicon atoms of the lattice and extra-
lattice aluminium) were detected by means of the diffuse reflectance IR-spectroscopy using low-temperature adsorption of molecular hydrogen. The lattice Lewis sites unlike the extralattice
centres were found to initiate C_2H_4 oligomerization via a cationic mechanism resulting in formation of branched products. The
reaction on the strongest Brønsted acidic sites of the high silica containing zeolites leads to linear oligomers. Ethoxy groups
were suggested to be intermediates in this reaction.

INTRODUCTION

Catalytic oligomerization of C_2-C_3 olefins is an important
stage of the fuel synthesis from non-oil raw materials (methanol,
ethanol, light olefins etc.) (ref.1). It has been established
that ethylene oligomerization proceeds on acidic OH-groups of ZSM
type zeolites even at 300K (ref.2-4). The reaction mechanism is
supposed to involve carbonium ion formation (ref.3). The less acidic HY zeolites are inactive in this reaction. On the other hand,
Liengme and Hall (ref.5) and Kubelková et al. (ref.6) have shown
that dehydroxylation of HY zeolite at high temperatures resulted
in the appearance of activity in ethylene oligomerization. On
this basis they supposed that this reaction may be also catalyzed
by Lewis acidic sites. However, the nature of these sites and the
conditions of their formation in zeolites are still not elucidated enough. Therefore, the question of the participation of these
centres in ethylene oligomerization on dehydroxylated samples
remains unsolved.

Recently we have developed a new method for the identification of aprotic acidic sites on the surface of adsorbents and catalysts based on IR spectroscopic study of low-temperature adsor-

ption of molecular hydrogen (ref.7). It allows us to distinguish experimentally different types of Lewis acidic sites arising during dehydroxylation. In the present work this technique together with spectroscopic study of ethylene adsorption was applied to elucidate the role of Lewis acidic sites in oligomerization of light olefins over highly dehydroxylated zeolites. In addition, we have also investigated C_2H_4 conversion over H-forms of zeolites containing only Brönsted acidic sites.

EXPERIMENTAL

HY zeolites (Si/Al=2.9) with a decationization degree of α=99.5 and 75%, H-mordenite (Si/Al=5.0), α=96% and H-ZSM-5 (Si/Al=35), α=99% were prepared from Na-forms by an ion exchange with a 0.2N NH_4Cl solution followed by a thermovacuum treatment at various temperatures. Dealuminated mordenite (Si/Al=12.5) was prepared by an acidic treatment of the Na-form with a 1N HCl.

IR-spectra of powdered samples in the 2000-4000cm^{-1} range were measured at 300K using a Perkin-Elmer 580B spectrophotometer supplied with a home-made diffuse reflectance adapter (ref.8). The measurements of IR-spectra of adsorbed hydrogen (P=4kPa) were carried out at 77K using a Beckman Acta M-YII spectrophotometer according to the procedure described in (ref.7).

Ethylene was purified from water traces by a prolonged storage over silica gel calcined at 970K in vacuum. The oligomerization reaction was studied at 300K spectroscopically directly in IR-cells under static conditions.

RESULTS AND DISCUSSION

Interaction of ethylene with Lewis acidic sites of zeolites.

Fig.1a represents the evolution of IR-spectra of HZSM-5 and HY zeolites, pretreated at 1270 and 970K, respectively, after C_2H_4 adsorption at 300K. With the increase of time the intensity of the bands at 2870, 2930 and 2970cm^{-1} which are attributed to the vibrations of CH_3 (2870 and 2970cm^{-1}) and CH_2 (2930cm^{-1}), are increasing. These groups definitely arise in the course of ethylene oligomerization leading to the formation of branched hydrocarbon chains. The absence of the bands in the region of 3550-3650cm^{-1} (Fig.1a) indicates that the samples under study do not contain any acidic OH-groups.

Dehydroxylation of zeolites is known to result in the formation of Lewis acidic sites. They can be detected IR- spectro-

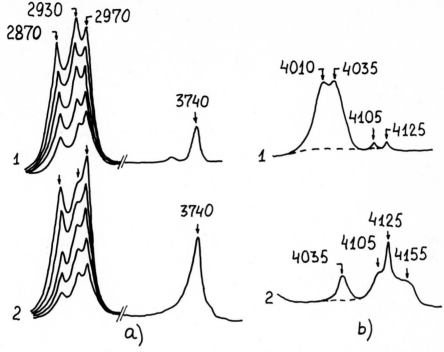

Fig.1. a) IR spectra of HZSM-5 (1) and HY (2) zeolites measured
in 1, 8, 20, 100 and 200 min after adsorption of ethylene. The samples were pretreated at 1270 and 970K, respectively, in vacuum.
b) IR spectra of H_2 adsorbed at 77K on the pretreated
HZSM (1) and HY (2) zeolites (solid lines) and after
ethylene adsorption at 300K (dotted lines).

scopically by the adsorption of molecular hydrogen. In Fig.1b IR
-spectra of hydrogen adsorbed at 77K on dehydroxylated HZSM and
HY zeolites are shown. According to our previous data (ref.9,10)
the bands at $4010 cm^{-1}$ and $4035 cm^{-1}$ belong to H_2 molecules adsorbed on the lattice three-coordinated aluminium atoms and silicon
ions, respectively. The lines in the region of $4100-4160 cm^{-1}$ correspond to the molecules interacting with residual Na^+ cations or
non-acidic SiOH-groups.

The previous oligomerization of C_2H_4 on these zeolites at
300K does not affect the intensity of the bands of adsorbed hydrogen at $4100-4160 cm^{-1}$. Consequently, the growing oligomer
chains do not block sodium cations and SiOH groups inside the cavities and channels of these zeolites. Ethylene preadsorption
leads, however, to a disappearance of the bands at $4010 cm^{-1}$ and
$4035 cm^{-1}$ (Fig.1b, dotted lines), which are connected with the

lattice Lewis acidic sites. These data show that active sites of ethylene oligomerization on dehydroxylated zeolites contain either three coordinated aluminium or silicon atoms in the lattice.

Oligomerization of C_2H_4 branched oligomers also takes place on the acidic Lewis sites of the lattice of decationized and dealuminated mordenites. The extent of the chain branching depends on the structure of a zeolite. The highest branching degree was observed for HY zeolites ($I^{CH}3/I^{CH}2>3$, where I are the intensities of the bands of CH_3 and CH_2 groups at 2970 and 2930cm^{-1}, respectively). On the contrary, with HZSM and HM zeolites which have narrow channels the oligomers exhibit comparable amounts of CH_3 and CH_2 fragments ($I^{CH}3/I^{CH}2=0.8-1.2$).

In addition to the lattice Lewis sites, dehydroxylation sometimes also results in the formation of extralattice aluminium atoms, which are formed because of the partial dealumination of the zeolitic framework. These sites are connected with the band at 4060cm^{-1} in the spectra of adsorbed hydrogen (see Fig.2). According to IR-spectra they do not, however, take part in ethylene oligomerization and form only weak complexes with ethylene molecules which can be removed by evacuation at 300K (see Fig.2).

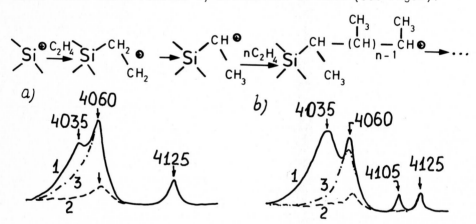

Fig.2. IR spectra of hydrogen adsorbed at 77K on H-forms of zeolites of Y (a) and mordenite (b) type pretreated at 970K under deep-bed conditions (1), after adsorption of ethylene (2) and after evacuation at 300K for 1h (3).

All these results about the formation of branched oligomers on the Lewis acidic sites of the lattice could be explained by the cationic mechanism given above.

Interaction of ethylene with Brönsted acidic sites.

In Fig.3 IR-spectra of ethylene adsorbed at room temperature

on HZSM-5 zeolite pretreated at 870K in vacuum are shown. The
sample contains only Brönsted acidic sites which are indicated by
the IR band at 3610cm^{-1}. Ethylene adsorption results in hydrogen
bonding between these hydroxyl groups and C_2H_4 molecules (the
broad band at 3250cm^{-1}) as well as in the increase of the intensity of the lines at 2860 and 2930cm^{-1} belonging to the vibration of CH_2 groups in oligomerization products. This is an evidence for the linear structure of the oligomer chains in agreement with the data presented in (ref.3). In the course of reaction the concentration of the hydrogen-bonded complexes of acidic
OH-groups with C_2H_4 molecules decreases whereas the concentration of the complexes formed between these hydroxyl groups and
saturated hydrocarbon chains of the growing oligomers increases,
resulting in the appearance of the band at 3470cm^{-1}. A similar
line was earlier observed in (ref.9) after the adsorption of
C_6H_{14} on HZSM-5 zeolite. During the reaction the release of some
part of acidic OH-groups also takes place. These transformations
are connected with the existence of two isobestic points in the
IR-spectra near 3400 and 3000cm^{-1}.

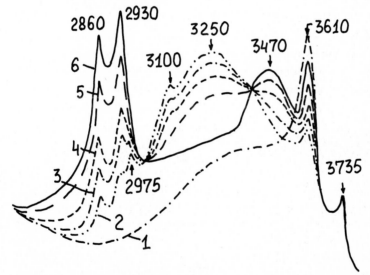

Fig.3. IR spectra of HZSM zeolite pretreated at 870K (1) and
  1, 8, 15, 120 and 240 min after adsorption of $2 \cdot 10^{20}$ ethylene molecules per gramm of zeolite (2-6).

Adsorption of ethylene on the deuterium form of ZSM-5 zeolite and of C_2D_4 on HZSM-5 shows that at 300K there is a fast proton exchange between hydroxyl groups and ethylene molecules occuring in hydrogen-bonded complexes. For example, in the case of

C_2D_4 adsorption on HZSM, the appearance of the intensive bands at 2660cm^{-1} and 2400cm^{-1} attributed to free OD-groups and those perturbed by ethylene, respectively, was detected even one minute after ethylene adsorption. The rate of proton exchange considerably exceeds that of oligomerization. It can proceed via formation of either ethoxy structures (I) or carbonium ions (II), depending on the ionic character of the C-O bond:

$$\begin{array}{c} \text{H} \\ \text{O} \\ \triangle \end{array} + \ C_2H_4 \ \rightleftharpoons \ \begin{array}{c} CH_2{\cdots}CH_2 \\ \text{H} \\ \text{O} \\ \triangle \end{array} \ \rightleftharpoons \ \begin{array}{c} CH_2\text{-}CH_3 \\ \text{O} \\ \triangle \end{array} \ (I)$$

$$\begin{array}{c} \oplus CH_2\text{-}CH_3 \\ \ominus \\ \text{O} \\ \triangle \end{array} \ (II)$$

As a result of such a fast exchange, C_2HD_3 or C_2H_3D molecules are formed, which are transformed into oligomers containing CHD methylene fragments. They are displayed in the IR-spectra by the lines in the regions of C-H or C-D vibrations at 2920cm^{-1} and 2143cm^{-1}, respectively.

Thus, there are two possible ways of C_2H_4 oligomerization with the participation of the strong Brönsted acidic sites of zeolites: the carbonium ion mechanism (a) suggested in (ref.3) and the concerted mechanism (b) with the structures I or II as active intermediates, respectively:

$$\begin{array}{c} \oplus CH_2CH_3 \\ \ominus \\ \text{O} \\ \triangle \end{array} \xrightarrow{C_2H_4} \begin{array}{c} \oplus OH_2CH_2CH_2CH_3 \\ \ominus \\ \text{O} \\ \triangle \end{array} \rightarrow \begin{array}{c} CH_3\overset{\oplus}{C}HCH_2CH_3 \\ \ominus \\ \text{O} \\ \triangle \end{array} \rightarrow \ \dots \ (a)$$

$$\dots (b)$$

We believe that the mechanism (b) which should lead to the formation of the linear polymethylene chains is more probable. Besides that, the carbonium ions should isomerize to more energetically preferable secondary or tertiary ions resulting in the formation of the branched oligomers. This does not agree with the experimental data. As it follows from the experiment on ethylene oligomerization with the participation of Lewis acidic sites, the branching is not restricted by the size of the cavities and channels in all zeolites under study. In principal, the possibility of formation of the branched chains in channels of ZSM zeolites is also confirmed by the data on propene oligomerization (ref.3).

The study of ethylene oligomerization over decationized and dealuminated mordenites which do not contain Lewis sites shows that this reaction proceeds only on the strongest acidic OH -groups which are characterized in the spectra by the band at $3610 cm^{-1}$.

Thus, the data obtained in this work demonstrate that the active sites of ethylene oligomerization at 300K on different types of zeolites could involve both strong Brönsted acidic sites and lattice Lewis acidic centres. The structure of hydrocarbon chains which are formed in this reaction depends on the nature of these active sites (Brönsted or Lewis). This could be used for the identification of the nature of active centres.

REFERENCES

1 J.H.C. van Hooff, in P.Prins and G.C.A. Schuit (Eds.), Chemistry and Chemical Engineering of Catalytic Processes, NATO Advanced Study Institute Series, Series E, Applied Sciences, 1980, No.39, p. 599.
2 V. Bolis, J.C. Vedrine, J.P. van den Berg, J.P. Wolthuizen and E.G. Derouane, J. Chem. Soc., Faraday Trans. I, 76(1980),1606.
3 J. Nováková, L. Kubelková, Z. Dolejšek and P. Jírů, Coll. Czechosl. Chem. Commun., 44 (1979), 3341.
4 J.P. Wolthuizen, J.P. van den Berg and J.H.C. van Hooff, in B. Imelik et al. (Eds.), Catalysis by Zeolites, Elsevier, Amsterdam, 1980, p.85.
5 B.V. Liengme and W.K. Hall, Trans. Faraday Soc., 62(1966),3229.
6 L. Kubelková, J. Nováková, B. Wichterlová and P. Jírů, Coll. Czechosl. Chem. Commun., 45 (1980), 2280.
7 L.M. Kustov, A.A. Alexeev, V. Yu.Borovkov and V.B. Kazansky, Doklady Akad. Nauk SSSR, 261 (1981), 1374.
8 L.M. Kustov, V. Yu. Borovkov and V.B. Kazansky, J. Catal., 72 (1981), 149.
9 V.B. Kazansky, L.M. Kustov and V.Yu. Borovkov, Zeolites, 3 (1983), 77.
10 V.B. Kazansky, V.Yu. Borovkov and L.M. Kustov, in press.

P.A. Jacobs et al. (Editors), *Structure and Reactivity of Modified Zeolites*
© 1984 Elsevier Science Publishers B.V., Amsterdam — Printed in The Netherlands

VARIOUS STATES OF Cr IONS IN Y AND ZSM-5 ZEOLITES AND THEIR CATALYTIC ACTIVITY IN ETHYLENE POLYMERIZATION [*]

B. WICHTERLOVÁ, Z. TVARŮŽKOVÁ, L. KRAJČÍKOVÁ AND J. NOVÁKOVÁ

J. Heyrovský Institute of Physical Chemistry and Electrochemistry, Czechoslovak Academy of Sciences, CS-121 38 Prague 2 (Czechoslovakia)

ABSTRACT

The state of Cr ions in Y and ZSM-5 zeolites is characterized by means of IR, ESR, XPS and TPR data. The considerable activity of intensively oxidized zeolites in ethylene polymerization compared with Cr(III) and Cr(II) zeolites is connected with the formation of Cr(V) finely dispersed within zeolite cavities most probably in chromate-like compounds. The catalytic effect of Cr in HZSM-5 is not observed because of high polymerization activity of this zeolite itself. While ethylene polymerization on CrHY and CrSY zeolites yields long linear chains of polymer on both the CrHZSM-5 and HZSM-5 zeolites shorter, slightly branched polymers are formed. Fast polymerization causes blocking of the ZSM channels and interruption of the polymerization reaction.

INTRODUCTION

The catalytic activity of finely dispersed Cr ions has iniciated a number of studies in the field of Cr zeolite chemistry (refs.1-3). Recently we have paid attention to the characterization of the state of Cr ions in Y zeolites and its relationship to the zeolite activity in ethylene polymerization (refs.4-7).

It has been found that the zeolite activity increases with increasing Cr valency from Cr(II)to Cr(V); Cr(VI) is inactive. This finding was related to the electron acceptor properties of the Cr ions and their free coordination sphere.

This paper describes an attempt to characterize in more detail the state of Cr ions in oxidized Y zeolites with the highest activity in ethylene polymerization. Further, the spectral characteristics of the Cr ions in CrHZSM-5 zeolite are considered, as well as the activity of this zeolite in ethylene polymerization.

EXPERIMENTAL

The zeolites used were HY, stabilized SY and HZSM-5 zeolites with Si/Al ratios of 2.5, 2.8 and 13.5, respectively. H forms of zeolites were prepared via NH_4^+ ion exchange followed by thermal decomposition of NH_4^+ ions. The Cr ions were introduced into the zeolites by ion exchange with $CrCl_3$ solution (see ref.4) or

---

[*]Part V.: Cr Ions within Zeolites (refs.4-7).

were supported on the SY zeolite using 1M $(NH_4)_2CrO_4$ solution; after thermal decomposition the zeolite is denoted CrO_nSY. The preservation of the zeolite structure was checked by the sorption capacity for argon (13.3 kPa at 77 K; the Table) and the IR spectra of skeletal vibrations. Although the zeolite with supported Cr ions exhibited slightly lower value of the Ar sorption capacity the IR spectra evidenced its high crystallinity. However, partial breakdown of the zeolite structures was observed in the IR spectra of the intensively oxidized CrHY and CrSY zeolites.

The state of Cr in zeolites was followed by the IR and ESR spectra of adsorbed NO (1.3 kPa at 298 K), the ESR spectra of Cr(V), the XPS spectra of the Cr 2p level and the TPR of zeolites using hydrogen. The catalytic activity for ethylene polymerization was studied in a static apparatus (5.3 kPa at 353 K) by the weight increase of a zeolite plate with a correction for the amount of reversibly adsorbed ethylene. The character of ethylene interaction with zeolites (5.3 kPa at 353 K for 30 minutes followed by evacuation at 353 K for 15 minutes) was followed by the IR spectra at 298 K. Various zeolite treatments before measuring the zeolite spectral characteristics and activity in ethylene polymerization are given in the Table. The IR spectra were recorded on a Nicolet MX-1 FT-IR spectrometer at 298 K. A total of 270 scans was carried out for each spectrum. The X-band ESR spectra were measured on an ERS-220 spectrometer(Acad.Sci., Berlin) with 100 kHz frequency. The XPS spectra were obtained using an ESCA 3 MK II VG spectrometer. The $AlK\alpha_{1,2}$ line was used with a power of 200 W. The BE were related to the Si 2p level at 102.8 eV (the C 1s level at 285.0 eV). For details see refs. 4 and 7.

RESULTS AND DISCUSSION

Cr ions in HY and stabilized SY zeolites. It has been demonstrated previously (ref.4) that the short-term oxidation(at 770 K for 3 hours) of CrHY and CrSY zeolites is accompanied by the formation of rather unstable Cr(VI) and Cr(V) ions, the latter being detected by the ESR signal at 1.976 with only a slight anisotropy (the Table, Fig. 1). The instability of the Cr(VI) and Cr(V) ions was evidenced, in addition to changes in the Cr(V) ESR signal intensity, by their self-reduction under X-ray irradiation and in high vacuo in the ESCA system. The BE values for these oxidized zeolites were found to be practically equal to those for Cr(III) zeolites (578.4 eV, the Table, ref.7).The prolongation of the oxidation time (up to 30 hours at 770 K) considerably increased the Cr(V) ESR signal intensity ($g_\perp$=1.976, $g_\parallel$ =1.923) in comparison with that for zeolites oxidized for only 3 hours. Similarly, the XPS spectra of the Cr 2p level of intensively oxidized zeolites exhibited a BE value of 580.5 eV (ΔE=8.8 eV) showing much stable higher Cr valency (the Table). Even though the partial breakdown of the zeolite structure was observed the oxidation process was not

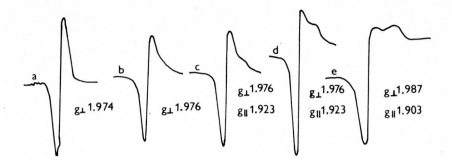

Fig. 1. ESR spectra of Cr(V) recorded at 298 K: in zeolites oxidized in O_2 at 770 K for 3 hours and evacuated at 620 K for 1 hour a) CrHZSM-5, b) CrSY, e) CrO_nSY; in zeolites oxidized in O_2 at 770 K for 30 hours and evacuated at 620 K for 1 hour, c) CrSY, d) CrHY.

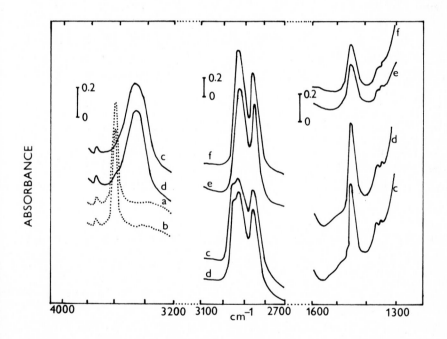

Fig. 2. IR spectra of ethylene interaction (1.3 kPa adsorbed at 353 K for 30 minutes, evacuation at 353 K for 15 minutes). Zeolites pretreated in vacuo at 620 K: a) HZSM-5, b) CrHZSM-5. Zeolites after interaction with ethylene: c) HZSM-5, d) CrHZSM-5, e) oxidized CrSY, f) CrO_nSY.

accompanied by the significant change of the surface concentration of Cr ions.

To bring more insight into this problem $(NH_4)_2CrO_4$ supported on SY was studied. The evacuation of CrO_nSY at elevated temperatures caused gradual reduction of Cr(VI) which was observed by TPR and by the intensity changes of the ESR

TABLE

Characteristics of various treated zeolites with exchanged and supported Cr ions.

Zeolite	Cr content wt.%	treatment	Ar capacity mmol/g^x	Cr2p level BE eV	Cr2p level dE eV	Cr(V) ESR $g_\perp$	Cr(V) ESR $g_\parallel$	Cr(V) ESR int. a.u.	IR cm^{-1}	Cr(III)(NO)$_2$ ESR $g_\perp$	Cr(III)(NO)$_2$ ESR $g_\parallel$	ethylene polymerization wt% after 30 minutes
CrSY	1.61	vacuo 293K	8.9	578.4	9.2							10.2
		vacuo 620K 18h							1900 1870	1.987	1.906	
Cr(III) exchange		ox. 770K 3h		578.4	9.1							26.6
		+vacuo 620K 1h				1.976	1.923	824				
		ox. 770K 30h		580.5	8.8							30.1
		+vacuo 620K 1h	8.8			1.976	1.923	1270				
CrHY	2.75	vacuo 293K	10.2	578.2	9.5							4.7
		vacuo 620K 18h							1900 1870	1.987	1.906	
Cr(III) exchange		ox. 770K 3h		578.3	9.2							13.7
		+vacuo 620K 1h										
		ox. 770K 30h		580.6	8.8							23.8
		+vacuo 620K 1h	9.1									
CrHZSM-5	0.08	vacuo 620K 18h	5.2									
Cr(III) exchange		ox. 770K 3h				1.974 sym.		184	1905 1875	1.991	1.914	see Fig.4
		+vacuo 620K 1h										
CrO$_n$ SY	3.80	vacuo 293K	8.7	580.0	9.0							6.0
		vacuo 620K 18h				1.987	1.903	157				5.4
(NH$_4$)$_2$CrO$_4$		vacuo 770K 18h				1.987	1.903	70				
supported		ox. 770K 3h										
		+vacuo 620K 1h				1.987	1.903	400				27.0

^{x}The sorption capacities of Ar for parent SY, HY and HZSM-5 zeolites are 8.9, 11.2 and 5.3, respectively.

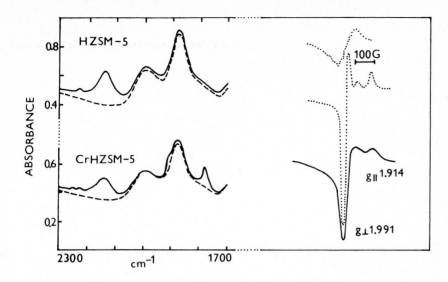

Fig. 3. IR and ESR spectra of NO adsorption on HZSM-5 and CrHZSM-5 zeolites pretreated in vacuo at 620 K for 18 hours: - - - background spectrum; adsorption of 1.3 kPa of NO at 298 K for 1 hour, ——— IR and ESR spectra recorded at 289 K, ... ESR spectrum recorded at 77 K.

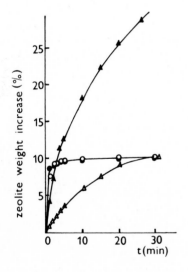

Fig. 4. Zeolite weight increase (wt.%) in time dependence of ethylene interaction with zeolite (at 353 K and 5.3 kPa). Zeolites e-vacuated at 620 K for 18 hours: HZSM-5, CrHZSM-5, CrSY, CrSY oxidized at 770 K for 30 hours and evacuated at 620 K for 1 hour.

signal at $g_\perp$=1.987, $g_\parallel$=1.903 (the Table, Fig. 1). The character of this Cr(V) signal with the axial symmetry implies distorted Td coordinated complexes (ref.8). In contrast, the ESR signal of Cr(V) in oxidized CrHZSM-5 is complete-ly symmetrical (g=1.974). For Cr(V) in CrY zeolites the axial symmetry of the signals is more pronounced for zeolite with a higher amount of Cr and that oxidized more intensively and, eventually, the higher intensity at $g_{\parallel}$ was found for CrO_nSY. This likely indicates highly isolated state of Cr ions in ZSM zeolite and formation of some chromates in intensively oxidized zeolites. The XPS spectra of the Cr 2p level of CrO_nSY exhibited a value of 580.0 eV, being again higher than the BE value for bulk chromates and CrO_3 oxide (cf. refs.7,9). This finding concerning the higher BE values is general for metal ions loaded in zeolites and was found also for Cr(III) ions in Y zeolites as well as on silica (refs.7,10,11). Bearing in mind this fact, the BE values of 580.5 and 580.0 eV for intensively oxidized CrY zeolites and CrO_nSY, respectively, cannot be simply compared with the BE values for bulk chromate compounds having BE values of 579.3, 577.3, 576.7 and 576.0 eV for Na_2CrO_4, Na_3CrO_4, Na_4CrO_4, $NaCrO_2$, respectively (ref.9). It should be noted that the spin-orbit splitting values cannot be also used for the determination of the Cr valence for reasons mentioned in (ref.7, cf. the Table). Thus, the estimation of the valence state of Cr in the surface layers of oxidized zeolites should be based rather on the BE differences between various valences of Cr in chromate-like compounds. As the difference in the BE values for Cr(III) in zeolites and that of oxidized zeolites (in O_2 at 770 K for 30 hours and briefly evacuated at 298 K) is 2.3 eV, it can be assumed that the surface layers of oxidized CrY zeolites contain Cr(V) and Cr(VI) ions. The similar Cr valence estimation is suggested for CrO_nSY.

The ethylene polymerization (performed at 353 K) on intensively oxidized CrSY and CrHY zeolites, as well as on CrO_nSY, proceeded with the formation of long linear polymer structures (the Table, Figs. 2,4). This is reflected in the appearance of IR bands at 2930 and 1460 cm^{-1} (vibrations of the $-CH_2-$, cf. refs. 2,12). The amount of polymer formed and thus also the polymerization rate on the intensively oxidized CrHY and CrSY zeolites were much higher than on these zeolites containing Cr(III) ions and roughly corresponded to the amount of Cr(V) in both oxidized CrSY and CrO_nSY zeolites.

It follows that the intensive oxidation of Cr zeolites, especially zeolites with higher Cr loading, results in the **release** of Cr ions from the cationic sites and the formation of much stable Cr(V) ions. These Cr(V) are finely dispersed within the zeolite structure likely in the form of supported chromates and are responsible for the high catalytic activity of zeolites in ethylene polymerization. The formation of long linear polymer structures (cf. ref.11) as well as a low activity of parent zeolites indicate that the main active sites for ethylene polymerization are Cr ions.

Cr in HZSM-5 zeolite. The adsorption of NO on Cr(III)HZSM-5 yielded the IR and ESR spectra given in the Table and Fig. 3. A couple of IR bands at 1785 and 1905 cm^{-1} with parameters very close to those for Cr(III)(NO)$_2$ complexes found in CrY zeolites (the Table, ref.7) can be ascribed to these complexes within the channel system of the ZSM zeolite. As the band at 2140 cm^{-1} was also found

after NO adsorption on the parent HZSM-5 zeolite, it cannot originate from the
NO adsorption on the Cr ions. The ESR spectrum of NO on CrHZSM-5 recorded at
298 K exhibited a signal at $g_\perp$ =1.991 and $g_\parallel$ =1.914 which can be similarly attributed to the dinitrosyl complexes. The same ESR signal was found at 77 K, however, superposition with an additional signal was observed. Comparing this spectrum with the ESR signal of adsorbed NO on HZSM-5 (g_{av}=1.98,ΔH=90G A=14G, monitored at 77 K) indicates that the additional signal originates from the NO adsorption on sites present on HZSM-5. This type of signal was observed after NO
adsorption on Lewis sites of HZSM-5 zeolite (ref.13) and on dehydroxylated HY
and AlHY zeolites (ref.14). We also observed a similar signal (of much higher
intensity) with well resolved hyperfine splitting from Al nuclei for another
sample of HZSM-5 which contained a substantial amount of amorphous material.
Therefore, the HZSM-5 zeolite considered here most probably contains a minor a-
mount of amorphous phase containing Al with electron acceptor properties. On
the basis of these findings, it can be suggested that the IR band at 2140 cm^{-1}
belongs to the NO adsorbed on Lewis sites of HZSM-5 and CrHZSM-5 zeolites.

The CrHZSM-5 zeolite weight increase due to the ethylene polymerization e-
quals that on the HZSM-5 zeolite. The initial very fast polymer formation was
stopped after 2 minutes of ethylene interaction, while for CrY zeolites their
weight increased continuously (Fig.4). As a result of ethylene interaction with
both the ZSM zeolites the IR band of OH groups (3610 cm^{-1}) was shifted to a lower frequency (3470 cm^{-1}) being caused likely by the interaction of hydroxyls
with the polymer formed (ref.2). Further, the IR spectra of ethylene interaction
with HZSM-5 and CrHZSM-5 zeolites exhibited bands at 2930 cm^{-1} and 1465 cm^{-1}
(vibrations of $-CH_2-$) and bands of lower intensity at 2950, 1380 and 1365 cm^{-1}
(vibrations of $-CH_3$).It follows that the shorter chains of linear polymers as
well as a small amount of branched polymers are formed on both the ZSM zeolites
in contrast to CrY zeolites. Formation of linear polymers on HZSM-5 has already
been reported (ref.12) while branched polymers were observed on dehydroxylated
HY and AlHY zeolites (ref.15), being ascribed to the simultaneous presence of
Brönsted and Lewis sites. Therefore, it seems that the formation of short, only
slightly branched polymers is mainly a result of the presence of acidic hydroxyls and minor amount of Lewis sites in CrHZSM-5 as well as HZSM-5 zeolite.

It can be concluded that the Cr ions incorporated into the HZSM-5 zeolite by
ion exchange exhibit similar properties as the Cr ions in the cationic sites of
Y zeolites. They form dinitrosyl complexes with NO and are able to be oxidized
to Cr(V). However, the Cr ions in CrHZSM-5 do not contribute significantly to
the zeolite activity in ethylene polymerization. The considerably high initial
rate of ethylene polymerization over HZSM-5 zeolite together with the constrain
effect of the ZSM structure causes fast blocking of zeolite channels leading to
the termination of polymer growth and inaccessibility of Cr ions for the ethy-

lene molecules.

ACKNOWLEDGEMENT

We are greatly indebted to Dr L. Kubelková for helpful discussion.

REFERENCES

1   C. Naccache and Y. Ben Taarit, JCS, Faraday Trans. 1, 69 (1973) 1475.
2   Z. Tvarůžková and V. Bosáček, Coll. Czech. Chem. Commun. 45 (1980) 2499.
3   J.R. Pearce, D.E. Sherwood, M.B. Hall and J.H. Lunsford, J. Phys. Chem.
    84 (1980) 3215.
4   B. Wichterlová, Z. Tvarůžková and J. Nováková, JCS, Faraday Trans. 1, 79
    (1983) 1573.
5   S. Beran, P. Jírů and B. Wichterlová, JCS, Faraday Trans. 1, 79 (1983) 1584.
6   Z. Tvarůžková and B. Wichterlová, JCS, Faraday Trans. 1, 79 (1983) 1591.
7   B. Wichterlová, L. Krajčíková, Z. Tvarůžková and S. Beran, submitted.
8   V.B. Kazansky and J. Turkevich, J. Catal., 8 (1967) 231.
9   L. Lavielle and H. Kessler, J. Electron Spectrosc. Relat. Phenom., 8 (1976)
    95.
10 Kh.M. Minachev, G.V.A. Antoshin and E.S. Shpiro, Izv. Akad. Nauk U.S.S.R.,
    Ser. Chim.(1974) 1974 1012.
11 R.N. Merryfield, M. McDaniel and G. Parks, J. Catal. 77 (1982) 348.
12 L. Kubelková, J. Nováková, Z. Dolejšek and P. Jírů, Coll. Czech. Chem.
    Commun., 44 (1979) 3341.
13 J.C. Vedrine, A. Auroux, V. Bolis, P. Dejaifve, C. Naccache, P. Wierzchowski,
    E.G. Derouane, J.B. Hagy, J.P. Gilson, J.H.C. van Hoof, J.P. van der Berg
    and J. Wolthuizen, J. Catal., 59 (1979) 248.
14 K.M. Wang and J. Lunsford, J. Catal., 24 (1972) 262.
15 B. Wichterlová, J. Nováková, L. Kubelková and P. Jírů, Proc. VIth Int. Symp.
    Zeolites, Ed. L.V.C. Rees London (1980), pp. 373.

P.A. Jacobs et al. (Editors), *Structure and Reactivity of Modified Zeolites*
© 1984 Elsevier Science Publishers B.V., Amsterdam — Printed in The Netherlands

METAL CARBONYL COMPOUNDS ENTRAPPED WITHIN ZEOLITE CAVITIES. PREPARATION, CHARACTERIZATION AND CATALYTIC PROPERTIES.

Frédéric LEFEBVRE[1], Patrick GELIN[1], Boubaker ELLEUCH[1], Youssef DIAB[2] and Younès BEN TAARIT[1]

[1]Institut de Recherches sur la Catalyse, CNRS, 2, avenue Albert Einstein 69626 Villeurbanne, Cédex (France)

[2]Université Libanaise - Faculté des Sciences II - Mansourieh BP/72/ (Liban).

ABSTRACT

Transition metal ions in zeolite show a distinct trend to coordinate CO. The stability of the carbonyl species strongly depended on the ability of transition metal ions to transfer d electrons into the Π^* orbital of CO. The highly electrophilic character of the noble metal carbonyl carbon accounted for most of the chemical reactivity. Mononuclear monovalent dicarbonyls resulted from the reduction of high valent carbonyls via attack of an OH^- group on the carbonyl carbon. The structure of monovalent carbonyls was determined using a combination of XPS, UV, IR, ^{13}C NMR and volumetric methods. The nature of the bonding of the monovalent carbonyls was investigated and displacement of the lattice oxide ligands could occur using donor ligands. Further reduction of the monovalent carbonyls to tetra- and hexa-nuclear complexes was performed using $H_2O:CO$ or $H_2:CO$ mixtures. Again the structure was determined using IR and NMR studies supplemented by volumetric measurements. The catalytic activity of both types (mono- and poly-nuclear) of carbonyls was investigated in CO insertion reactions into organic substrates and in the water gas shift reaction.

INTRODUCTION

When exchanged with protons or with transition metal ions and complexes, the behaviour of zeolites was rather pertaining to solution chemistry than solid state. When hosting metal particles they could be considered as solid matrices as well as giant micelles. The solid solvent behaviour of zeolites was in fact unveiled a number of years ago by several authors (1-3). The lattice oxide ions behaved as interchangeable labile ligands providing for the translational motion within the porous structure of the zeolite. It also soon appeared that numerous molecules (H_2O, amines, olefins, NO, NO_2, CO, aromatics, etc...) could compete more or less effectively with lattice oxide ions to coordinate the exchanged transition metal ions and form well defined complexes within the zeolite cavities.

One of the newest features, besides the conversion of methanol to gasoline, resides in the ability of zeolites to host the entire synthesis of complex-structure molecules involving metal-metal bond formation such as polynuclear as well as mononuclear carbonyl complexes. These appear particularly important not only from the academic point of view but also because of their potential as CO carriers in Carbon Monoxide Chemistry.

In heterogeneous catalysis, the CO molecule moved rapidly from the status of a simple probe molecule merely used to characterize particular surfaces (metal and oxide surfaces) into a fashionable and almost precious reagent used to upgrade cheap organic substrates into high cost chemicals.

The metal -to- CO bonding scheme

Because of its long known and wide spread use as a probe molecule and because transition metal carbonyls have been known for many years, the nature of the CO bonding to metal ions and/or atoms have been thoroughly studied. The knowledge of this bonding scheme is so important in order to predict the reactivity of this molecule, and that of the carbonyls which are the potential transfer agents of this molecule to other substrates.

The CO molecule appeared to behave, in these complexes and surface species, as a poor σ donor (the 3σ carbon lone pair) ligand towards metal d orbitals. This poor donor character would not result in stable enough complexes without a reverse scheme involving electron transfer from the metal (ion) filled d orbitals of Π symmetry towards the Π^* (antibonding) molecular orbital of the CO molecule.

This latter interaction known as back-bonding or back donation would be significantly favoured only when the metal is at a low valent state (formally -1, 0, +1). Under such circumstances, a strong metal-to-carbon bond is established upon populating the Π^* orbital of CO, which then would weaken the C-O bond. Thus depending on the valence state of the metal, its ability to back-bonding will vary decisively affecting several factors : the electron flow into the M-C bond, the electron density at the carbon atom, the force constant of the C-O bond. These three factors would, in turn, determine the properties of the carbonyls : the stability of the carbonyl, the reactivity of the carbon towards electrophiles and nucleophiles, the possibility of C-O bond cleavage respectively.

THE MONONUCLEAR CARBONYLS

In spite of the fact that a prerequisite condition to the formation of a carbonyl, is the metal ion being at a low valence state, a few examples, in zeolites show, that indeed, similar complexes although with specific properties do form with high valent transition metal ions. Even fewer examples are known in solution (4).

In the faujasite type zeolite, it is particularly easy to isolate transition metal ions (including noble metal ions) in a high oxidation state. Suitable complexes, usually introduced according to a conventional ion exchange procedure,

could be easily converted into highly coordinatively unsaturated transition
metal ions bound to lattice oxide ions by simple heat treatment in oxygen at
rather moderate temperatures (250-350°C). Under these conditions, the coordinatively unsaturated ions would more willingly interact with carbon monoxide,
introduced in the vacated cavities of the zeolite, than their solution analogues
which coordination sphere would be more or less crowed by at least solvent
molecules. Thus simply because zeolites provide the means to coordinatively
unsaturated cations, formation of high valent carbonyl is made possible.

Most of the evidence of this interaction came from infrared data (5). The
presence of potential ligands (water, amines) in the medium may preclude the
formation of these weak complexes or decompose them since σ donor ligands such
as NH_3 would compete more effectively than CO for vacant coordination site of a
high valent cation.

By contrast addition of such ligands in appropriate amounts providing for
Co-adsorption of the base and the CO molecule in a non competitive way significantly increases the back bonding (M → CO) therefore enhancing the stability of
the carbonyl at the expense of the C-O bond strength as evidenced by a low
frequency shift of the ν_{CO} absorption with respect of the pure zeolite-cation
carbonyl (6).

As far as the electron density at the carbon is concerned, ^{13}C NMR studies
(7) showed, as the 3σ electron transfer would suggest, an electron deficient
carbon with a high affinity towards nucleophiles. This electron deficiency
reflected by a strong upfield shift in the ^{13}C resonance with respect to that
of free CO was attenuated as electron donor ligands entered the coordination sphere of the carbonyl (8).

First and second series transition metal carbonyls were essentially similar
in their characteristics : conditions of formation and structure. However they
essentially differ in subsequent interaction with water and possibly other
ligands.

While first series transition metal carbonyls were reversibly destroyed by
water vapour according to the following equilibrium scheme :

$$
\begin{array}{c}
O_L \diagdown \\
O_L \rightleftharpoons M - CO \ + \ H_2O \\
O_L \diagup
\end{array}
\longrightarrow
\begin{array}{c}
O_L \diagdown \\
O_L \rightleftharpoons M \ \leftarrow OH_2 \ + \ CO \\
O_L \diagup \\
\qquad \downarrow\uparrow H_2O
\end{array}
$$

$$
\begin{array}{c}
O_L \diagdown \qquad \diagup OH \\
\qquad M \qquad\qquad + \ O_L H \\
O_L \diagup \qquad \diagdown OH_2
\end{array}
$$

Transition noble metal carbonyls may suffer addition of a neutral or ionized water molecule into the coordination sphere without decomposing but rather undergoing a rearrangement induced by the electrophilic character of the carbonyl carbon, subsequently resulting in the reduction of the central ion and leading possibly to a catalytic reaction in presence of excess CO and water.

The formation of monovalent carbonyls would appear as a result of a subsequent reaction of the high valent carbonyl precursor with CO or CO and water. This reduction step could be envisionned as a simple Redox reaction affecting both the cation and the zeolite lattice as in other instances. In such a case CO would act as a reducing agent first and second as a ligand. This formation should rather be considered as a result of the unstable character of the high valent carbonyl because of the extreme reactivity of this carbonyl carbon towards nucleophiles and especially ^-OH originating from trace amounts of water (or hydroxyl groups) often present within the zeolite matrix.

In fact transition metal ions are usually exchanged into the zeolite using an aqueous ammine solution of these complexes : $|Pt(NH_3)_4|^{2+}$, $|Pd(NH_3)_4|^{2+}$, $|Rh(NH_3)_5Cl|^{2+}$ and $|Ir(NH_3)_5Cl|^{2+}$. These ammine and chloro-ammine complexes were generally decomposed by heating in an oxygen flow at temperature within the range 250-350°C. The result is usually a divalent (Pt, Pd) or trivalent (Rh, Ir) oxo or hydroxo complex bound at least partly to the zeolite lattice and evenly distributed within this lattice. No surface enrichment in transition metal could be observed by XPS nor did X-rays diffraction give a hint to the formation of any oxide lattice. UV spectroscopy however gave evidence for the presence of divalent platinum and palladium (9) and trivalent rhodium (10). XPS measurements were also consistent with essentially trivalent rhodium and iridium.

Upon addition of CO to transition metal zeolites activated in oxygen as already mentioned then evacuated at similar temperatures gave rise to two types of species.

First the high valent carbonyl type characterized by a high frequency ν_{CO} single absorption (table 1). This species essentially similar to those reported for the first series transition metal ion-zeolites, formed reversibly.

TABLE 1

ν_{CO} absorption frequency of the high valent transition metal carbonyls in NaY zeolite.

Cation	Pt^{II}	Pd^{II}	Rh^{III}	Ir^{III}
ν_{CO} (cm^{-1})	2118	2135	2138	2130

Simultaneously another species developped and was accompanied by the production of carbon dioxide as apparent from the IR band at 2360 cm^{-1}. In addition to the high frequency ν_{CO} band, other absorptions developped to lower frequencies.

The oxidation state of the transition metal, the stoechiometry of the carbonyl, the nature of the other ligands within the coordination sphere, the structure of the carbonyl and last but not least its chemical reactivity must be determined using various but quite often complementary methods and techniques. Only combination of all these data could reliably be used to identify the carbonyl.

As an example, combination of XPS, UV and volumetric measurements may help determine the actual oxidation state of the central transition metal within the complex together with the possible number of carbonyl ligands in the coordination sphere, provided a single species in dealt with at a time. Mixtures are far more difficult to characterize and it is perhaps the most tedious part of the work to delineate experimental conditions where a pure species is isolated.

Infrared and NMR investigation of the carbonyl ligands must be used in order to probe the structure of the complex and also the number of carbonyl ligands together with the nature of remaining ligands and the chemical reactivity of the complex with respect to various molecules.

The structure of the carbonyl

The UV spectrum exhibited by rhodium zeolites activated in oxygen was similar to that reported for various Rh^{III} compounds (10). XPS studies gaves $3d_{3/2}$ and $3d_{5/2}$ binding energies at 315.7 and 310.8 eV respectively, in the range of Rh^{III} complexes. Upon reaction with CO, the UV bands characteristic of trivalent rhodium were washed out, indicating reduction of this ion. No EPR evidence was found for mononuclear d^9 (Rh(o)) species, while limited concentrations of Rh^{II}(CO) were observed. By contrast, XPS measurements showed that the resulting rhodium carbonyl compound exhibits a $3d_{3/2}$ binding energy of 308.1 eV which compares favourably with that observed for $Rh(CO)_2Cl_2$ supported on NaY (308.3 eV) (11). Thus essentially a monovalent rhodium carbonyl is formed. Similarly when the Ir^{III}-zeolite precursor was oxygen treated at 250°C so as to be stripped of its initial ligands again XPS studies showed that the trivalent oxidation state was preserved. The $4f_{5/2}$ and $4f_{7/2}$ binding energies of Ir^{III}-zeolite treated with carbon monoxide appear to fit into the range of Ir^I binding energies.

In both the rhodium and iridium cases, the compound appeared to be evenly distributed within the bulk and the surface of the zeolite cristallites.

Volumetric measurements were also used to determine the number of carbonyl ligands in the complex and to confirm the oxidation state of the central metal atom. Indeed the total consumption of CO accounted for two distinct, simultaneous and/or consecutive processes : (i) reduction of the transition metal ion *to produce CO_2*, (ii) coordination of a number of CO molecules to the reduced transition metal ion to form the carbonyl complex. Thus the volume of CO_2 formed is indicative of the Redox process stoechiometry, assuming a quantitative reduction.

In all volumetric experiments, the amount of CO_2 formed was, within experimental errors, in a good agreement with a two electron-reduction of trivalent ions to monovalent metal complexes.

As the structure of the trivalent ion bound to the zeolite may vary depending on the cation and the activation conditions, the following reduction schemes are proposed to account for such 2 electron-reduction.

$$\left[Al \diagup^{O}\diagdown Si \right]^- + M^{3+} + (n+1)CO \rightarrow CO_2 + |M^I(CO)_n|^+ + \left[Al \quad ^+Si \right]$$

in the case of anhydrous zeolites

$$Al \diagup^{O}\diagdown Si^- + (MOH)^{2+} + (n+1)CO \rightarrow CO_2 + |M^I(CO)_n|^+ + \overset{\overset{H}{|}}{Al}\diagdown^{O}\diagup Si$$

in the case of partially hydrated zeolite

$$\left\{ M \diagup^{O}_{\diagdown O}\diagup M \right\}^{2+} + 2(n+1)CO \rightarrow 2 CO_2 + 2 |M^I(CO)_n|^+$$

in the case of oxobridges.

The total CO consumption may contribute to the determination of the mean number of CO ligands attached to each metal centre. However such measurements must be considered with an extreme care for the following reasons :
(i) the coordinated CO is obtained as a difference of the total CO and that involved in the reduction and determined by measuring CO_2 formed,
(ii) the assumption of quantitative reduction and the formation of a single

species may be valid only to a certain extent and may depend on the CO pressure.
(iii) metallic impurities (Fe^{3+}) in the zeolite may affect notably the total
CO uptake,
(iv) further CO consumption may proceed depending on the amount of moisture
present.

Infrared spectroscopy would appear far more reliable in order to determine
the symmetry and therefore the structure of the carbonyl. Indeed the symmetry
of the carbonyl was shown to determine the IR active vibration modes of the
molecule. As an example a dicarbonyl or a tricarbonyl (C_{2v} and C_{3v} respectively)
would exhibit two IR active ν_{CO} vibration modes each : the A_1, B_1 or E modes.

TABLE 2

Comparison of expected and calculated ν_{CO} frequencies for isotopic mixtures of
rhodium and iridium dicarbonyls

Metal	Isotopic isomer		$L_nM(CO)_2$	L_nM^*COCO	$L_nM(CO)_2$
	Symmetry group		C_{2v}	C_s	C_{2v}
	Active ν_{CO} modes		A_1 B_1	A' A'	A_1 B_1
Rhodium	Calculated ν_{CO}	Species I	2101 2022	2084 1993	2054 1977
		Species II	2116 2048	2099 2018	2069 2002
	Experimental ν_{CO}		2100 2018 2116 2050	2086 1990 2100 2018	2050 1976 2066 2000
Iridium	Calculated ν_{CO}	Species I	2088 2004	2066 1981	2041 1959
		Species II	2102 2030	2085 2001	2055 1985
	Experimental ν_{CO}		2088 2004 2102 2030	2062 1980 2086 2000	2038 1957 2056 1980

In each case two ν_{CO} absorptions would be expected. Therefore simple IR data
cannot provide the means to distinguish di- and tri-carbonyl. However upon
partial substitution of ^{12}CO by ^{13}CO ligands the local symmetry is broken down
to C_s in both cases and whereas the dicarbonyl would give rise to three major
species with relative concentrations depending on the $^{12}CO/^{13}CO$ and two infra-

red active A' modes, the tricarbonyl would give rise to four major species and
three IR active vibration modes (two A' and one A" mode). Therefore the combination of IR spectroscopy with partial isotopic exchange would make it possible
to discriminate di- and tri-carbonyl complexes. Furthermore following the method
of Cotton and coll (12) it is possible to derive the ν_{CO} frequencies   in
each   case. In the   case of the rhodium I and iridium I carbonyls comparison
of predicted and observed frequencies showed an excellent agreement (table 2)
indicating that the compounds are dicarbonyl in nature.

The bonding of the carbonyl to the zeolite

By contrast to the abundance of data demonstrating the dicarbonyl nature of
these complexes, little data deal with the nature of the other ligands to the
metal centre. Rhodium dicarbonyls are essentially 16 electron square planar
complexes. The assumption of a C_{2v} symmetry which has been confirmed seems to
indicate that the metal centre was bound to two lattice oxide ions (X and L type
ligands). Additional evidence to the reality of this bond stems from ^{13}C NMR
measurements. Indeed ^{13}C NMR has been successfully applied to the investigation
of the interaction of zerovalent carbonyls with the zeolite matrix. Even bulky
molecules such as $Fe(CO)_5$ showed sharp resonances indicating a reasonable
degree of motional freedom in the zeolite supercages (13). In the case of monovalent rhodium and iridium carbonyls no signal could be observed under conventional conditions while upon spinning the rhodium dicarbonyl sample around an
axis at an angle of 54°44'off the magnetic field at frequencies around 3000 Hz,
a narrow line was observed at $\sim$ 183 ppm downfield the TMS. Spinning the sample
at the magic angle (54°44') undermines the effect of line broadening chief
contributors the dipolar interaction (negligible in this instance) and the
chemical shift anisotropy, which in the solid state results from a restricted
motion. This restricted motion is blamable on the rhodium-to-lattice bond. On
the other hand, structural studies of zeolites by infrared spectroscopy showed
that the assymmetric elongation of the TO_4 tetrahedra (T = Si, Al) occuring at
$\sim$ 985 cm^{-1} (in Y zeolite) was shifted to lower frequencies as the relevant
tetrahedra interacted with protons or cations, thus weakening the T-O bond. The
formation of the monovalent dicarbonyls was indeed accompanied by a shift of
this asymmetric elongation to 870 cm^{-1} indicating that indeed the metal centre
was interacting with the zeolite lattice as well.

Further evidence for this metal-to-lattice bond was also evidenced through
the steric constraints imposed on the carbonyl structure. As the OTO bond angle
must be restricted to a fixed value within slight distortions compatible with
the zeolite structure, a complex where the metal centre would be bound to two
such oxide ions would be subjected to this imposed structure, influencing the
bond angle of the two carbonyls. Indeed, in the case of an anhydrous zeolite, the

OCMCO bond angle determined from the relative intensities of the asymmetric and symmetric ν_{CO} absorptions showed an important departure from ideal $\Pi/2$ angle in square planar complexes (112°).

Substitution reactions of the zeolite ligands

The influence of the zeolite on the complex structure could be also revealed in a paradoxical way : by its removal, water molecules added into the zeolite affect the infrared spectrum of the carbonyl. Contrary to high valent carbonyls, it does not decompose the monovalent ones. Table 3 shows the ν_{CO} frequency changes and bond angle variations as more and more water was added.

TABLE 3

Variation of ν_{CO} frequencies and the OCMCO angles on the water content

Complex	ν_{CO} frequencies	bond angle
O$_L$\M/CO, O$_L$/M\CO	2216 2048	112°
O$_L$\M/CO, H$_2$O/M\CO	2101 2022	104°
HO\M/CO, H$_2$O/M\CO or |H$_2$O\M/CO, H$_2$O/M\CO|$^+$	2090 2030	95°

Other potential ligands such as ethylene and ammonia could also substitute lattice oxide ions, the latter amine ultimatly decomposes the carbonyl.

THE POLYNUCLEAR CARBONYLS

The tetranuclear carbonyls

Monovalent dicarbonyls react with $CO:H_2O$ mixture to form CO_2 at mild temperatures. As CO_2 was formed the ν_{CO} bands due to the dicarbonyl decreased while an other complex structure appeared in lieu. Simultaneously - in the case of rhodium zeolite - two new bands developped at 1870 and 1832 cm^{-1} due to bridging carbonyls. Their appearance indicated the formation of rhodium-rhodium bonds demonstrating the reduction of the monovalent metal centre to the zerovalent state as already suggested by the formation of CO_2 as shown by the infrared spectrum.

Similar behaviour was observed in the case of zeolite bound monovalent iridium dicarbonyl. However only new ν_{CO} bands characteristic of linearly bound carbonyls developped simultaneously to the CO_2 absorption band under the action of $CO:H_2O$ mixture. No bridging carbonyls appeared. However upon removing excess

H_2O two bands appeared in the bridging CO region appeared again suggesting the reduction of the monovalent iridium and establishment of Ir-Ir bonds. This phenomenon was reversible and water readdition removed the bridging CO absorptions.

Apart from the fact that the infrared spectra of these two carbonyls closely paralleled the infrared spectra of $Rh_4(CO)_{12}$ and $Ir_4(CO)_{12}$ respectively, additional data were necessary in order to ascertain the identification of the tetranuclear carbonyls.

The structure of the tetranuclear carbonyls

The M/CO ratio could be easily determined upon simple combustion of the carbonyls to form CO_2 which amount could be determined by volumetric methods. Raw data were all within the CO/M ratio of 3. Again infrared spectroscopy may be beneficial in ascertaining the structure of these carbonyls.

The frame of $Rh_4(CO)_{12}$ for instance is made up with four tetrahedrally bound rhodium atoms. The basal rhodium atoms are linked by three edge bridging carbonyls in addition to 2 linearly bound CO's. This basal $M_3(CO)_3$ structure assumes a D_{3h} local symmetry neglecting the apical $Rh(CO)_3$ group or a C_{3v} local symmetry taking into account this latter group. Thus again partial substitution of ^{12}CO by ^{13}CO ligands was used to confirm this structure. The same procedure was applied to $Ir_4(CO)_{12}$ which assumed a similar structure when excess water was removed from the zeolite.

Additional data in favour of the tetranuclear structure was provided by ^{13}C NMR used under Magic Angle Spinning. While hydrated $Ir_4(CO)_{12}$ ^{13}CO enriched exhibited a single sharp line at $\delta=160$ ppm consistent with expected chemical shift for this compound should it be soluble, $Rh_4(CO)_{12}$ in NaY zeolite exhibited two resonances at 180 and 243 ppm dowfield the TMS respectively in an approximate ratio of 3:1. These resonances are characteristic of linear and edge bridging carbonyls in close agreement with the ^{13}C NMR spectrum of this compound in solution. A slight shift of the edge bridging carbonyls resonance is probably indicative of solvent like effect of the zeolite matrix. In one example $Ir_4(CO)_{12}$ has been extracted in refluxing toluene and subsequently characterized by infrared and elemental analysis. Thus, the CO/M ratio, the specific $M_3(CO)_3$ frame structure and finally the overall structural data from IR and NMR techniques convincingly reveal the formation of tetranuclear iridium and rhodium dodecarbonyls.

As to the mechanism of formation of these clusters no decisive evidence was reported yet. They were reported to form either under $H_2:CO$ or $H_2O:CO$ atmosphere. The latter circumstances are quite similar to those prevailing in solution. Indeed, $Rh_4(CO)_{12}$ was prepared by bubling CO through a solution of the $(Rh(CO)_2Cl)_2$ dimer in presence of water which was recognized as a necessary ingredient. We feel that the monovalent carbonyl $H_2ORhOH(CO)_2$ obained in presence of excess water would undergo a hydroxide shift to form formate species followed by decarboxylation giving rise to a very reactive hydride, this will, in turn react with a monovalent species to form a dinuclear octocarbonyl complex *in presence of CO*. This latter would easily dimerise to form a zerovalent tetra-metal dodecacarbonyl according to the following overall scheme

The hexanuclear carbonyls

Heating $M_4(CO)_{12}$ in H_2:CO atmosphere at temperature within the range 100-300°C resulted in sharpening of the linear ν_{CO} absorptions and a significant shift to lower frequencies of the bridging ν_{CO} carbonyls (1800 or 1760 cm^{-1} for M = Rh depending on the hydration state of the zeolite, 1765 - 1730 cm^{-1} for M = Ir). The infrared spectrum as a whole was consistent with that of $M_6(CO)_{16}$ in a nujol mull or in cyclohexane or sublimed into the zeolite (14).

The oxidative decarbonylation under O_2 at 100°C produced CO_2 amounts in fair agreement with the general CO:M ratio of 2.6 - 2.7 consistent with hexanuclear carbonyl stoechiometry. Furthermore recent radial distribution studies by means of Fourier transform of X-rays diffraction pattern showed upon substracting the zeolite background the appearance of M-M distances in excellent agreement with M-M distances in these compounds.

The bonding of polynuclear carbonyls to the zeolite

The amazing feature of the synthesis of metal carbonyl clusters is that zeolites appear to provide the right pot for relatively easy selective synthesis whereas, in solution, whatever the conditions, an intricate mixture is usually obtained leading to tedious separation procedures. This might be interpreted as a specific solvent effect. Indeed it would appear that bonding (loose) of the metal cluster to the zeolite occurs preferentially via interaction of the bridging carbonyls, which absorption frequencies decreased by 30 - 50 cm^{-1} indicating a significant interaction of the carbonyl oxygen with lattice species strongly enough to result in a weakening of the C-O bond. It is also possible that particular arrangement of the zeolite, the cations and the cluster may be thermodynamically favoured, thus accounting for the discrete bridging ν_{CO} absorptions observed depending on the hydration state and the nature of the major charge compensating cation (15). Such specific atomic order would appear more clearly as the driving force in the case of $Ir_4(CO)_{12}$: this latter cluster assumes in the solid state, and in solution, a tetrahedral arrangement with four tetrahedrally symmetric $Rh(CO)_3$ groups ; in anhydrous zeolites, it assumes the $Rh_4(CO)_{12}$ like configuration, the very one it assumed in argon matrix at low temperature where a well defined ordering at the atomic scale prevails.

CATALYTIC PROPERTIES OF ZEOLITE-HOSTED METAL CARBONYLS

Only those properties involving CO activation and transfer will be dealt with, even through these compounds appear to be active in such processes as isomerization, dimerization and oligomerization of olefins, hydrogenation reaction etc.

The water shift reaction

Hydrogen is needed free from organic impurities. The water gas shift reaction $CO + H_2O \rightarrow CO_2 + H_2$ may provide such high purity hydrogen when this reaction was carried out selectively. It has been reported (16) that this reaction proceeded selectively on monovalent Ru bis- and triscarbonyl generated upon mild reduction of Ru(III) exchanged faujasite type zeolites using carbon monoxide.

It was concluded that a specific geometry was needed for the complex to catalyze this Redox Cycle type reaction. Only supercages of the faujasite type could accomodate such a complex.

On the other hand, the Si/Al ratio, the nature and number of major charge compensating cations as well as the presence of chloride ions within the coordination sphere of the catalyst appeared to influence the activity of the complex. These observations were rationalized in terms of the electronegativity concept and in terms of favourable geometrical arrangement depending on the size of cavities and cations.

Other transition metal complexes including monovalent carbonyls would catalyze this water gas shift reaction, though with variable effeciencies. The probable scheme would involve ionization of a water molecule to give hydroxide ions (hence the electronegativity concept) which would effect a nucleophilic attack on a carbonyl carbon (usually electrophilic) to form an intermediate formate which decarboxylates spontaneously to a very reactive metal-hydride. The latter would react with coordinated water to leave a coordinate hydroxide and remove molecular hydrogen.

The nature of the zeolite $(-O_L-)$ and that of the major charge balancing cation would, of course, determine the ease of ionization of the water molecule, while the nature of the metal centre and the surrounding ligands would indeed determine the electrophilic character of the carbonyl carbon and the ease of the internal rearrangement that produces the formate species and later influences the reactivity of the hydride towards coordinated water protons.

Carbonylation of methanol

This reaction, improperly named, proceeds in fact via the actual carbonylation of the methyl iodide which is directly introduced in the reaction mixture or generated in situ by reaction of methanol with an iodide promoter usually HI. The detailed reaction obeys the following equations

$$CH_3I + CO + Cat \rightarrow CH_3COI\text{-Catalyst}$$
$$CH_3COI\text{-Cat} + CH_3OH \rightarrow CH_3COOCH_3 + Cat + HI$$
$$HI + CH_3OH \rightarrow CH_3I + H_2O$$

Under normal pressure this reaction has a significant rate on catalysts in the temperature range 100-180°C. The practical catalyst is a Monsanto designed Rh(I) dicarbonyl associated with HI.

Zeolites, as well as other inorganic carriers have been successfully used to host the monovalent carbonyl complex ; the reaction always needed the iodide promoter to proceed. Both $L_2Rh^I(CO)_2$ and $L_2Ir^I(CO)_2$ bound to zeolites exhibited reasonable activities and high selectivities to methyl acetate. Presumably because of its practical and academic importance as a model for a first C-C bond formation, this simple reaction prompted numerous studies (17).

Kinetic studies showed both homogeneous and heterogeneous reactions to obey the same rate laws : first order in MeI and zero order in MeOH for rhodium catalysts, first order in MeOH and zero order in MeI for iridium catalysts and a general zero order in CO.

The rates encountered in homogeneous media appeared to be an order of magnitude higher than those reported for zeolite (18). This should be probably due to diffusion limitations preventing most of the metal load from acting catalytically. Zeolites, in turn exhibited better and more stable activities than other inorganic and polymer type carriers or carbon presumably because of a better dispersion and better bonding preventing aggregation and leaching.

Mechanistic considerations

Most of the mechanistic studies involved addition of MeI to the catalyst precursor : the monovalent dicarbonyl, investigated by UV, IR and ^{13}C NMR spectroscopy. MeI was shown to add oxidatively to the dicarbonyl precursor.

UV spectroscopy showed that indeed trivalent rhodium was generated upon addition of the methyl iodide. IR spectroscopy confirmed the oxidative addition as the ν_{CO} absorptions were shifted to higher frequencies. ^{13}C NMR unveiled the highly electrophilic character of the carbonyl carbons following addition of the methyl iodide as a high field shift of about 50 ppm was observed. On the other hand while in the case of iridium this oxidative addition proceeded rapidly, it proceeded more slowly in the case of rhodium, which is in agreement with the kinetic laws. Moreover in the latter case a spontaneous rearrangement of the coordination sphere occured via migration of the methyl to react with electrophilic carbonyl carbon resulting in an acetyl ligand as suggested by the CO absorption at 1680-1700 cm^{-1}. Methanol reacted immediatly with the rearranged complex to induce a reductive elimination of the acetyl halide and form the methyl acetate.

In the case of iridium no spontaneous rearrangement could be observed. However introduction of methanol leads to a rearrangement or perhaps to the formation of an alkoxy-carbonyl by direct nucleophilic attack of the methoxy on the electrophilic carbonyl carbon.

Reductive elimination restores the precursor dicarbonyl in presence of CO. Hence the Redox cycle and the intramolecular rearrangement whether spontaneous or induced are the key features of this reaction. Therefore all factors influencing the course of any of these two features will doubdtless affect the catalytic activity of the zeolite bound carbonyl. Such factors are the type of the zeolite, the nature of the central atom, the nature of the major charge compensating cation and that of the alcohol. The effect of zeolites may be to modify for instance the relative ease of Redox cycles with respect to other solvent. Such a modification would account for such an amazing observation that iridium zeolites are almost one order of magnitude more active than the rhodium zeolite. Other factors determining the dispersion and location of the active metal centre may also modify relative activities.

Carbonylation of nitroaromatics

   This reaction depicted by the following equation :

$$Ar - NO_2 + 3CO \rightarrow 2CO_2 + AR - N = CO$$

It has been also demonstrated that the reaction proceeds via a Redox cycle of
the transition metal carbonyl catalyst (19). The carbonyl undergoes an oxidative
addition of a nitrene which later affects a nucleophilic migration to insert a
coordinated carbonyl ligand. This migration may be assisted by donor ligands
within the coordination sphere and the presence of donor substituents on the
aromatic ring and adversely affected by electron withdrawing substituents, in
line with the behaviour encountered in methanol carbonylation.

CONCLUSIONS. THE ZEOLITE SPECIFICITY

   The zeolite material showed a propensity throughout numerous studies to
behave as a polydentate ligand to a number of carbonyl complexes both high
valent and low valent. The zeolite lattice influenced chiefly the structure
and the chemical reactivity of these species : the structure was subjected to
the expected lattice constraints. The reactivity of the complex towards ligands
and potential reagents was modified by the charge distribution within the
lattice depending not only on the Si/Al ratio and the zeolite structure but also
on the cations (nature and distribution) within the same type of structure. A
number of complexes (with hydroxo ligands) were stabilized whereas they were
never isolated in other media.

   It would seem that part of this stabilization of odd species was due to a
zeolite induced expansion of the thermodynamic domain of stability and higher
resolution (separation) of the different domains. This would be the major cause
that made the zeolite a sort of an ideal pot where various complexes, including
zerovalent carbonyl clusters, were selectively prepared under given temperature and pressure whereas in other media an intricate mixture would be
produced.

   The bonding between zeolites and zerovalent clusters seems to involve precise
and discrete atomic arrangements where the total energy of the binary system
would be significantly lower than in random arrangements.

   CO insertion reactions typical of the carbonyls hosted by the zeolite were
also affected by these matrices as they influence the major two steps which
determine the activity and the selectivity in such reactions. the Redox cycle
and the substrate migration to insert the carbonyl. The electrostatic fields
which depend both on the zeolite structure and the cation distribution may also
induce an effective polarization of the organic substrate and assist the first
reaction step :the oxidative addition or the water ionization in the water gas

shift reaction. Thus via its geometrical structure and its electronic configuration zeolites act as unique ligands and matrices.

REFERENCES

1 J.H. Lunsford, A.C.S.Symposium Series, 40 (1977) 473.
2 C. Naccache and Y. Ben Taârit in J.P. Bonnelle, B. Delmon and E. Derouane (Eds.), Surface Properties and Catalysis by non-metals, Reidel, Dordrecht 1983, pp. 405-431.
3 M. Che and Y. Ben Taârit in K.L. Mittal and E.J. Fendler (Eds.), Solution Behavior of Surfactants : Theoretical and Applied Aspects, Plenum Press, New York, 1982, pp. 189-214.
4 W.J. Cherwinski, B.F.G. Johnson, J. Lewis and J.R. Norton, J.C.S., Dalton, (1975) 1156.
5 J.W. Ward in J.A. Rabo (Ed.), Zeolite Chemistry and Catalysis, A.C.S. Monograph, 171 (1976) 118.
6 Y.Y. Huang, J. Catal., 32 (1974) 482.
7 Y. Ben Taârit, Chem. Phys. Letters, 62 (1979) 211.
8 Y. Ben Taârit, H. Tzehoval, C. Naccache and B. Imelik, in L.V.C. Rees (Ed.), Proceedings of the Fifth International Conference on zeolites, Heyden, London, 1980, pp. 723-731.
9 E. Garbowski, M. Primet, M.V. Mathieu, J. Chim. Phys., 75 (1978) 329.
10 M. Primet and E. Garbowski, Chem. Phys. Letters, 72 (1980) 472.
11 M. Primet, J.C. Vedrine and C. Naccache, J. Mol. Cat. 4 (1978) 411.
12 F.A. Cotton and C.S. Kraihanzel, J.A.C.S., 84 (1962) 4132.
13 J. B Nagy, M. Van Eenoo and E.G. Derouane, J. Catal., 58 (1979) 230.
14 P. Gelin, Y. Ben Taârit and C. Naccache, J. Catal., 59 (1979) 357.
15 F. Lefebvre et al., to be published.
16 P.A. Jacobs, R. Chantillon, P. De Laet, J. Verdonck and M. Tielen, A.C.S. Symposium Series, 218 (1983) 439.
17 P. Gelin, F. Lefebvre, B. Elleuch, C. Naccache and Y. Ben Taârit, A.C.S. Symposium series, 218 (1983) 455.
18 B. Christensen and M.S. Scurrell, J.C.S. Faraday I, 73 (1977) 2036.
19 B. Elleuch, Y. Ben Taârit, J. Basset and J. Kervehnal, Angew. Chemie, Int. Ed. Engl., 21 (1982) 687.

P.A. Jacobs et al. (Editors), *Structure and Reactivity of Modified Zeolites*
© 1984 Elsevier Science Publishers B.V., Amsterdam — Printed in The Netherlands

AROMATIZATION OF ETHANE ON METAL — ZEOLITE CATALYSTS

O.V.BRAGIN[1], T.V.VASINA[1], Ya.I.ISAKOV[1], N.V.PALISHKINA[1],
A.V.PREOBRAZHENSKY[1], B.K.NEFEDOV[2], Kh.M.MINACHEV[1]

[1]N.D.Zelinsky Institute of Organic Chemistry Academy of Sciences
of the USSR, Leninsky Prospekt 47, Moscow (USSR)

[2]The All-Union Institute for Petroleum Industry, Moscow (USSR)

ABSTRACT

It was demonstrated that the M/H-ZVM (M = Pt, Pd or Rh) system
was active in the ethane aromatization reaction. The H-form of
high-silica ZVM zeolites also exhibited a considerable catalytic
activity in the ethane aromatization and a high activity in the
ethylene aromatization. The activity was found to increase in the
following order: Na-ZVM HNa-ZVM H-ZVM. In the process of
ethane or ethylene aromatization the development of M/H-ZVM catalytic systems occurred accompanied by the activity improvement
with time.

INTRODUCTION

Dehydrocyclooligomerization of ethane to benzene and other
aromatics described earlier (ref.1) for metal-alumina catalysts,
M/Al_2O_3 (M = Pt, Pd, Ir, Rh or Ru), is of a considerable interest
as a method for producing valuable products from a readily available lowcost feedstock. The yields of aromatics, however, did not
exceed 5-7% even on the most active samples.

This paper presents some results obtained by studying the behaviour of ZVM-based zeolite catalysts in the conversion of ethane and, for comparison, ethylene. Catalytic properties of high-
silica ZVM (ЦВМ) zeolites that are analogs of ZSM-5 and belong to
pentasils, were studied. Particular attention was paid to metal-
containing systems (M/H-ZVM) exhibiting hydrogenating-dehydroge-
nating and acidic functions.

METHODS

The zeolites were synthesized by a new method using no organic
reagents. The metal catalysts were prepared by the method of ionic
exchange. The experiments were performed in a flow unit and in a
pulse microcatalytic system at normal pressure described pre-

viously (ref.1).

RESULTS and DISCUSSION

Activity and selectivity data obtained for the studied catalysts are summarized in Tables 1 and 2.

Of particular interest are the results obtained with an H-form
of ZVM Sample(III) which gave, without additional incorporation
of dehydrogenating components, up to 10% yield of aromatics, mainly C_6-C_8 (95%), from ethane. It is also noteworthy that the H-ZVM
catalyst showed the highest activity in the ethylene aromatization; the aromatics yield at 500°C was 75% (Table 1).

TABLE 1
Results of ethane and ethylene catalytic conversions on
ZVM zeolites[x]

Catalyst	Feed-stock	T ($^\circ$C)	Yield of liquid product (%)	Composition of liquid product (%) [xx]		Composition of gas (%) [xxx]			
				aromatics					
				C_6-C_8	C_9-C_{12}	CH_4	C_2H_4	C_2H_6	$C_3;C_4$
I	ethane	600	0	–	–	2.5	7.5	90	traces
Na-ZVM	ethylene	500	0	–	–	0	96.5	1.5	2
	ethylene	600	traces	–	–	0	94.5	1.5	4
II	ethane	600	10	91	9	17	4	79	traces
HNa-ZVM	ethylene	600	69	67	33	12	6	22	60
III	ethane	600	10	95	5	3	1	96	traces
H-ZVM	ethylene	500	75	55	45	3	3	48	46

x A flow system, 120 hr^{-1} space velocity, catalyst treatment in
air (5 hrs at 550°C).
xx No C_{5+} aliphatic hydrocarbons were found.
xxx Here and in Table 2, only the hydrocarbon portion of the gas
product is shown (H_2 was also present in addition to hydrocarbons).

It was interesting to study the ethane and ethylene reactions
to examine the catalytic activity of zeolites with the acidity
varying in the following order: Na-ZVM(I) ≪ HNa-ZVM(II) < H-ZVM(III).

Sample I exhibited no appreciable activity in the studied reactions; even when the temperature was raised to $600^{\circ}C$ the reaction products contained essentially no aromatics and only negligible amounts of C_1-C_4 hydrocarbons (Table 1). Based on the composition of the three catalysts it is reasonable to expect that in case of Na-ZVM the acid sites necessary for the catalytic activity were in fact completely "neutralized" by Na^+ ions.

The HNa-ZVM zeolite in which a considerable part of Na^+ (43%) was exchanged with H^+ exhibited some lower (but still sufficiently high) activity in the ethylene aromatization as compared to Sample III, but did not differ much from H-ZVM in the ethane reactions.

TABLE 2

Results of ethane and ethylene catalytic conversions on metal-zeolite catalysts[x]

Catalyst	Feed-stock	T ($^{\circ}C$)	Yield of liquid product (%)	Composition of liquid product (%) aromatics		Composition of gas (%)			
				C_6-C_8	C_9-C_{12}	CH_4	C_2H_4	C_2H_6	C_3,C_4
IV 0.5% Pt/ /Na-ZVM	ethane	600	traces	–	–	6.5	11.5	82	traces
	ethane	600[xx]	0	–	–	38	16	46	traces
V 0.5% Pt/ /H-ZVM	ethane	500	10	74	26	18	4	75	3
	ethane	600	20	83	17	19	3	78	traces
	ethylene	500	40	54	46	35	2.5	62	0.5
VI 0.5% Rh/ /H-ZVM	ethane	500	10	85	15	6	2	92	traces
	ethane	600	2.5	–	–	29	4	67	traces
	ethylene	450	35	71	29	25	3	37	35
VII 0.5% Pd/ /H-ZVM	ethane	600	2.5	–	–	12	3	85	traces
	ethane	600[xx]	12.5	54	46	7.5	2.5	90	traces
	ethylene	500	56	67	33	35	5	42	18

[x] A flow system, 120 hr^{-1} space velocity, catalyst treatment in air (5 hrs at $550^{\circ}C$).

[xx] Successive catalyst treatments in air (5 hrs at $550^{\circ}C$) and in H_2 (5 hrs at $500^{\circ}C$).

The above results bear a strong evidence of an important role of protonic acid sites in catalytic reactions of C_2H_4 and C_2H_6 on ZVM zeolites.

The analysis of C_2H_6 and C_2H_4 aromatization data obtained with H-ZVM and HNa-ZVM suggested that the incorporation of dehydrogenating components in zeolites should contribute to their catalytic activity in the ethane aromatization. A series of metal-containing H-ZVM - based catalysts, both single-function (Pt/Na-ZVM) and bifunctional (M/H-ZVM), was synthesized and tested to study this effect (Table 2).

Indeed, as shown in Table 2, the addition of Pt to the zeolite markedly improved the catalyst dehydrogenating activity which resulted in a 1.5-2.0 times higher activity in the ethane aromatization compared to H-ZVM (Sample III). In the ethylene aromatization, however, the activity of metal-containinig catalysts (Samples V through VII) was considerably lower which is supposed to be due to the higher rates of hydrogenation and hydrogenolysis reactions of hydrocarbons on metal-zeolite systems.

In spite of an intensive methane formation the aromatics yield on Pt/H-ZVM was as high as 20% of the converted ethane (Table 2).

For a Pd-containing sample (VII) the catalytic activity was found to be a strong function of the pretreatment conditions. For example, the highest activity in the ethane aromatization at $600^{\circ}C$ was shown by a Pd-containing sample treated successively in air (5 hrs at $550^{\circ}C$) and in H_2 (5 hrs at $500^{\circ}C$). The aromatics yield on this sample was 12.5%. The catalysts treated in air or in hydrogen only gave lower aromatics yields, not exceeding 2.5% at $600^{\circ}C$ (Fig.1). This substantial difference was apparently due to the different palladium state in the catalysts formed during the oxidation, oxidation-reduction treatment of the parent metal-zeolite system. This view is in agreement with published data (ref.2).

This is in agreement with the fact that Sample VII which had a low activity in the C_2H_6 reaction after the air treatment, demonstrated a relatively high activity (56% aromatics, Table 2) in the C_2H_6 reaction which can occur without Pd°. The tested Pd-containing sample apparently had a sufficient amount of acid sites necessary for the ethylene aromatization. Unlike the palladium catalyst, the tested Pt-catalysts were less sensitive to the pretreatment.

A characteristic feature of the catalytic behaviour of metal-
containing zeolites in the C_2H_6 aromatization was their "development" with time in the course of the reaction. This development effect was particularly pronounced in the ethylene reactions.
In the case of ethane a qualitatively similar effect was observed
(Fig. 2a). Experiments carried out with H-ZVM demonstrated that
the above effect was not displayed in the absence of metal (Pt,
Pd) (Fig. 2b, curve 3).

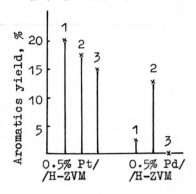

Fig. 1. The effect of pretreatment
conditions on the activity of metal-
zeolite catalysts in the ethane aromatization at $600^{\circ}C$ in flow system;
1 - air (method A);
2 - air-hydrogen (method B);
3 - hydrogen (method C).

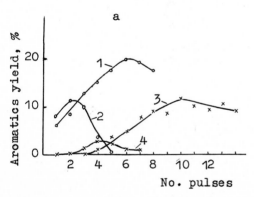 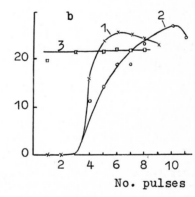

Fig. 2. The development effect observed for the Pt/H-ZVM and
Pd/H-ZVM catalysts in the aromatization of ethane and ethylene
at $550^{\circ}C$ by a pulse method.
(a) Ethane: 1 and 2 - Pt/H-ZVM and Pd/H-ZVM, respectively,
treated by method A; 3 and 4 - Pt/H-ZVM and Pd/H-ZVM, respectively,
treated by method B.
(b) Ethylene: 1,2 and 3 - Pt/H-ZVM, Pd/H-ZVM and H-ZVM, respectively, treated by method A.

This suggested that the effect of the reaction mixture which
caused a modification of the catalytic system was one of the factors responsible for the variation of the activity and selectivity of the studied contact materials in operation. Thus, the valence

and physical states of the transition element could be changed
(final reduction of M^{n+} to M^o, aggregation and migration of M^o)
which could result in a different metal distribution in the channels and on the surface of zeolite crystals. The relative contribution of each factor is supposed to depend on the metal nature,
its content in the catalyst, the level of Na^+ exchange with M^{n+}
in the zeolite, and the reaction temperature.

Thus, the experimental studies demonstrated a high catalytic
activity of ZVM zeolites in the ethylene aromatization; the aromatics yield from C_2H_4 on an H-ZVM zeolite was as high as 75%. At
the same time ZVM-based catalysts exhibited an appreciable activity in the aromatization of ethane, the modified samples yielding
up to 20% aromatics. Experiments carried out to study the effect
of the high temperature pretreatment on the catalytic properties
of metal-containing ZVM zeolites revealed a bifuncitonal nature
of their behaviour in the ethane aromatization reaction.

REFERENCES

1 O.V.Bragin, A.V.Preobrazhensky and A.L.Liberman, Izvestia
   Acad. Nauk SSSR, ser.khim., 2751-2757 (1974).
2 W.I.Reagan, A.W.Chester and G.T.Kerr, J.Catal., 69 (1981)
   89-100.

P.A. Jacobs et al. (Editors), *Structure and Reactivity of Modified Zeolites*
© 1984 Elsevier Science Publishers B.V., Amsterdam — Printed in The Netherlands

SHAPE SELECTIVE ISOMERIZATION AND HYDROCRACKING OF NAPHTHENES OVER Pt/HZSM-5 ZEOLITE

J. WEITKAMP[1,2], P.A. JACOBS[3], and S. ERNST[1]

[1]Engler-Bunte-Institute, Division of Gas, Oil, and Coal, University of Karlsruhe, D-7500 Karlsruhe 1 (Federal Republic of Germany)

[2]Author for correspondence

[3]Centrum voor Oppervlaktescheikunde en Colloidale Scheikunde, Katholieke Universiteit Leuven, B-3030 Leuven (Belgium)

ABSTRACT

The pure naphthenes methylcyclopentane, methylcyclohexane, and ethylcyclohexane were converted over a bifunctional Pt/HZSM-5 zeolite catalyst under hydrogen pressure. For comparison the reactions of ethylcyclohexane were investigated over two large pore zeolites, viz. Pd/LaY and Pt/CaY. A variety of shape selectivity effects are encountered over Pt/HZSM-5, especially in the isomerization of methylcyclohexane and ethylcyclohexane.

INTRODUCTION

In the past, considerable efforts were undertaken to elucidate the mechanisms of catalytic hydrocarbon conversion via carbocations (ref. 1). Particular attention was devoted to large pore zeolite catalysts, especially faujasites, due to their widespread use in the petroleum refining industry. It has been shown that model studies with alkanes of reasonably long carbon chains and suitably selected bifunctional zeolite catalysts furnish most valuable insight into the mechanisms of rearrangement (ref. 2,3) and β-scission (ref. 4,5) of alkylcarbenium ions. Much less work has been done with naphthenes even though they represent another major class of petroleum hydrocarbons. Voorhies et al. reported on the kinetics of cyclohexane (ref. 6,7) and decalin (ref. 8) hydroconversion over acidic mordenites and faujasites loaded with palladium. Schulz et al. (ref. 9) found that the isomerization of C_6-naphthenes on Pt/CaY is accompanied by a disproportionation reaction leading to C_7-/C_8-naphthenes and C_5-/C_4-alkanes. The isomerization of ethylcyclohexane and other naphthenes over non-zeolitic catalysts such as Ni/SiO_2-Al_2O_3 was thoroughly studied by Pines et al. (ref. 10-12).

More recently, zeolite catalysts of medium pore width were introduced on a commercial scale, e.g., in M-forming (ref. 13) and dewaxing (ref. 14). In these refinery processes catalysts of the ZSM-5 type are applied (ref. 13).

It is most likely that, inside the pores of such zeolites, the conversion of hydrocarbons is again determined by the chemistry of carbocations which, however, is modified significantly by shape selectivity effects. It is common practice to classify the latter into (i) reactant shape selectivity, (ii) product shape selectivity, and (iii) restricted transition state shape selectivity, according to Csicsery's proposal (ref. 15,16). A variety of product and restricted transition state shape selectivity effects were recently shown to occur in the hydroconversion of long-chain n-alkanes over Pt/HZSM-5 zeolites (ref. 17,18). It is the intention of the present paper to extend these investigations to naphthenic model hydrocarbons with 6 to 8 carbon atoms. For comparison, some few data will be added which were obtained on large pore zeolite catalysts, viz. Pd/LaY and Pt/CaY.

EXPERIMENTAL

The pure naphthenes methylcyclopentane (M-CPn, purity 99.95 wt.-%, 0.05 % CHx), methylcyclohexane (M-CHx, no detectable impurities), and ethylcyclohexane (E-CHx, purity 99.06 wt.-%, 0.08 % heptanes, 0.21 % M-CHx, 0.48 % other C_8-naphthenes, 0.17 % unidentified impurities, presumably ethylcyclohexenes) were catalytically converted in the gas phase under hydrogen pressure in a fixed bed consisting of 0.47 g of Pt/HZSM-5. The flow type apparatus has been described in detail (ref. 19,20). The content of noble metal and the molar ratio Si/Al of the catalyst were 0.5 wt.-% and 60, respectively. Its preparation has been described previously (ref. 17,18). The 0.5 Pt/CaY-86 zeolite was a commercial sample (Union Carbide, SK 200) which has been used extensively in prior studies by the Karlsruhe group (e.g., ref. 3). The 0.27 Pd/LaY-72 zeolite was prepared from NaY by ion exchange with an aqueous solution of $La(NO_3)_3$ at 80 °C followed by ion exchange with $[Pd(NH_3)_4]Cl_2$. The Pd content, the molar ratio Si/Al, the total content of La and Na cations per Al, and the La ion exchange level amounted to 0.27 wt.-%, 2.46, 1.0 equiv./mol, and 72 equiv.-%, respectively. All catalysts were used in a particle size between 0.2 and 0.3 mm and pretreated successively in a purge of oxygen at 300 °C, a purge of nitrogen at 350 °C, and a purge of hydrogen at 300 °C.

High resolution capillary GLC with a flame ionization detector was employed for product analysis. During a run, at least two product samples were analyzed in the on-line mode, usually with polypropylene glycol (PPG) as stationary phase. In addition, liquid product samples were collected at ca. -190 °C which were later analyzed off-line using a different stationary GLC phase. In the experiments with ethylcyclohexane the combined application of a relatively polar (PPG) and a non-polar phase (e.g., OV-1) was a prerequisite for achieving satisfactory separation of feed isomers, especially when octanes were present. It was found that the variation of the temperature program in the

GLC oven is a valuable tool for separating unresolved peaks. The assignment of product peaks was based on the experience acquired in preceding studies on hydrocarbon conversion (e.g., ref. 3,9,19), on the boiling points of isomeric C_8-naphthenes, and ancillary GC/MS analyses (Hewlett-Packard 5987 A, ionization by 70 eV electron impact). All commercially available C_8-naphthenes were used as reference substances. Particular care was paid to the identification of methylcycloheptane.

Unless otherwise stated the partial pressures of the feed hydrocarbon and hydrogen at the reactor inlet were 19.4 kPa and 2.0 MPa, respectively, and W/F was 135 g·h/mol. The reaction temperature was varied between 200 and 360 °C. Between two successive runs the catalyst was purged in pure hydrogen at 2.0 MPa and 300 °C.

RESULTS AND DISCUSSION
Conversion and types of reaction

It was ascertained in preliminary experiments and by repeated control runs that there was no catalyst deactivation under the conditions applied in this study. In Figure 1 the conversions of the three naphthenes on Pt/HZSM-5 are plotted versus the reaction temperature. For comparison the conversion of ethylcyclohexane on the Pd/LaY zeolite is also given. As expected, the reactivity of the hydrocarbons increases with increasing carbon number. The Y type

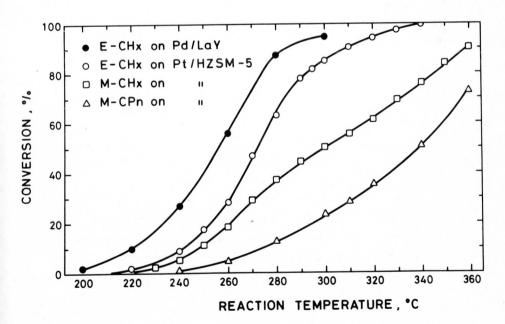

Fig. 1. Conversion of naphthenes over bifunctional zeolite catalysts.

zeolite is slightly more active than Pt/HZSM-5 which is in all probability due to the higher *number* of Brønsted acid sites in Pd/LaY. Earlier TPD experiments with NH_3 revealed that HZSM-5 possesses very strong acidic sites (ref. 21) whereas it is a reasonable assumption that sites of intermediate acid strength predominate in Pd/LaY with its moderate exchange level of 72 %.

The following types of reaction of the naphthenes will be distinguished: (i) Isomerization to other naphthenes, (ii) ring opening to alkanes with the same carbon number as the feed, (iii) hydrocracking to alkanes or cycloalkanes containing less carbon atoms than the feed, and (iv) formation of aromatics with the carbon number of the feed. It is evident from Figures 2 and 3 that under mild conditions isomerization is the sole reaction, regardless of the catalyst and the feed. Under more severe conditions on Pt/HZSM-5 methylcyclo-pentane undergoes mainly ring opening while hydrocracking predominates with methylcyclohexane and ethylcyclohexane. On Pd/LaY ca. 90 % of the ethylcyclo-hexane feed can be isomerized without substantial carbon-carbon bond rupture.

It is noteworthy that on Pt/HZSM-5 methylcyclopentane does not dispropor-tionate to any measurable extent as it does on, e.g., Pt/CaY (ref. 9). The potential products of such a disproportionation, i.e., C_4- and C_5-alkanes as well as C_8- and C_7-naphthenes, could at least in part diffuse out of the intra-crystalline framework since they are formed by hydrocracking and isomerization of M-CHx and E-CHx (vide infra). Therefore, it is concluded that the lack of disproportionation reactions in Pt/HZSM-5 is due to restricted transition state shape selectivity. Indeed, a bimolecular alkylation step is involved in the disproportionation reaction (ref. 9) which requires a bulky transition state. Evidence for some contribution of bimolecular reactions of ethylcyclo-hexane on Pd/LaY will be presented below.

Isomerization

Both on Pd/LaY and Pt/HZSM-5 isomerization proceeds in a stepwise manner. E.g., the primary products from ethylcyclohexane are the other monobranched C_8-naphthenes (Fig. 4), viz. mainly propylcyclopentane and little methyl-cycloheptane. Dibranched and tribranched isomers are formed in consecutive reactions. Presumably, the isomer distribution found on Pd/LaY around 300 °C is close to thermodynamic equilibrium. On Pt/HZSM-5 much higher selectivities of dibranched isomers are encountered. Probably, diffusion of most, if not all tribranched isomers (trimethylcyclopentanes) out of the ZSM-5 intracrystalline framework is severely hindered. It is reasonable to assume that, for the most part, these bulky isomers are formed on the *external* HZSM-5 surface or on platinum clusters located at the *external* zeolite surface.

The rapid formation of monobranched isomers from ethylcyclohexane is easily understood since this is a so-called type A isomerization (ref. 2,3,23). By definition, the number of ramifications remains constant in type A

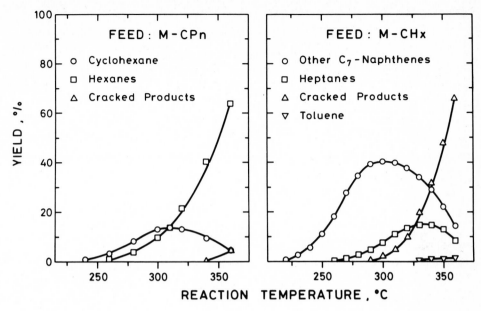

Fig. 2. Yields of products from methylcyclopentane and methylcyclohexane over Pt/HZSM-5.

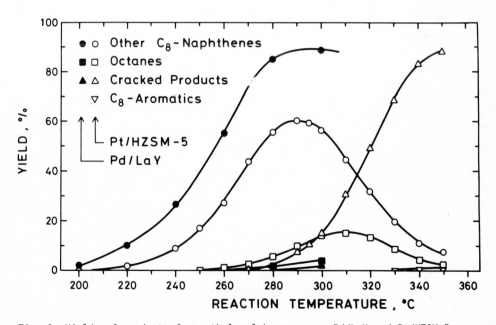

Fig. 3. Yields of products from ethylcyclohexane over Pd/LaY and Pt/HZSM-5.

TABLE 1

Selectivity of isomerization of methylcyclohexane on Pt/HZSM-5 (expressed as moles of corresponding C_7-naphthene formed per moles of M-CHx isomerized) and thermodynamic equilibrium distribution [1] of C_7-naphthenes.

T, °C		220	250	300	360	227	327
X_{M-CHx}, %		1.0	11.2	50.4	90.5	Equilibrium	
$Y_{Isomers}$, %		1.0	11.0	40.3	14.4	Distribution,	
			Selectivity, %			mol-%	
M-CHx		-	-	-	-	57.9	38.8
E-CPn		68	38.4	21.8	17.5	4.8	9.2
1,1-DM-CPn		0	0.1	1.8	9.0	7.2	9.0
1c2-DM-CPn		0	0	0.2	4.3	2.0	3.6
1t2-DM-CPn		0	0.1	0.5	15.9	12.3	16.5
1c3-DM-CPn		17	30.7	37.8	26.6	10.0	14.0
1t3-DM-CPn		15	30.7	37.9	26.7	5.8	8.9

[1] Calculated from ref. 22

isomerizations whereas in type B isomerizations a new branching forms or an existing one disappears. It has been found previously with aliphatic substrates and a Pt/CaY zeolite (ref. 2) that type A isomerizations are somewhat faster

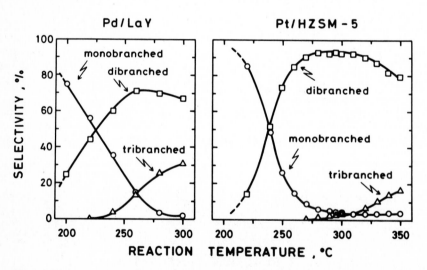

Fig. 4. Isomers formed from ethylcyclohexane (selectivity is defined as in Table 2).

TABLE 2

Selectivity of isomerization of ethylcyclohexane (expressed as moles of corresponding C_8-naphthene formed per moles of E-CHx isomerized in %).

Catalyst	Pt/HZSM-5				Pt/CaY	Pd/LaY	
T, °C	220	240	290	330	220	200	280
P_{E-CHx}, kPa	19.4	19.4	19.4	19.4	25.1	19.4	19.4
P_{H_2}, MPa	2.0	2.0	2.0	2.0	3.9	2.0	2.0
W/F, g·h/mol	135	135	135	135	250	280	135
X, %	1.9	9.0	78.1	97.6	3.9	4.9	87.6
$Y_{Isomers}$, %	1.9	8.9	60.4	19.5	3.9	4.9	85.7
P-CPn	85.7	48.5	5.3	4.1	59.9	58.8	4.3
M-CHp	0	0	0.1	0.1	6.1	4.0	0.1
(1-M-E)-CPn	0	0.5	2.8	2.1	0	0	0.8
1-E-1-M-CPn	0	0	0.3	1.4	1.1	1.7	1.3
1-E-c-2-M-CPn	0	0	0	1.1	2.8	1.9	0.7
1-E-t-2-M-CPn	0	0	0	1.5	14.7	13.8	4.7
1-E-c-3-M-CPn	6.0	15.4	17.5	17.0	2.2	2.1	4.2
1-E-t-3-M-CPn	6.6	15.3	17.4	12.5			
1,1-DM-CHx	0	0	0.1		3.3	3.7	8.8
1c2-DM-CHx	0	0	0	0.4	1.6	1.8	1.2
1t2-DM-CHx	0	0	0.3	4.0	3.3	5.4	8.5
1c3-DM-CHx	0	0	0		4.1	5.4	
1t4-DM-CHx	1.7	16.0	40.1	32.4	0	0	30.0
1t3-DM-CHx	0	0	0	5.8			
1c4-DM-CHx	0	4.3	13.6	6.5	0.9	1.4	9.7
1,1,2-TM-CPn	0	0	0	1.0	0	0	2.7
1,1,3-TM-CPn	0	0	1.7	3.3	0	0	6.8
1c2c3-TM-CPn	0	0	0	0	0	0	0.1
1c2t3-TM-CPn	0	0	0.1	1.0	0	0	1.9
1t2c3-TM-CPn	0	0	0.2	2.2	0	0	5.1
1c2c4-TM-CPn	0	0	0.1	0.6	0	0	1.5
1c2t4-TM-CPn	0	0			0	0	
1t2c4-TM-CPn	0	0	0.4	3.0	0	0	7.6

than type B isomerizations. Mechanistically, type A rearrangements via carbocations are best accounted for by consecutive hydride and alkyl shifts while type B rearrangements are nowadays (ref. 3,23,24) believed to proceed via protonated cyclopropanes (PCPs).

More detailed information on the selectivities of isomerization is presented in Tables 1 and 2. The type B isomerization of methylcyclohexane (Table 1) exhibits a pronounced shape selectivity: Up to at least 50 % conversion virtually no 1,1- and 1,2-dimethylcyclopentane are formed while both 1,3-isomers occur even at very low conversion. A clear discrimination between product and restricted transition state shape selectivity cannot be made as long as the diffusion coefficients of the isomers are unknown. At this stage of the investigation, we believe that product shape selectivity is mainly responsible for the absence of 1,1-, 1cis2-, and 1trans2-dimethylcyclopentane.

Even more pronounced shape selectivity effects are observed in the isomerization of ethylcyclohexane (Table 2). Whereas on Pt/CaY and Pd/LaY almost all possible isomers with two branchings occur at low conversion 1-ethyl-3-methyl-cyclopentane and 1,4-dimethylcyclohexane are the preferred products on Pt/HZSM-5. A comparison of the low conversion data for the three zeolites clearly reveals that the selectivity is determined by the type of zeolite rather than by the nature of the noble metal. This rules out that isomerization on Pt/HZSM-5 occurs on external Pt clusters, except perhaps for the formation of trimethyl-pentanes. It is also interesting to notice that small amounts of methylcyclo-heptane are formed from ethylcyclohexane, especially on the large pore zeolites.

Probable routes from ethylcyclohexyl cations to the products observed at low conversions are depicted in Figures 5 and 6. Based on earlier results obtained with aliphatic substrates (ref. 2,25) it was assumed that formation of PCPs (A,B,D in Fig. 5) starts from the set of secondary cations rather than from the tertiary cation. It is evident that the latter can only react to one particular PCP designated A in Fig. 5. However, such a pathway can be neglected at least for Pt/HZSM-5 since the predicted isomers formed via A are absent.

The formation of dimethylcyclohexanes can be explained by a type A rearrangement starting from ethylmethylcyclopentyl cations with the positive charge in the side chain (Fig. 6). This way, the fast formation of 1,4-dimethylcyclo-hexane from ethylcyclohexane on Pt/HZSM-5 is readily accounted for. Alternatively, a pathway to dimethylcyclohexanes can be considered in which the anticipated sequence of type B and type A rearrangements is changed: Type A rearrangements of ethylcyclohexyl cations lead to both propylcyclopentyl and methylcycloheptyl cations. Type B rearrangements of the latter give dimethyl-cyclohexyl cations with all possible carbon skeletons.

It cannot be decided from the results of this study whether the interconversion of cycloalkanes and cycloalkylcarbenium ions proceeds via olefins or not. In principle, the noble metal opens the route via olefins. Another major function of the noble metal is to avoid catalyst deactivation by the formation of coke deposits.

Ring opening

Carbon-carbon bond rupture within the ring of cycloalkylcarbenium ions is a slow reaction. This has been explained in terms of unfavorable orbital orientation during β-scission (ref. 26). Hence, the low rate of ring opening even at high conversions of ethylcyclohexane on Pd/LaY is readily understood. The relatively high selectivities for trimethyl isomers (cf. Table 2 for Pd/LaY at 280 °C) are in sharp contrast to isomerization of aliphatic hydrocarbons, e.g.,

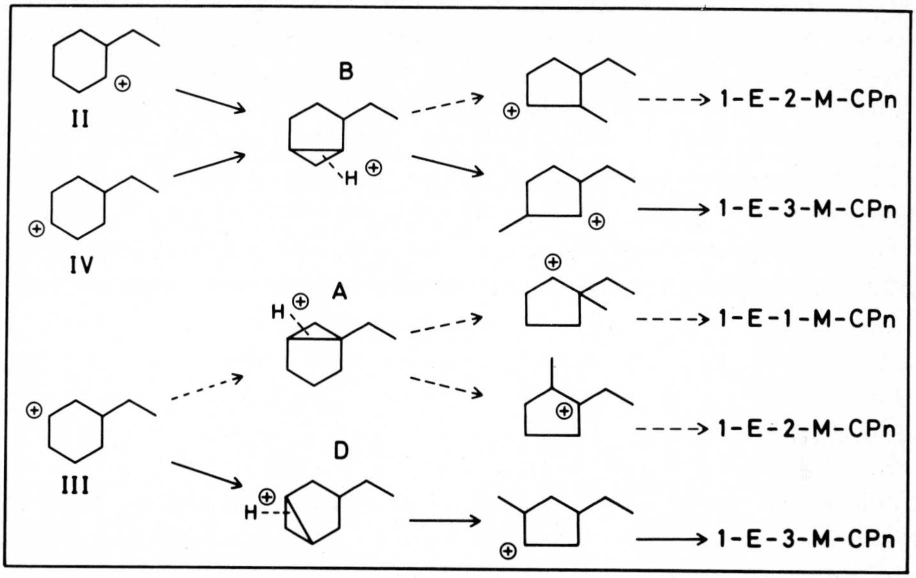

Fig. 5. Type B rearrangements of secondary, ethylcyclohexyl cations. Full arrows lead to products which are strongly favored over Pt/HZSM-5.

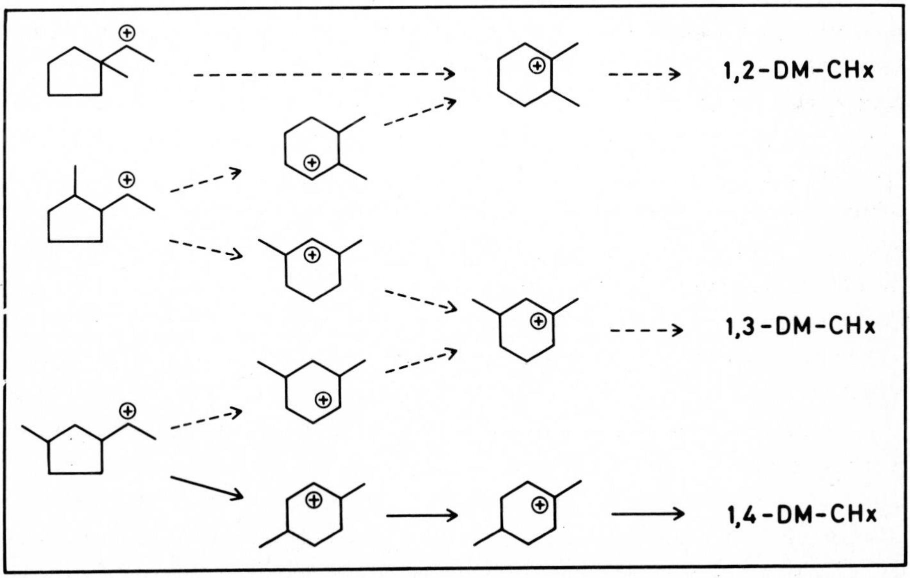

Fig. 6. Formation of dimethylcyclohexanes by ring enlargement of ethylmethyl-cyclopentyl cations via tpye A rearrangements.

of n-octane over similar bifunctional catalysts (ref. 27,28). In particular, the occurrence of 1,1,3-trimethylcyclopentane is noteworthy since this isomer possesses an α,α,γ-tribranched carbon skeleton and, hence, it is capable to undergo the energetically favored type A β-scission as defined in a recent paper (ref. 18).

On Pt/HZSM-5 the maximum yield of isomers from ethylcyclohexane is considerably lower than on Pd/LaY (Fig. 3). Further work is needed to decide whether this is due to the higher strength of acidic sites in the ZSM-5 zeolite or to geometric constraints or to other reasons. It is interesting to note that on Pt/HZSM-5 the maximum yields of octanes from ethylcyclohexane (Fig. 3) and of heptanes from methylcyclohexane (Fig. 2) are relatively low. This means that there is no fast mechanism of desorption of the alkenyl cations formed by cleavage of the naphthenic rings. Rather, the cations formed by ring opening undergo consecutive β-scissions into smaller fragments.

The alkanes formed by ring opening mainly consist of monomethyl isomers irrespective of the nature of the feed and the catalyst. Besides, the n-alkanes are present but there is no indication whatsoever that n-alkanes are primary products of ionic ring opening. Hence, there is no indication for a so-called direct ring opening via non-classical carbonium ions as proposed in the literature (ref. 29). With ethylcyclohexane as feed 2,4- and especially 2,5-dimethylhexane were always present in the products of ring opening. The latter isomer has the carbon skeleton which one would expect to result by a type A β-scission of the tertiary 1,1,3-trimethylcyclopentyl cation.

Hydrocracking

Typical distributions of the cracked products from ethylcyclohexane are presented in Figure 7. Over Pt/HZSM-5 the carbon number distribution is essentially symmetrical with very low selectivities for C_1 and C_2 and, correspondingly, C_7 and C_6. This curve is best interpreted in terms of an ionic mechanism of hydrocracking leading to C_3 through C_5 and a superimposed hydrogenolysis on Pt clusters. Relatively large amounts of $C_3 + C_5$ are formed on Pt/HZSM-5 compared to hydrocracking of C_8-hydrocarbons on large pore zeolites, e.g., ethylcyclohexane on Pd/LaY (Fig. 7, left side) or n-octane on Pt/CaY (ref. 27).

On Pd/LaY neither methane nor ethane are formed at 300 °C. Nevertheless, substantial amounts of C_7- and C_6-hydrocarbons occur under the same conditions, and slightly more C_5 than C_3 is found. Most of the C_7- and C_6-products are naphthenes. To account for all these findings some contribution of a bimolecular mechanism must be invoked: A C_8 species is added to a second one which results in C_{16} unit. The latter undergoes rearrangements until a favorable structure for abstraction of C_4 or C_5 exists. This way, 2 C_8 can be converted into, e.g., $C_4 + C_5 + C_7$. According to such a mechanism one could as well expect products with a carbon number above C_8, i.e., the occurrence of a disproportio-

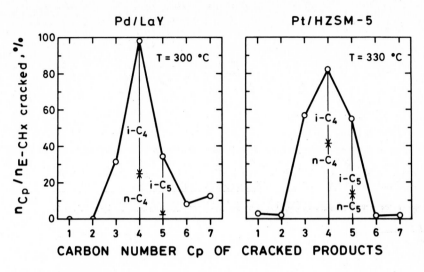

Fig. 7. Hydrocracking of ethylcyclohexane. Distribution of the cracked products.

nation type of reaction. However, no hydrocarbons with 9 or more carbon atoms were ever observed during this study. This indicates that the desorption of such intermediates from the acidic sites is slow compared to their cleavage.

Hydrocracking of methylcyclohexane over Pt/HZSM-5 gives mainly propane and i-butane beside little n-butane and small amounts of hydrocarbons formed by hydrogenolysis. In contrast, the cracked products from methylcyclopentane, e.g., at 360 °C contain large amounts of methane, ethane, butanes, and pentanes, i.e., the contribution of hydrogenolysis is high. This is another example for the fact that model hydrocarbons with six (or less) carbon atoms are often excluded from the favored pathways of ionic cracking and escape into different mechanisms (cf. ref. 27). The use of such small hydrocarbons for the investigation of ionic mechanisms is, therefore, discouraged.

ACKNOWLEDGEMENTS

The authors thank Mr. W. Stober and Mr. A. Dietl for valuable technical assistance and Mr. S. Maixner who carried out the mass spectroscopy work. Financial support by Deutsche Forschungsgemeinschaft and Fonds der Chemischen Industrie is gratefully acknowledged.

REFERENCES

1 M.L. Poutsma, in J.A. Rabo (Ed.), Zeolite Chemistry and Catalysis, ACS Monograph 171, American Chemical Society, Washington, D.C., 1976, pp. 437-528.
2 J. Weitkamp and H. Farag, Acta Universitatis Szegediensis, Acta Physica et

Chemica, 24 (1978) 327-333.

3   J. Weitkamp, Ind. Eng. Chem., Prod. Res. Dev., 21 (1982) 550-558.

4   J. Weitkamp, Erdöl, Kohle - Erdgas - Petrochem., 31 (1978) 13-22.

5   M. Steijns, G. Froment, P. Jacobs, J. Uytterhoeven and J. Weitkamp, Ind. Eng. Chem., Prod. Res. Dev., 20 (1981) 654-660.

6   A. Voorhies, Jr. and J.R. Hopper, in E.M. Flanigen and L.B. Sand (Eds.), Molecular Sieve Zeolites II, Adv. Chem. Ser. 102, American Chemical Society, Washington, D.C., 1971, pp. 410-416.

7   M.G. Luzarraga and A. Voorhies, Jr., Ind. Eng. Chem., Prod. Res. Dev., 12 (1973) 194-198.

8   R. Beecher, A. Voorhies, Jr. and P. Eberly, Jr., Ind. Eng. Chem., Prod. Res. Dev., 7 (1968) 203-209.

9   H. Schulz, J. Weitkamp and H. Eberth, Proc. 5th Intern. Congr. Catalysis, J.W. Hightower (Ed.), Vol. 2, North-Holland Publishing Co., Amsterdam, 1973, pp. 1229-1239.

10 H. Pines and A.W. Shaw, J. Am. Chem. Soc., 79 (1957) 1474-1482.

11 H. Pines and A.W. Shaw, Adv. Catal. 9 (1957) 569-574.

12 H. Pines and N.E. Hoffman, in G.A. Olah (Ed.), Friedel-Crafts and Related Reactions, Vol. II, Part 2, Interscience Publishers, New York, London, Sydney, 1964, pp. 1211-1252.

13 H. Heinemann, in J.R. Anderson and M. Boudart (Eds.), Catalysis - Science and Technology, Vol. 1, Springer-Verlag, Berlin, Heidelberg, New York, 1981, pp. 1-41.

14 H.R. Ireland, C. Redini, A.S. Raff and L. Fava, Hydrocarbon Process., 58 (No. 5, 1979) 119-122.

15 S.M. Csicsery, in J.A. Rabo (Ed.), Zeolite Chemistry and Catalysis, ACS Monograph 171, American Chemical Society, Washington, D.C., 1976, pp. 680-713.

16 S.M. Csicsery, Preprints, Div. Fuel Chem., Am. Chem. Soc., 28 (No. 2, 1983) 116-126.

17 P.A. Jacobs, J.A. Martens, J. Weitkamp and H.K. Beyer, Faraday Discuss. Chem. Soc. 72 (1982) 353-369.

18 J. Weitkamp, P.A. Jacobs and J.A. Martens, Appl. Catal. 8 (1983) 123-141.

19 H. Schulz and J. Weitkamp, Ind. Eng. Chem., Prod. Res. Dev., 11 (1972) 46-53.

20 J. Weitkamp and K. Hedden, Chem.-Ing.-Tech., 47 (1975) 537.

21 P.A. Jacobs, J.B. Uytterhoeven, M. Steijns, G. Froment and J. Weitkamp, in L.V.C. Rees (Ed.), Proc. 5th Intern. Conference Zeolites, Heyden, London, Philadelphia, Rheine, 1980, pp. 607-615.

22 D.R. Stull, E.F. Westrum, Jr. and G.C. Sinke, The Chemical Thermodynamics of Organic Compounds, John Wiley & Sons, New York, London, Sydney, Toronto, 1969, pp. 348-350 and 357.

23 D.M. Brouwer and H. Hogeveen, in R.W. Taft and A. Streitwieser, Jr. (Eds.), Progr. Phys. Org. Chem., Vol. 9, Interscience Publishers, New York, London, Sydney, Toronto, 1972, pp. 179-240.

24 J. Weitkamp, in T. Seiyama and K. Tanabe (Eds.), Proc. 7th Intern. Congr. Catal., Elsevier Scientific Publishing Co., Amsterdam, Oxford, New York, 1981, pp. 1404-1405.

25 J. Weitkamp and P.A. Jacobs, Preprints, Div. Petr. Chem., Am. Chem. Soc. 26 (1981) 9-13.

26 D.M. Brouwer and H. Hogeveen, Recl. Trav. Chim. Pays-Bas, 89 (1970) 211-224.

27 J. Weitkamp, in J.W. Ward and S.A. Qader (Eds.), Hydrocracking and Hydrotreating, Am. Chem. Soc. Symp. Ser. 20, American Chemical Society, Washington, D.C., 1979, pp.1-27.

28 H. Vansina, M.A. Baltanas and G.F. Froment, Ind. Eng. Chem., Prod. Res. Dev., 22 (1983) 526-531.

29 E.G. Christoffel and K.-H. Röbschläger, Ind. Eng. Chem., Prod. Res. Dev., 17 (1978) 331-334.

P.A. Jacobs et al. (Editors), *Structure and Reactivity of Modified Zeolites*
© 1984 Elsevier Science Publishers B.V., Amsterdam — Printed in The Netherlands

LIQUID PHASE SYNTHESIS OF AROMATES AND ISOMERS ON POLYFUNCTIONAL ZEOLITIC CATALYST MIXTURES

H. NGUYEN-NGOC[+], K. MÜLLER and M. RALEK

Institut für Technische Chemie, Technische Universität Berlin,

Straße des 17. Juni 135, D 1000 Berlin 12 (Germany)

[+]Present address: Institute of Chemistry, National Center for Scientific
Research of Vietnam
Ho Chi Minh City, Vietnam

ABSTRACT

In order to carry out the one-stage synthesis of aromates and isomers on a suspension of the ZSM-5 zeolite and a Fischer-Tropsch catalyst, the application of the Kölbel-reactor is taken into account. The Kölbel-reactor, which had been developed for the Fischer-Tropsch synthesis, is well suited to convert synthesis gases containing a high concentration of CO, which are generated by the economical second generation gasifiers.

The manganese-iron precipitation catalyst was used as the Fischer-Tropsch component. This catalyst and the ZSM-5 zeolite were suspended in tetralin, which showed sufficient stability in the presence of the zeolite. Although there was no direct contact between the catalyst particles in the suspension, the primary products of the Fischer-Tropsch catalyst were converted into aromates and isomers by the zeolite. A yield of 127 g gasolin fraction, which mainly consisted of aromates and isomers, was obtained per m^3 of converted synthesis gas.

INTRODUCTION

The products of the conventional Fischer-Tropsch (FT) synthesis consist mainly of straight-chain hydrocarbons (1) and are therefore only directly suitable for the field of diesel engine fuel. The FT gasolin fraction has a RON-number of approximately 70 and therefore requires further processing before it can be used as super gasoline. The Mobil Oil Corporations "MTG process" offers an alternative way of producing super gasoline. Here, methanol is first of all produced from synthesis gas and in a second stage this methanol is converted on a zeolite of the ZSM-5 variety (2).

Recently, the possibility of producing anti-knock gasoline by combining the FT synthesis with a conversion on a ZSM-5 zeolite, has aroused growing interest. The olefines and the compounds containing oxygen, which are produced by the FT synthesis, react on ZSM-5 to produce hydrocarbon mixtures that are rich in isomers and aromates (3), which can then be refined to super gasoline or to raw materials for the chemical industry. Compared to the methanol synthesis, the FT synthesis in the liquid phase is remarkably advantageous; the CO-rich synthesis gases, which are generated by economical second generation coal gasifiers, can be used without the additional stage of shift conversion. A study by the MITRE Corporation for the US Department of Energy (4), which compares the economy of different FT reactors in combination with modern second generation coal gasifiers, concludes that the Kölbel-reactor is the most favourable type.

In a two-stage process, Mobil Oil investigates the suitability of the Kölbel-reactor for converting the synthesis gas. In the second stage, where the ZSM-5 catalyzed reaction takes place, conventional reactor types such as fixed-bed or fluized-bed are used (5).

Synthesis runs using a physical mixture of an FT catalyst and an H-ZSM-5 zeolite placed in a fixed-bed reactor, demonstrated the possibility of synergistic interactions between both catalyst components (6, 7). The liquid products derived from this one-stage synthesis contained aromatic fractions three times greater than those of the liquid products, which were synthesized in two-stage synthesis runs with separated arrangement of the catalyst components. In contrast to the two-stage process, where every step can be carried out under individually optimized reaction conditions, the components used in the one-stage process have to work under the same reaction conditions (7).

Bearing these factors in mind, an attempt was made to conduct the one-stage synthesis of aromates and isomers in the liquid phase with a suspension of a ZSM-5 zeolite and a FT catalyst. The selective manganese-iron (MnFe) precipitation catalyst was used as the FT component. In order to avoid accelerated coking, the MnFe catalyst should not be operated above 300°C. Therefore, special care was taken as regards the effect of the synthesis temperature on the activity of the ZSM-5 zeolite. Furthermore, the dependence of the synthesis on the composition and concentration of the catalyst suspension, as well as on the size of the catalyst particles was investigated.

EXPERIMENTAL

All synthesis runs were previously carried out with a continuously stirred tank reactor to avoid any difficulties arising between those parameters of the Kölbel-reactor which can be directly adjusted and those which cannot. The reactor had a volume of one litr and was equipped with a magnetically actuated stirrer, which could be operated to a maximum speed of 1500 RPM.

The usage of zeolites in the presence of hydrocarbons obviously causes the danger of cracking and can result in a loss of liquid phase. Therefore the stability of the following hydrocarbons were tested against this activity of ZSM-5: $C_{60}-C_{70}$ paraffin, squalene, diphenyl, naphthalin and tetralin (8).

At 300°C, all aliphatic hydrocarbons cracked very quickly and only the aromatic hydrocarbons offered satisfactory stability. Because of its low price, tetralin was used as liquid phase in all synthesis runs. Before entering the reactor, the synthesis gases were saturated with liquid phase under the conditions of the reaction, by passing them through a saturator. This, together with a relatively high pressure of approximately 50 bars, were necessary to prevent a continuous loss of liquid phase from the reactor.

The MnFe catalyst was prepared by the continuous precipitation method optimized by Lehmann et al. (9). Catalyst activation was carried out at 260°C by carburation with carbonmonoxide and reduction with hydrogen, for 24 h in both cases. To this end, a fraction of the MnFe catalyst was taken, whose average diameter is 50 μm. The diameter of the FT catalyst particles in the synthesis can be smaller, due to decay caused by mechanical stress occuring in the stirred reactor.

The synthesis of the ZSM-5 zeolite is based on Argauer and Landolt's US patent (10). After crystallization, the zeolite was calcined at 540°C in an airflow and was finally treated with 1 n HCl to obtain the protonic form. To examine the effect of the zeolite particle size on the synthesis, two sieve fractions of differing particle diameters (d) were used: d < 50 μm and 50 μm < d < 200 μm.

Without the presence of air, the activated FT component and the zeolite were dispersed in a specific quantity of tetralin and then poured into the reactor. The carbonmonoxide-to-hydrogen ratio of the synthesis gas was 1.4:1. The catalyst load refers merely to the mass of the MnFe catalyst, as only this catalyst is responsible for the primary reaction of synthesis gas. The synthesis temperatures were varied between 265°C and 300°C. At 240°C, the catalyst mixtures produced only negligible amounts of aromatic compounds. The educt and product gases as well as the liquid products were analysed by gas chromatography.

RESULTS AND DISCUSSION

The synthesis conditions and results, obtained with a slurry of MnFe catalyst and H-ZSM-5 zeolite (MnFe-H-ZSM-5) after a run of 240 h, are summarized in table 1. For comparison, the table also gives the data achieved for the MnFe catalyst alone. Compared to the MnFe catalyst, the application of the MnFe-H-ZSM-5 catalyst leads to a greater yield of C_{5+} phase and to a drastic decrease in the yields of long and short chain olefins. As in the fixed-bed synthesis, the synthesis gas conversion in the liquid phase was not affected by the addition of zeolite to the MnFe catalyst. If the yields obtained in the fixed-bed and in the liquid phase reactor are based on the volume synthesis

TABLE 1

Conditions, yields (Y) and carbon selectivities (C) on a MnFe catalyst and MnFe-ZSM-5 catalyst.

Catalyst concentration		10 % wt. MnFe		10 % wt. MnFe 10 % wt. H-ZSM-5	
T	(°C)	265		265	
p	(bar)	50		50	
CO/H_2		1.4		1.4	
Catalyst load ($m_N^3 \, h^{-1} kg^{-1}$)		0.396		0.402	
Conversion	($CO+H_2$)	0.52		0.48	
run time	(h)	192		240	
		Y $(g/m_N^3)_{feed}$	C (%)	Y $(g/m_N^3)_{feed}$	C (%)
CH_4		15.2	12.7	11.1	9.9
C_2-C_4-Olefins		23.9	22.7	0.2	0.2
C_2-C_4-Paraffins		26.9	24.3	28.5	27.8
Liquid phase, C_{5+}		42.5	40.4	61.0	62.1
Liquid phase, C_{5+}					
Structure isomers		13.9	13.2	19.2	17.0
Nonbranched HC		28.6	27.2	5.8	5.1
Aromates		0	0	36	40.0
Olefins		19.8	18.8	0	0
Paraffins		22.7	21.6	25	22.1
	Distribution of the aromatic product fraction % wt.				
	Benzene			6	
	Toluene			6	
	Xylene, Ethyltoluene			21	
	C_9-aromates			39	
	C_{10}-aromates			28	

gas converted, there are no significant differences between them.

Whereas the Schulz-Flory equation fits the product distribution of the MnFe catalyst, it fails to describe the product distribution of the MnFe-H-ZSM-5 catalyst (Fig. 1).

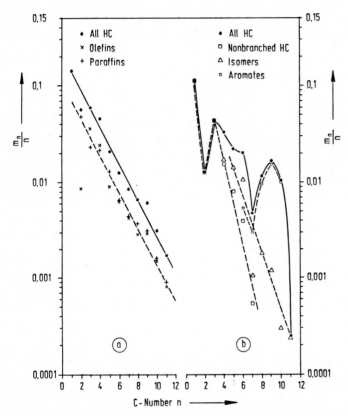

Fig. 1. Product distribution according to Schulz-Flory
a) MnFe catalyst
b) MnFe-H-ZSM-5 catalyst

Short-chain olefins, which are primary products of the FT catalyst, are immediately converted by the zeolite into aromates of 6 to 10 carbon atoms per molecule. Aromatic hydrocarbons of more than 10 carbon atoms are not produced by the H-ZSM-5 zeolite because of the limited size of its pores. Only a small amount of the primary olefins are converted into linear-chained paraffins by hydrogenation.

As already stated, the conversion of synthesis gas on a mixed MnFe-H-ZSM-5 catalyst is a complex system of reactions which are bound together by olefin intermediates (7). The olefin intermediates can react in two separate ways. This depends upon the activity of both catalysts and the average distance be tween the different catalyst particles:

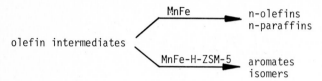

Further proof of the synergistic action was that the yield of aromates decreased as the zeolite particle size increased (8). This decrease was caused by a separation of the reactions occurring on the different catalyst components. The effective reaction rate of olefin intermediates on the zeolite was limited by diffusion (11).

The catalyst concentration of the slurry was varied between 10 % and 30 % wt. When the catalyst load and conversion were kept constant, an increase of the catalyst concentration led to a significant increase in aromates (7). An optimum yield of aromates was achieved with equal mass portions of both the catalyst components suspended in the liquid phase. Using a 20 % concentration of such a catalyst in the slurry in a synthesis run at 265°C and at the conversion level of 48 %, 101 g hydrocarbons were produced per m_N^3 feed. The liquid product contained 36 g aromates and 19 g isomers per m_N^3 feed. When an excess of zeolite was used, the cracking of long-chain hydrocarbons was accelerated. This results in an increased yield of C_2-C_4 hydrocarbons and a decreased yield of liquid phase C_{5+}. The yield of aromates however, does not diminish if zeolite is used in excess. Similar to the MTG process, the aromatic hydrocarbons, which are yielded by using ZSM-5 zeolite and MnFe catalyst in liquid phase, are predominantly alkylated benzol derivatives.

It is common in petrochemistry to regenerate zeolites using oxidative treatment. An attempt was made to simulate this oxidative regeneration by oxidizing the catalyst mixture with a diluted stream of oxygen at 350°C. After activation with carbonmonoxide and hydrogen, the initial behaviour of the MnFe catalyst in synthesis was attained.

The medium oil phases (200° - 300°C), which are obtained from coal hydrogenation, consist mainly of polycyclic aromatic hydrocarbons and can possibly be used as an alternative to tetralin as liquid phase. Integration of the one-stage synthesis of aromatic hydrocarbon into direct coal liquification, is also

an interesting way of enhancing the flexibility of the whole process, with regard to the quality of the coal to be converted and the market development.

ACKNOWLEDGEMENT

Thanks are due to the Ministry of Research and Technology of the Federal
Republic Germany for financial support.

REFERENCES

1 C.D. Frohning, H. Kölbel, M. Ralek, W. Rottig, F. Schnur und H. Schulz in
   J. Falbe (Ed.), Chemical Feedstocks from Coal, John Wiley, 1982, p 309
2 C.O. Chang, J.C.W. Kuo, W.H. Lang, S.M. Jacob, J.J. Wise und A.J. Silvestri,
   Ind.Eng.Chem.Process Des.Dev., 17 (1978) 255
3 R.A. Stowe und C.B. Murchison, Hydrocarbon Processing, 1 (1982) 147
4 D. Gray, M. Lytton und M. Nenworth, DOE Contract No.: EF-77-C-01-2783
   (Nov. 1980), MITRE Corp. McLean, Virginia
5 W.O. Haag, Mobil Research and Development Corporation, Princeton, USA,
   lecture given in the Institut für Technische Chemie, Technische Universität
   Berlin, 14.10.1983, DOE contract No. DE-AC22-80 PC 30022, Oktober 1983
6 P.D. Caesar, J.A. Brennan, W.E. Garwood, J. Civic, J. Catal., 50
   (1979) 274
7 K. Müller, W.-D. Deckwer und M. Ralek in P.A. Jacobs et al. (Ed.),
   Metal Microstructures in Zeolites, Elsevier, Amsterdam, 1982, p 267
8 H. Nguyen-Ngoc, Thesis, Technische Universität Berlin, 1983
9 H.J. Lehmann, Thesis, Technische Universität Berlin, 1981
10 US Pat. 3,702,866 (1972), R.J. Argauer und G.R. Landolt
11 P.B. Weisz, Adv. Catal., 13 (1962) 137

P.A. Jacobs et al. (Editors), *Structure and Reactivity of Modified Zeolites*
© 1984 Elsevier Science Publishers B.V., Amsterdam — Printed in The Netherlands

CONCEPTS OF REDUCTION AND DISPERSION OF METALS IN ZEOLITES

N.I. JAEGER, P. RYDER[1] and G. SCHULZ-EKLOFF
Angewandte Katalyse, Fachbereich 2 - Chemie
[1]Werkstoffphysik und Strukturforschung, Fachbereich 1 - Physik
Universität Bremen, D-2800 Bremen 33 (FRG)

ABSTRACT
   Experimental observations concerning the formation, dispersion and stability
of metal phases in stoichiometrically ion-exchanged zeolites are reconsidered
on the basis of (i) redox equilibria, taking into account ion activities and
zeolite lattice rearrangements, and (ii) possible nucleation mechanisms of
metal clusters. New experimental evidence for the existence of an orientation
relationship between platinum crystals and the surrounding zeolite matrix is
presented and discussed.

INTRODUCTION
   The redox and dispersion processes in metal loaded zeolites have been reviewed repeatedly (refs.1-3). The number of publications on this subject is
rapidly increasing. The observed variety of phenomena connected with the preparation, agglomeration and stabilization of zeolite supported metal phases is
so confusing, that related principles of reduction and dispersion of metals
should be reconsidered. In the following paper, selected examples are used to
illustrate the application of basic concepts to complex phenomena. The discussion of reducibility will focus on nickel ion-exchanged faujasites, in view
of the attention nickel supported on zeolites has received and due to conflicting interpretations regarding the results on the reducibility of nickel exchanged faujasites. The mechanisms of reduction and agglomeration of the reduced metal will be discussed and new results on the growth of a Pt phase within a zeolite X matrix will be presented.

REDUCIBILITY OF TRANSITION METAL IONS IN STOICHIOMETRICALLY EXCHANGED
ZEOLITES
   With respect to the thermodynamics of reduction, Riekert (ref. 4) successfully applied the general concept of a solid electrolyte to the ion-exchanged
zeolite. The existence of redox equilibria has been established repeatedly
(ref.2). This approach has been a guideline in assessing the properties of
acidic and ion-exchanged zeolites with respect to their catalytic activity
(refs.5-7). The reducibility of transition metal ions introduced into the

zeolite by stoichiometric ion-exchange is the fraction of the metal ions which can be reduced according to the redox equilibrium. The solid electrolyte or crystalline liquid concept has been refined by Barthomeuf (refs.8-10) who defined a coefficient of activity depending on the ionic strength, in analogy to strong electrolytes in aqueous solutions. The ionic strength depends on the Si:Al ratio and the valence state of the charge compensating cation. For a given exchanged cation, e.g. protons, the coefficient of activity should increase in the order A → X → Y,Mordenite. A more quantitative approach towards activity was developed by Mortier and by Jacobs et al. (refs.11-13) who applied the Sanderson electronegativity principle to the interaction of the charge compensating metal ions with the zeolite framework. The model works well in explaining trends within a homologous series. The correlation between Sanderson electronegativity and initial rate of reduction can be understood on the basis of a linear free energy relationship for the reduction kinetics (ref.14).

The following discussion will be based on the activities of the ions, despite the fact that the activity coefficients cannot be determined quantitatively for the solid electrolyte (refs. 4,15). In the case of nickel, the Nernst equation for acidic solvents like zeolites can be written as

$$\mu = \mu_{o,Ni} + RT\ln\{ (1-n)a_{Ni^{2+}} / (a_{H^+} + 2na_{Ni^{2+}})^2 \} \tag{1}$$

where a_{H^+} is the proton activity prior to reduction, $a_{Ni^{2+}}$ is the nickel ion activity and n is the fraction of the reduced metal ions. $\mu_{o,Ni}$ contains a given constant partial pressure of hydrogen. The activity of the metallic nickel is assumed to be $a_{Ni} = 1$ which is certainly not correct for very small metal clusters which are probably charged. Equation (1) indicates the effect of the initial proton activity a_{H^+} on the reducibility of the metal ion. Samples with high initial proton activity are more difficult to reduce (refs. 16,17). Initial proton activity can be due to proton exchange or hydrolysis involving metal ions. The increase in reducibility of a given transition metal in the order Y → X → A (refs. 2,3,18) can be explained by assuming an enhancement of the removal of protons in the same direction. Protons created during reduction may be removed by dealumination of the zeolite according to

$$(\equiv SiO)_4Al^- + H^+ + 3H_2O \rightarrow 4\equiv SiOH + Al(OH)_3 \tag{2}$$

The associated improvement of the reducibility should be more pronounced for lower Si:Al ratios. Both, the increasing instability towards acids and the decrease in thermal stability of the zeolite lattice in the order Y → X → A (ref. 19) are in support of this assumption. Effects of the activity coefficient, which depends on the Si:Al ratio (ref. 10), can be neglected in this context. Evidence for the dealumination process during reduction can be obtained from nitrogen physisorption capacities rather than from X-ray diffraction (ref.18). For samples dehydrated at 723K the capacity drops from 80 ml N_2/ml zeolite for NaX to 55 ml N_2/ml zeolite for NiCaNaX ($Ni_{9,6}Ca_{10,6}Na_{45,5}X$). Patzelova et al. (ref.20) report improved reducibility for stabilized NiY-zeolites and conclude that the number of nickel ions in "open" positions has increased. An alternative interpretation would be the removal of proton activity in the course of the stabilization of the $NaNH_4$Y-zeolite used in the experiments.

While these general concepts work well in explaining trends, their authors agree that the local environment within the zeolite lattice must be taken into account. The models do not contribute to resolving conflicting experimental evidence regarding the reducibility of Ni^{2+} ions in different lattice sites and the influence of co-exchanged non-reducible cations. In equation(1), local activity coefficients and especially local proton activity may have to be considered (refs.21,22). The experimental evidence obtained for Ni^{2+} exchanged faujasites points towards easier reduction of Ni^{2+} in S_I positions (refs.22-24). These observations and the influence of co-exchanged non-reducible cations on the reducibility of the nickel in certain positions can be interpreted with the aid of the local activity concept. For zeolite X Briend Faure et al. (ref.25) generally find that the Ni^{2+} sites are equivalent with respect to reduction. However, in the presence of La^{3+} a more difficult reduction for Ni^{2+} in supercage positions is reported (ref.26).

The presence of Ce^{3+} in Ni exchanged X zeolites was found to increase the proportion of Ni^{2+} in S_I positions and to improve the reducibility (refs.16,27). It is suggested that the highly charged cations La^{3+} and Ce^{3+} lower the local activity of the reducible ions in supercage positions, where at the same time the local proton activity might be considerable. A relative or even absolute increase of the activity of nickel ions in S_I positions combined with low local proton activity may lead to the observed improved reducibilities.

METAL NUCLEATION AND DISPERSION

Metal atoms (Me^0) formed inside the zeolite framework by a reduction process will migrate through the channels and cages more rapidly than metal ions due to smaller electrostatic interactions resulting in lower energy barriers in the paths on the potential hypersurface. For the metal atoms reaching the outside of the zeolite crystal, the probability of migration along the outer surface will be higher compared to the probability for reentering the zeolite lattice for geometrical reasons. The frequently observed enrichment of the Me^0

outside the zeolite framework by agglomeration and crystallization, can thus be readily understood. Evaporation, which has been observed for a few metals, e.g. Hg, Zn and Cd (ref.28), can be excluded for most of the metals under the reduction conditions generally used, i.e. temperatures below 700K.

Nevertheless, metal phases in reduced ion-exchanged zeolites can be formed preferentially or even exclusively inside the zeolite matrix as was revealed by X-ray photoelectron spectroscopy (refs.29,30). This indicates a high probability for the formation of nuclei and agglomerates in the channels and cages. The nucleation should result from the collision of migrating species and the following two alternatives could be the initiating steps:

$$Me^0 + Me^0 \rightleftarrows Me_2^0 \tag{3}$$

$$Me^0 + Me^{n+} \rightleftarrows Me_2^{n+} \tag{4}$$

Even though a high supersaturation with respect to Me^0 can be assumed to exist within the zeolite during the process of reduction, nucleation following step (4) should be favoured. The free enthalpy of nucleation will be less for charged as compared to uncharged nuclei. The difference in the free enthalpies of nucleation corresponds to the difference in the enthalpies of formation for the two nuclei. The enthalpy of formation for nucleation step (4) will roughly correspond to the difference of the first and the nth ionization potential, i.e. 5-15 eV for n=2. The formation enthalpy for process (3) will be close to the evaporation enthalpy, i.e. 2-7 eV and is thus smaller. The nucleation process should therefore be promoted by a high density of transition metal ions populating the migration paths of the metal atoms as collision partners. A nucleus Me_2^{n+}, stabilized in the zeolite framework by electrostatic forces, can grow by further collisions with Me^0 forming charged clusters which have repeatedly been found in zeolites (refs.31,32). Furthermore, finer dispersion of the metal phase is often observed if non-reducible transition metal ions, e.g. Cr^{3+} (ref.33), Fe^{3+} (ref.34), Ce^{3+} (ref.27) are present in the zeolite framework during the reduction process, operating as preferred nucleation sites.

This model needs refinement when it is applied to real zeolites, such as faujasites, in which open and hidden sites participate in the nucleation process. The autoreduction of platinum tetrammine exchanged faujasites during temperature programmed dehydration leads to a high growth rate of Pt-nuclei in the interconnected supercages and a relatively coarse metal dispersion, since all the platinum tetrammine complexes are located there due to their size (ref.35). Calcination procedures prior to reduction direct the platinum ions into the sodalite cages and the nucleation process has to take place in the separated positions (ref.36). The higher number of centers for nucleation according to equation (4), leads to a finer dispersion of the Pt as a consequence.

Nickel loaded faujasites, however, exhibit high occupancies of the most hidden sites, i.e. S_I sites, for the nickel ions (ref.37). The generally observed bidispersities, i.e. large fractions of reduced nickel at the outer surface and another fraction of smaller clusters inside the zeolite crystal, could be due to a relatively low ion density in the migration path of Ni^0. Monodispersed nickel phases, located exclusively inside the faujasite matrix, have been found with NiCaX samples prepared by simultaneous exchange of nickel and calcium ions (ref.38).

Further parameters, such as zeolite crystal size, reduction temperature, degree of dehydration or formation of mobile species, e.g. hydrides, contribute to the migration process and the variety of observed dispersion phenomena.

STRUCTURE AND GROWTH OF METAL CRYSTALS

The agglomeration of transition metals in faujasites can result in metal particles of about 1nm in diameter which fit into the supercages. This was established for platinum (refs.39,40) and nickel (ref.41). Platinum particles of this size can be imaged readily in the transmission electron microscope (Fig.1). They show very weak, broad lines in X-ray diffraction and also in electron diffraction patterns (Fig.2).

Metal particles exceeding supercage dimensions have been found repeatedly embedded in a faujasite matrix. Aggregate sizes ranging from 3 nm for Pd (ref.36) or 4 nm for Ru (ref.42) to 8 nm for Pt (ref.40) and Ni (ref.38) were determined mainly by electron transmission microscopy. It was shown by electron diffraction from selected areas and single particles that such aggregates are single crystals (refs.22,43). These single crystals grow in an intact faujasite matrix, as revealed by direct imaging in transmission electron micrographs (ref.38) or by electron diffraction (ref.22).

Fig. 3 shows an electron transmission micrograph of reduced Pt faujasite with Pt particles of 5 to 10 nm size. Fig. 4 depicts the electron diffraction pattern taken from a selected area of the specimen shown in Fig. 3. A clear ring pattern is observed and can be indexed on the basis of the normal structure of metallic Pt (fcc, a = 0.392 nm). The dark field micrograph of Fig. 3b was taken using the two innermost rings, i.e. (111) and (200).

It was proposed that metal crystals encapsulated in the faujasite matrix grow by a mechanism involving atomic rearrangements within the zeolite lattice (ref.22). Since the energy of formation of the metal lattice is gained, the

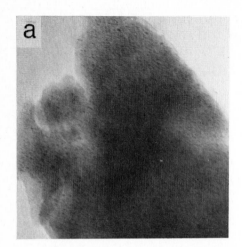
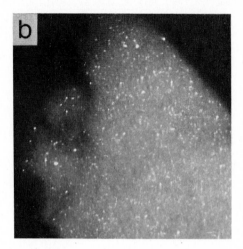

Fig. 1. a) Electron transmission micrograph of reduced Pt faujasite showing Pt particles $\leq$ 2 nm. The preparation conditions are given elsewhere (ref.40). b) Corresponding dark field micrograph (conical beam illumination) from the (111) ring of the diffraction pattern (Fig. 2). Magnification: 200,000.

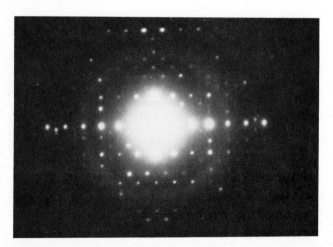

Fig. 2. Electron diffraction pattern of the specimen shown in Fig. 1. The spot pattern is from the zeolite and is near the [110] orientation. The diffuse ring is from the Pt particles.

recrystallization of the zeolite can be achieved at temperatures < 600 K, i.e. much lower than that of the ultrastabilization (ref. 44).

The recrystallization can be recognized experimentally via the change of temperature for lattice destruction and the capacity of nitrogen physisorption, as is demonstrated in the following example. The reduction of NiCaX ($Ni_{9,6} Ca_{10,6} Na_{45,5}X$) samples to a degree of approx. 50% led to monodispersed nickel with 8 nm crystals located exclusively inside the zeolite matrix (refs. 22,30,38). The temperature for lattice breakdown is increased from 1000 K for the ion exchanged samples to 1050 K for the reduced specimens, indicating the stabilizing effect of the framework recrystallization. The capacity for nitrogen physisorption is increased from 55 to 65 ml N_2/ml zeolite, indicating the partial removal of lattice defects which decrease the nitrogen uptake. After growth nickel crystals occupy 0,25% of the zeolite volume. The zeolite lattice fragments originating from this volume must presumably migrate to defect sites, thus increasing the thermal stability of the lattice and its capacity for nitrogen uptake. Comparing the 55 ml N_2/ml zeolite with the maximum value of 110 ml N_2/ml zeolite found for NaX (ref.45), it follows that 20% of the zeolite voids blocked for the N_2 uptake are opened by the recrystallization. From these values, it can be estimated that the fragments will migrate into and stabilize a volume having 10 times the diameter of the nickel crystals.

Bright field and dark field electron micrographs and electron diffraction patterns revealed a random location and a random orientation for faujasite embedded crystals of palladium (ref. 43) and nickel (ref.22). In the case of platinum, however, a preferred orientation with respect to the zeolite lattice can be observed (Fig.4).

The fact that continuous rings are observed in the Pt diffraction pattern shows that all possible orientations are present, i.e. there is no single, exact orientation relationship between the Pt and the zeolite matrix. However, the intensity is not evenly distributed around the rings, indicating that the Pt orientations are not completely random. The concentration of diffracted intensity in certain directions can only be explained by the existence of one or more preferred orientation relationships.

Diffraction patterns from smaller areas within a single zeolite crystal showed single crystal spot patterns from the zeolite and ring patterns with preferred orientation from the Pt particles. An example is shown in Fig. 5. In this pattern the zeolite spots index as a [110] zone, and the regions of high intensity on the Pt rings correspond to zeolite directions of the same crystallographic type (Fig. 5 b). Thus the (111) ring shows increased intensity near the points of intersection of the rings with two { 111 }-type directions of the zeolite reciprocal lattice. Similar intensity maxima are also seen at

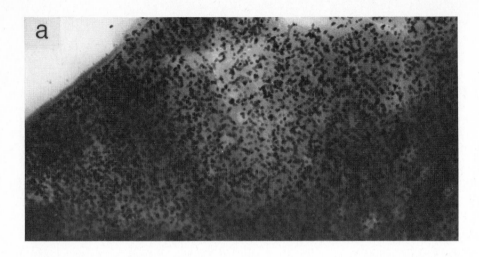

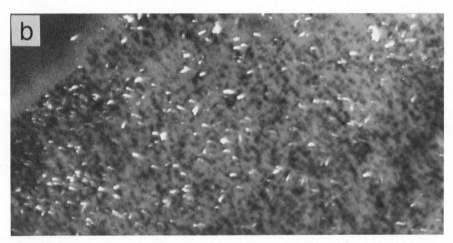

Fig. 3 a) Electron transmission micrograph of reduced Pt faujasite showing Pt particles of 5 to 10 nm size. The preparation conditions are given in ref. 49. b) Corresponding dark field micrograph of the (111) and (200) rings of the electron diffraction pattern (Fig. 4). Magnification : 150 000.

the intersection of the (200) and (022) zeolite direction with the corresponding Pt rings. The intensity maxima of the Pt diffraction pattern may therefore be indexed also as a [110] zone, and the corresponding orientation relationship is simply $<100>_{Pt} \parallel <100>_{zeolite}$. The same orientation relationship was also observed in crystals with different orientations relative to the electron beam, e.g. [332] and [112].

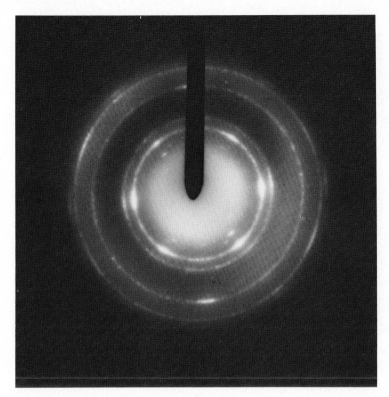

Fig. 4. Electron diffraction pattern of the specimen shown in Fig. 3 with Pt rings and a few zeolite spots.

The orientation relationship is observed even though there is no simple lattice matching between the two phases, so that coherent precipitation must be ruled out. The existence of a preferred orientation of the Pt crystallites can be explained by assuming an orientated growth of the fcc platinum within the cubic faujasite structure. This requires that the intrinsic voids of the zeolite matrix exhibit equilibrium shapes fitting the morphology of the metal crystals. The observed orientation of the platinum crystals can in turn be taken as additional proof, that the metal crystals must have been grown in an intact zeolite matrix and not in a priori intrinsic voids.

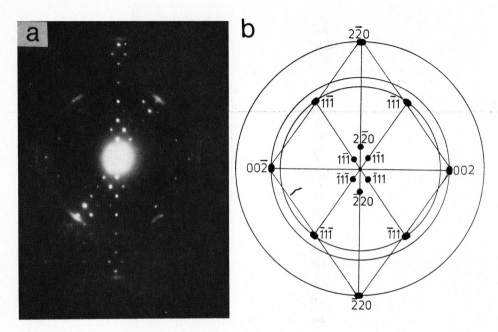

Fig. 5 a) Electron diffraction of a selected area of the specimen shown in Fig. 3. The selected area is from a faujasite single crystal. The figure shows an orientation relationship between the Pt crystals (5-10 nm size) and the zeolite lattice (b).

SINTERING AND REDISPERSION

Zeolite embedded metal particles far exceeding the apertures of the cages cannot agglomerate by a migration mechanism of the particles themselves. Mobile, uncharged species formed by reacting gases, e.g. hydrides, oxides or carbonyls, could serve as vehicles for the transport of the metal to the outer surface of the zeolite crystals. Such a coarsening process can be retarded by measures, which considerably decrease the rate of diffusion of the mobile species. For example, the coarsening of a monodispersed nickel phase in a faujasite matrix during the methanation reaction (350°C, 1 bar, CO/H_2 = 3/7, 1 week time-on -stream) via mobile carbonyls is strongly inhibited by calcium ions randomly distributed in the zeolite lattice (ref.46).

Extensive redispersion by transport from the surface back into the bulk zeolite should, however, be limited to those metals forming mobile, unstable compounds with low vapour pressures, e.g. oxides or sulfides, which readily migrate to cation positions and may be decomposed easily e.g. by reactions of the type

$$Pt_xO_y + 2\ Z-O-H \rightarrow H_2O + Pt^{2+} + Pt^o_{x-1} + \frac{y-1}{2}\ O_2 + 2\ ZO^- \tag{5}$$

Metals forming highly stable oxides, such as ruthenium, cannot be redispersed in oxygen (ref.47). A generalized scheme of these possibilities was given by Jacobs (ref.14).

OUTLOOK

The preparation of monodispersed metal phases with narrow particle size distributions in zeolite matrices offers the chance to study the chemical properties of metal crystals as a function of their size. Crystal size effects respectively surface structure sensitivity have been assumed repeatedly to be operative in catalysis by metals (refs.48,49). Improvements of the selectivity in the carbon monoxide hydrogenation were claimed for metals supported by zeolites instead of silica or alumina (refs.50,51). In this case, however, it would be even more difficult to detect unambiguous correlations between catalytic selectivities and catalyst structure. Difficulties would arise because (i) metal loaded zeolites will exhibit a bifunctional and shape selective behaviour, (ii) sintering and redispersion will occur, (iii) during formation and deactivation, physical and chemical properties of both the surface and the bulk of the metal crystals will change. Furthermore, electron transfer processes between metal particle and carrier have to be taken into account. In spite of the complexity of these systems it can be stated that the potentials of the zeolite matrices for a better control of metallic particle sizes are far from being exhausted.

ACKNOWLEDGMENTS

We thank Mr. D. Exner for the preparation of the platinum samples, Mr. A. Kleine for the electron micrographs and Mr. H. Kompa for the zeolite lattice stability test. We are indebted to Drs. H. Karge and A. J. Melmed for their thorough review of the manuscript, and to the Deutsche Forschungsgemeinschaft for financial support. Critical comments by Prof. L. Riekert are gratefully acknowledged.

REFERENCES

1   Kh.M. Minachev and Ya.I. Isakov, in "Zeolite Chemistry and Catalysis", (J.A. Rabo, Ed.) ACS Monograph 171, Washington, D.C. 1976, p. 552.
2   J.B. Uytterhoeven, Acta Phys. et Chem. (Szeged) 24 (1978) 53.
3   P.A. Jacobs, "Carboniogenic Activity of Zeolites", Elsevier, Amsterdam 1977, p. 183.
4   L. Riekert, Ber. Bunsenges. Phys.Chem. 73 (1969) 331.

310

5   P.H. Kasai and R.J. Bishop, J. Phys. Chem. 77 (1973) 2308.

6   J.A. Rabo, R.D. Bezman and M.L. Poutsma, Acta Phys. et Chem. (Szeged) 24 (1978) 39.

7   J.A. Rabo, Catal. Rev. - Sci. Eng. 23 (1981) 293.

8   D. Barthomeuf, Acta Phys. et Chem. (Szeged) 24 (1978) 71.

9   D. Barthomeuf, J. Phys. Chem. 83 (1979) 249.

10   D. Barthomeuf in "Catalysis by Zeolites" (B. Imelik et al., Eds.) Elsevier, Amsterdam 1980, Stud. Surf. Sci. Catal., vol. 5, p. 55.

11   W.J. Mortier, J. Catal. 55 (1978) 138.

12   P.A. Jacobs, W.J. Mortier and J.B. Uytterhoeven, J. Inorg. Nucl. Chem. 40 (1978) 1919.

13   P.A. Jacobs, Catal. Rev. - Sci. Eng. 24 (1982) 415.

14   P.A. Jacobs, in "Metal Microstructures in Zeolites", (P.A. Jacobs et al., Eds.) Elsevier, Amsterdam 1982, Stud. Surf. Sci. Catal., vol. 12, p. 71.

15   J.T. Richardson, J. Catal. 21 (1971) 122.

16   M. Briend Faure, M.F. Guilleux, J. Jeanjean, D. Delafosse, G. Djega Maria-dassou and M. Bureau-Tardy, Acta Phys. et Chem. (Szeged) 24 (1978) 99.

17   E.D. Garbowski, C. Mirodatos and M. Primet, in "Metal Microstructures in Zeolites" (P.A. Jacobs et al., Eds.) Elsevier, Amsterdam 1982, Stud. Surf. Sci. Catal., vol. 12 p. 235.

18   W. Romanowski, Polish J. Chem. 54 (1980) 1515.

19   Ref. 3, p. 9.

20   V. Patzelova, Z. Tvaruzkova, K. Mach and A. Zukal, in "Metal Microstructures in Zeolites" (P.A. Jacobs et al., Eds.) Elsevier, Amsterdam,1982 Stud. Surf. Sci. Catal. vol. 12, p. 151.

21   Ref. 3, p. 45.

22   D. Exner, N.I. Jaeger, R. Nowak, G. Schulz-Ekloff and P. Ryder, in "Proceedings of the 6th International Zeolite Conference" (A.Bisio and D.H. Olson, Eds.) Butterworth Scientific, Guildford 1983, p.

23   T.A. Egerton and J.C. Vickerman, J. Chem. Soc., Faraday Trans. I 69 (1973) 39.

24   Ch. Minchev, V. Kanazirev, L. Kosova, V. Penchev, W. Gunsser and F. Schmidt, in "Proceedings of the 5th Int. Conf. on Zeolites" (L.V.C. Rees, Ed.). Heyden London 1980 p. 355.

25   M. Briend Faure, J. Jeanjean, M. Kermarec and D. Delafosse, J. Chem. Soc., Faraday Trans. I 74 (1978) 1538.

26   M. Briend Faure, J. Jeanjean, D. Delafosse and P. Gallezot, J. Phys. Chem. 84 (1980) 875.

27   D. Delafosse, in "Catalysis by Zeolites", (B. Imelik et al., Eds.) Elsevier, Amsterdam 1980, Stud. Surf. Sci. Catal., vol. 5, p. 235.

28   D.J.C. Yates, J. Phys. Chem. 69 (1965) 1676.

29   Kh. M. Minachev, G.V. Antoshin and E.S. Shpiro, Commun. Acad. Sci. USSR, Chemical Series (1974), p. 972 (Engl. transl.).

30   G. Schulz-Ekloff, D. Wright and M. Grunze, Zeolites 2 (1982) 70.

31   H.K. Beyer and P.A. Jacobs, in "Metal Microstructures and Zeolites", (P.A. Jacobs et al., Eds.) Elsevier, Amsterdam 1982, Stud. Surf. Sci. Catal., vol. 12, p. 95.

32   P. Gallezot and G. Bergeret, in "Metal Microstructures in Zeolites", (P.A. Jacobs et al., Eds.) Elsevier, Amsterdam 1982, Stud. Surf. Sci. Catal., vol. 12, p. 167.

33   J.D. Lawson and H.F. Rase, Ind. Eng. Chem., Prod. Res. Develop. 9 (1970) 317.

34   H. Winkler, A. Ebert, W. Ebert and E. Riedel, Surface Sci. 50 (1975) 565.

35   D. Exner, N. Jaeger, K. Möller and G. Schulz-Ekloff, J. Chem. Soc., Faraday Trans. I 78 (1982) 3537.

36   P. Gallezot, in "Catalysis by Zeolites", (B. Imelik et al., Eds.) Elsevier, Amsterdam 1980, Stud. Surf. Sci. Catal., vol. 5, p. 227.

37   J. Jeanjean, D. Delafosse and P. Gallezot, J. Phys. Chem. 83 (1979) 2761.

38   D. Exner, N. Jaeger, R. Nowak, H. Schrübbers and G. Schulz-Ekloff, J. Catal. 74 (1982) 188.

39 P. Gallezot, A. Alarcon-Diaz, J.A. Dalmon, A.J. Renouprez and B. Imelik, J. Catal. 39 (1975) 334.
40 D. Exner, N. Jaeger and G. Schulz-Ekloff, Chem. Ing. Techn. 52 (1980) 734.
41 M. Che, M. Richard and D. Olivier, J. Chem. Soc., Faraday Trans. I 76 (1980) 1526.
42 H.H. Nijs, P.A. Jacobs, J.J. Verdonck and J.B. Uytterhoeven, in "Growth and Properties of Metal Clusters" (J. Bourdon, Ed.) Elsevier, Amsterdam 1980, Stud. Surf. Sci. Catal., vol. 4, p. 479.
43 D. Exner, N. Jaeger, K. Möller, R. Nowak, H. Schrübbers, G. Schulz-Ekloff and P. Ryder, in "Metal Microstructures in Zeolites", (P.A. Jacobs et al., Eds.) Elsevier, Amsterdam 1982, Stud. Surf. Sci. Catal., vol. 12, p. 205.
44 D.W. Breck, "Zeolite Molecular Sieves", Wiley, New York 1974, p. 507.
45 Ref. 44, p. 611.
46 N. Jaeger, U. Melville, R. Nowak, H. Schrübbers and G. Schulz-Ekloff, in "Catalysis by Zeolites", (B. Imelik et al., Eds.) Elsevier, Amsterdam 1980, Stud. Surf. Sci. Catal., vol. 5, p. 335.
47 J.H. Lunsford, in "Metal Microstructures in Zeolites", (P.A. Jacobs et al., Eds.) Elsevier, Amsterdam 1982, Stud. Surf. Sci. Catal., vol. 12, p. 1.
48 M. Boudard, Avd. Catal. 20 (1969) 153.
49 R. van Hardeveld and F. Hartog, Surf. Sci. 15 (1969) 189.
50 H.H. Nijs, P.A. Jacobs and J.B. Uytterhoeven, J. Chem. Soc., Chem. Commun. (1979) 180.
51 H. Schrübbers, P. Plath and G. Schulz-Ekloff, React. Kinet. Catal. Lett. 13 (1980) 255.

P.A. Jacobs et al. (Editors), *Structure and Reactivity of Modified Zeolites*
© 1984 Elsevier Science Publishers B.V., Amsterdam — Printed in The Netherlands

LOCAL STRUCTURE AND BONDING IN ZEOLITES BY MEANS OF QUANTUM
CHEMICAL AB INITIO CALCULATIONS: METAL CATIONS, METAL ATOMS AND
FRAMEWORK MODIFICATION

J. SAUER, H. HABERLANDT and W. SCHIRMER
Central Institute of Physical Chemistry, Academy of Sciences,
DDR-1199 Berlin, Rudower Chaussee 5 (German Democratic Republic)

ABSTRACT
    Results on molecular models are reported which allow conclu-
sions about the nature of bonding of metal cations and atoms in
zeolites. The bonding of Ni(2+) is found to be basically ionic
and shows no substantial differences compared with Mg(2+). The
interaction of Mg, Ni and Cu atoms with models of aluminosilicate
framework sites is very weak (van-der-Waals forces only). Net
charges calculated for aluminophosphate suggest that the electric
field inside the pores is not stronger than that in microporous
silica.

INTRODUCTION
    Quantum chemical ab initio calculations can provide both a
deeper insight into bonding in minerals and predictions of local
structures and properties which sometimes are not accessible by
experiments. Since these methods are computationally highly demanding their application is limited to relatively small systems.
It is therefore crucial for our approach to make a careful choice
of appropriate models for active sites in zeolites (refs.1-4).
    Usually, zeolites contain alkaline or alkaline earth metal cat-
ions which are coordinated by $\equiv$Si-O-Al$^-\equiv$ sites of the framework.
This bonding is basically ionic with a very small degree of charge transfer from the anionic framework to the cation (refs.1,3).
It is the first aim of this paper to see whether this conclusion
holds true for transition metal cations (Ni^{2+}) as well.
    These cations may be easily reduced to form transition metal
atoms whereby systems are formed which are well-suited for studying metal support interactions (ref.5). Since little is known
from theoretical point of view about the interaction of transition metal atoms with oxygen sites in general and with aluminosilicates in particular, calculations were made which throw some

light on the nature of this type of interactions.

Recently several <u>microporous aluminophosphates</u> were synthesized (ref.6) which, similarly to microporous silica, do not contain
extra-framework metal cations. The third part of this paper deals
with <u>atomic charges</u> on such frameworks and consequences for their
adsorption properties.

MODELS AND METHODS

Models adopted for studying the interaction of metal cations
and metal atoms with oxygen sites of aluminosilicate frameworks
are displayed in Fig. 1. The water molecule is the simplest system appropriate for this purpose. For Ni^{2+} and Ni (d^8 and d^8s^2
configurations, respectively) the ^3F states are considered.

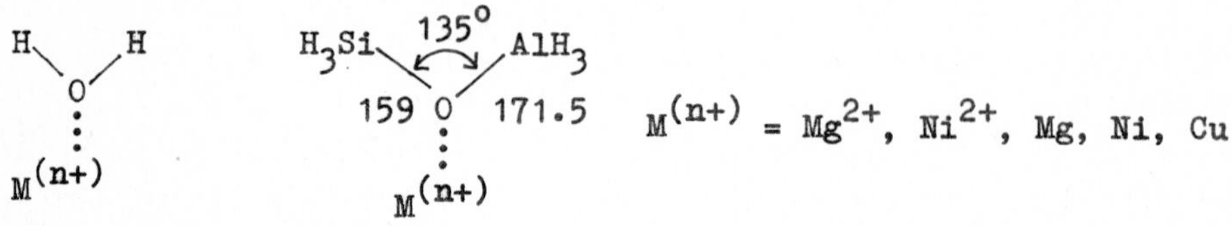

H \ / H H_3Si $\overset{135^\circ}{\frown}$ AlH_3

O   159 O 171.5 $M^{(n+)}$ = Mg^{2+}, Ni^{2+}, Mg, Ni, Cu

$M^{(n+)}$ $M^{(n+)}$

Fig. 1. Model complexes studied. For H2O the experimental geometry (R(OH)=95.72 pm, ∠ HOH=104.52°) and for H3SiOAlH3⁻ a standard geometry (R(SiH)=148, R(AlH)=158 pm, tetrahedral angles on
Si and Al) are adopted.

Quantum chemical SCF calculations are performed employing two
types of basis sets: the minimal basis set MINI-1 (ref.7) which
proved particularly useful for interaction problems (ref.8) and
split valence (SV) basis sets, namely the 4-31 G set for H_2O and
basis sets with split 3d, 4s and 4p shells for Ni and Cu (ref.9
gives details, cf. ref.10). In general, SV basis sets yield more
reliable results. Substantial improvements can be achieved
(refs.1,8) by making corrections for the basis set superposition
error, BSSE (ref.11).

INTERACTION OF MG^{2+} AND NI^{2+} WITH OXYGEN SITES

SCF calculations at the SV level yield equilibrium distances
of 190 and 191 pm, stabilization energies of 382 and 409 kJ/mol
and ionic charges of 1.96 and 1.85 for $H_2O...Mg^{2+}$ and $H_2O...Ni^{2+}$
complexes. These data as well as the MINI-1 results for complexes
of H_2O and $H_3SiOAlH_3^-(A^-)$ (Table 1) suggest that only minor differences exist for the bonding of Mg^{2+} and Ni^{2+} ions. This is al-

so true for other transition metal ions as calculations on
$H_2O...Zn^{2+}$ and $H_2O...Cu^{2+}$ (refs.8,12) show. The (corrected) total interaction energies of -1094 and -1110 kJ/mol for $A^-...Mg^{2+}$
and $A^-...Ni^{2+}$, respectively, may be compared with electrostatic
and induction energies of -1050 and -125 kJ/mol obtained for the
interaction of A^- with a point charge ($+2$) at $R=185$ pm. This
energy analysis and the small extent of charge transfer prove
that the interaction is basically ionic both for Mg^{2+} and Ni^{2+}.
However, the cation charge is smaller for Ni^{2+} than for Mg^{2+},
indicating that there is a higher covalence character in the bonding of Ni^{2+} (which is also reflected by slightly larger binding
energies).

Increasing the number of coordination sites when passing from
$H_2O...Ni^{2+}$ to $Ni^{2+}(H_2O)_6$ decreases the cation charge from 1.85
(SV result, vide supra) to 1.52 (ref.10). Assuming that Ni^{2+} is
coordinated by three $\equiv Si-O-Al^- \equiv$ sites we do not expect that the
cation charge for Ni in zeolites is smaller than 1.5. In contrast, charges as low as 0.3 were reported from CNDO/2 calculations (ref.13). For comparison, cation charges predicted by the
semiempirical CNDO/2 method (CLACK parametrization) are included
in Table 1. They are by far too small, in particular for the complexes containing Ni^{2+}. Hence, the influence of the zeolite
framework on the redox properties of cations should be re-consid-
ered assuming that the electron donation of the framework is
smaller than it was inferred from CNDO/2 calculations (ref.13).

TABLE 1
Equilibrium distances, R(O...M) in pm, interaction energies, ΔE
(kJ/mol) and cation charges, q (atomic units), for complexes of
Mg(2+) and Ni(2+) with various ligands. The subscript c refers
to values corrected for the BSSE.

Ligand	Cation	R	$-\Delta E$	$-\Delta E_c$	q	q_c	q(CNDO)
$H_3SiOSiH_3$	Mg^{2+}	200	236	207	1.91	1.92	1.32
H_2O	Mg^{2+} [a]	187	356	335	1.93	–	1.70 [c]
	Ni^{2+} [b]	188	366	–	1.89	–	1.41 [c]
$H_3SiOAlH_3^-$	Mg^{2+}	185	1138	1094	1.82	1.87	1.05
	Ni^{2+}	(185) [d]	–	1110	1.75	1.79	0.23 [c]

[a] Ref.8   [b] Ref.12   [c] At R=190 pm   [d] Not optimized

INTERACTION OF METAL ATOMS WITH OXYGEN SITES

Reference calculations on the $H_2O...Mg$ system

Nonempirical calculations on $H_2O...Mg$ revealed that there is
only a very small (ref.14) if any (ref.15) attractive well in the
SCF potential curve. Adding polarization functions to the basis
set of ref.15 and making corrections for the BSSE (ref.11) yield
an even more repulsive potential curve (see Fig. 2a). The decisive attractive energy contribution is the dispersion energy
which is not included in the SCF results. If the (nonexpanded) ab
initio values of ref.15 are added to our BSSE corrected SCF
result:

$$\Delta E \simeq \Delta E_c^{SCF} + DISP, \tag{1}$$

a curve with a minimum ($\Delta E = -28.2$ kJ/mol) at R=210 pm is ob-

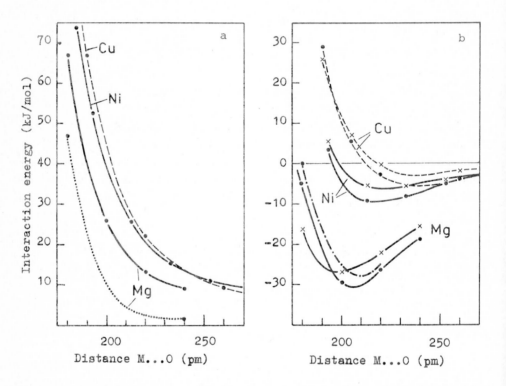

Fig. 2. Potential curves for H2O...M interactions (M=Mg, Ni, Cu).
(a) SCF results corrected for basis set superposition error (Cu,
Ni, Mg); Mg (dotted line): SCF results from ref.15 without
correction.
(b) Total interaction energy, i.e. SCF plus dispersion energy.
The curve -.-.- includes the ab initio values (ref.15) while
crosses and dots refer to the semiempirical dispersion energy
estimates DISP-A and DISP-B.

tained (see Fig. 2b, line -.-.-) which is in close agreement with the results of ref.14 (ΔE= -29.7 kJ/mol, R=212 pm).

Since ab initio results of the dispersion energy are rarely available two semiempirical procedures, DISP-A and DISP-B (ref.9), were tested which are based on the C_6-term of the multipole series (LONDON formula) and damped at shorter distances. While DISP-A adopts the damping function of ref.16 and does not make any reference to ab initio data, DISP-B refers to ab initio values for some scaling on C_6 and makes use of the damping function of ref.17. For details we refer to ref.9. If the dispersion energies obtained by these procedures are added to our SCF energies, potential curves result which are in reasonable agreement with the nonempirical curve mentioned above (see Fig. 2b).

From the calculations on H_2O...Mg we learn that in order to obtain reliable potential curves corrections for the BSSE should be made on the SCF energy and some estimate of the dispersion energy should be added.

Results for H_2O...Ni and H_2O...Cu interactions

The SCF potential curves (SV basis set, corrected for the BSSE) for both transition metal atoms are very similar to each other and purely repulsive as in the case of Mg (see Fig. 2a). The dispersion energy causes slightly attractive potentials (ΔE is about -5...-10 kJ/mol for H_2O...Ni and -3...-6 kJ/mol for H_2O...Cu) with minima at about 215...225 pm and 235...245 pm (see Fig. 2b). As it is anticipated for such weak interactions the calculated charge transfer is very small. After corrections for the BSSE, in the complexes with H_2O, Mg looses 0.02 e (at R=200 pm), while the transition metals gain 0.05 e (H_2O...Ni at R=213 pm) and 0.04 e (H_2O...Cu at R=220 pm).

Interactions of Mg and Ni with $H_3SiOAlH_3^-$

Finally we may ask how these results will change if we pass to larger models ($H_3SiOAlH_3^-$) which requires to employ a simpler method (MINI-1 basis set). For comparison, the complexes with water are reinvestigated by the same method. The results (Table 2) show that the SCF potential curve at larger distances becomes less repulsive (probably due to enhanced polarization of the metal atom), while in the range where the minimum is expected (see Fig. 2b) the difference is very small and the SCF potential for the interaction with the larger model can be even more repul-

TABLE 2

Comparison of H2O...M with H3SiOAlH3⁻...M (A⁻...M) complexes: Corrected SCF interaction energies (kJ/mol) and charge transfer to the cation, Δq (atomic units).

R (pm)	M = Mg			M = Ni	
	220	240	265	220	250
$\Delta(\Delta E_c^{SCF})$ [a]	-3.0	-6.3	-7.5	0.65	-5.2
$\Delta q(H_2O...M)$	0.030	0.024	0.017	-	0.022
$\Delta q(A^-...M)$	-0.014	-	-0.009	-0.011	-

[a] $\Delta(\Delta E_c^{SCF}) = \Delta E_c^{SCF}(M...A^-) - \Delta E_c^{SCF}(H_2O...M)$

sive as the results for Ni indicate. On the other hand the dispersion energy will be certainly larger, for the polarizability of Si-O bonds (and similarly of Al⁻-O bonds) is about 25% larger than the O-H bond polarizability. Moreover, in contact with the real aluminosilicate framework, the metal atom can interact with more Si-O and Al⁻-O bonds. Hence, the value of -5...-10 kJ/mol for H_2O...Ni is considered as lower limit for aluminosilicate..Ni interactions in zeolites, while we guess that the upper limit could be between -30 and -50 kJ/mol.

Furthermore, a very small amount of charge (some 0.01 e) is transferred onto the metal atom from the negatively charged framework (see Table 2).

Conclusion and discussion

From the model calculations follows that the interaction of simple metal and transition metal atoms with aluminosilicate frameworks or similar supports is generally very weak (van-der-Waals forces only). GEUS (ref.18) came to the same conclusion considering the work of adhesion of Ni on ceramics. His estimate of the heat of adsorption of Ni on oxidic supports is 20±2 kJ/mol. The binding energy of a metal atom is comparable to the energy of adsorption of hydrocarbons (about 25 kJ/mol for C_2H_6 in NaX). The weakness of this interaction explains that metal atoms will nearly freely move in zeolites until they are trapped by a site of stronger interaction (e.g. another metal atom, a cation or a defect site).

Moreover, we conclude that due to the weakness of the interaction no pronounced electronic effects should occur. Accordingly, a very small charge transfer (some 0.01 e) onto the metal atom is predicted. Conclusions from experiments about positively charged Pt clusters in zeolites are debatable (ref.19). Hence, the catalytic properties of single metal atoms and metal atom clusters within zeolite cavities should be understandable by considering the electronic properties of these species alone. The role played by the support is "non-electronic" and limited to offering the medium for distributing the metal species. This discussion does not apply to the effect of defect sites or other strong interaction sites on the properties of metal species, e.g. on terminal surfaces of crystallites.

FRAMEWORK MODIFICATION

One might expect that aluminophosphate frameworks are more polar than pure silica because of the electronegativity difference between aluminium and phosphorus. As the polarity of the framework determines the electrostatic potential and field inside the micropores, knowledge of atomic charges would permit to conclude on the adsorption properties of these novel materials. Therefore, atomic net charges q are calculated for a molecular model representing cornersharing AlO_4^- and PO_4^+ tetrahedra (T1=P, T2=Al):

T_1	T_2	$q(T_1)$	$q(T_2)$	$q(O_b)$
Si	Si	1.36	1.36	-0.71
P	Al	1.34	1.13	-0.57

This model is similar to our models which were previously adopted for describing bonding in pure silica (T1=T2=Si) and aluminosilicates (T1=Si,T2=Al$^-$) (refs.2,3). The STO-3G SCF method is employed. The T-O_b (b-bridging) and T-O_n (n-nonbridging) distances as well as the T1-O_b-T2 angle were optimized yielding R(P-O)=158 pm, R(Al-O)=177 pm and angle (P-O-Al)=125°.

The net charges reported above show that the framework possesses polarity even in the case of pure silica (T1=T2=Si). From adsorption experiments and theoretical investigations (ref.4) it is evident that this polarity is not sufficient to adsorb polar molecules, e.g. H_2O, to an appreciable extent. However, as our calculations show, the polarity is not increased if we pass to aluminophosphate. Contrary to expectations, the oxygen charge is reduced by 20% and the net charges both on aluminium and phosphorus are smaller than the silicon charge in pure silica. In conclusion, the adsorption properties (molecular shape effects are excluded from our considerations) of aluminophosphates should be similar to those of microporous silica. In particular, they may be hydrophobic.

REFERENCES

1   J. Sauer, P. Hobza and R. Zahradník, J. Phys. Chem., 84 (1980) 3318-3326.
2   J. Sauer and G. Engelhardt, Z. Naturforsch., 37a (1982) 277-279.
3   J. Sauer, R. Zahradník and W. Schirmer, in Adsorption of Hydrocarbons in Microporous Adsorbents - II, Preprints of the Workshop, Academy of Sciences of the GDR, Berlin, 1982, pp. 44-52.
4   P. Hobza, J. Sauer, C. Morgeneyer, J. Hurych and R. Zahradník, J. Phys. Chem., 85 (1981) 4061-4067.
5   P.A. Jacobs, N.I. Jaeger, P. Jírů and G. Schulz-Ekloff (Eds.), Metal Microstructures in Zeolites, Studies in Surface Science and Catalysis, Vol. 12, Elsevier, Amsterdam, 1982.
6   S.T. Wilson, B.M. Lok, C.A. Messina, T.R. Cannan and E.M. Flanigan, J. Am. Chem. Soc., 104 (1982) 1146-1147.
7   H. Tatewaki, Y. Sakai and S. Huzinaga, J. Comput. Chem., 2 (1981) 96-99;
    Y. Sakai, H. Tatewaki and S. Huzinaga, J. Comput. Chem., 2 (1981) 100-107.
8   J. Sauer and P. Hobza, Theoret. Chim. Acta, submitted for publication.
9   H. Haberlandt and J. Sauer, in preparation.
10 J. Kowalewski, A. Laaksonen, L. Nordenskiöld and M. Blomberg, J. Chem. Phys. 74 (1981) 2927-2930.
11 S.F. Boys and F. Bernardi, Mol. Phys., 19 (1970) 553-566.
12 H.J. Hofmann and P. Hobza, personal communication (1983).
13 S. Beran and P. Jírů, in ref. 5, pp. 53-59.
14 J. Bentley, J. Am. Chem. Soc., 104 (1982) 2754-2759.
15 E. Kochanski and J. Prissette, Chem. Phys. Letters, 80 (1981) 564-568.
16 C. Douketis, G. Scoles, S. Marchetti, M. Zen and A.J. Thakkar, J. Chem.Phys., 76 (1982) 3057-3063.
17 R. Ahlrichs, R. Penco and G. Scoles, Chem. Phys., 19 (1977) 119-130.
18 J.W. Geus, in J.R. Anderson (Ed.), Chemisorption and Reactions on Metallic Films, Vol. 1, Academic Press, London, 1971, Ch. 3, pp. 129-224.
19 P.H. Lewis, J. Catalysis, 69 (1981) 511-513;
    R.A. Dalla Betta, M. Boudart, P. Gallezot and R.S. Weber, J. Catalysis, 69 (1981) 514-515.

P.A. Jacobs et al. (Editors), *Structure and Reactivity of Modified Zeolites*
© 1984 Elsevier Science Publishers B.V., Amsterdam — Printed in The Netherlands

ON THE BIFUNCTIONAL ACTION OF MODIFIED ZEOLITES Y CONTAINING NICKEL

H. BREMER, W. P. RESCHETILOWSKI, F. VOGT, and K.-P. WENDLANDT
Technical University "Carl Schorlemmer" Leuna-Merseburg,
GDR-4200 Merseburg, Otto-Nuschke-Straße

ABSTRACT

Catalytic properties of modified zeolites Y containing nickel were characterized by the conversion of n-hexane, the reaction of cyclohexane, and the hydrogenolysis of ethane, respectively. The changes in the catalytic properties resulting from the modification of zeolite acidity suggest that there exists a direct interaction between the metal and the support, which is particularly evident in samples prepared by impregnation. Increasing acidity of the support leads to a decrease and finally to the suppression of both dehydrogenation and hydrogenolysis activity of the supported Ni/HNa-Y catalyst. At the same time the selectivity in skeletal isomerization is improved.

INTRODUCTION

Modified acidic zeolites containing transition metals (especially those of Group VIII) are widely used as efficient catalysts, e. g. in the hydrocracking of heavy gas oil and in the hydroisomerization of light paraffins (C_4 to C_6). The catalytic behaviour of zeolites Y containing nickel in a complex manner depends on several factors: Degree of reduction, dispersion and nature of the reduced metal, acidity of the support and nature of direct and indirect interactions between metal and zeolite. The existence of an electronic interaction between Group VIII metals (Pt, Pd) and acidic supports has recently been demonstrated (refs. 1,2). Such interactions modify the energy of the Fermi level and hence the catalytic properties of the supported metal.

The aim of the present work is to distinguish between the different factors in hydrocarbon reactions which control catalytic activity and selectivity of modified zeolites Y containing nickel. To this purpose, the catalytic activity and selectivity of zeolites Ni-Y of different acidity has been investigated in the n-hexane isomerization, the cyclohexane conversion, and the ethane hydrogenolysis. Furthermore, the reduction degree and the metal dispersion in the catalysts used were determined. Ni-containing zeolites were prepared by ion exchange as well as by impregnation with aqueous solutions of $Ni(NO_3)_2$.

EXPERIMENTAL

Modified Ni-containing zeolite samples were prepared from a zeolite Na-Y synthesized in the VEB Chemiekombinat Bitterfeld (SiO_2/Al_2O_3 mole ratio = 5.2) applying different methods: (i) One-step (in the case of the NiNa-Y samples) or two-step (in the case of the $(NH_4,Al)NiNa-Y$ samples) ion exchange (370 K, 2 h) with 0.1 n solutions of the nitrates; (ii) quantitative impregnation of the original zeolite with constant amounts of the aqueous solutions of NH_4^+ and Ni^{2+} nitrate in a vacuum rotation evaporator (samples $NH_4,Ni/Na-Y$); (iii) ion exchange with solutions of NH_4^+ and Ca^{2+} nitrate, resp., and subsequent impregnation with a solution of $Ni(NO_3)_2$ (samples Ni/NH_4Na-Y and $Ni/CaNa-Y$).

Reduction degrees are calculated from H_2-uptake during temperature programmed reduction (TPR). For this purpose the samples were exposed in situ to flowing air at 720 K for 2 h. The TPR measurements were carried out at linear increasing temperature (8 $K \cdot min^{-1}$) from 300 K to 920 K in a stream of 7.5 vol% H_2 in argon.

Data on the dispersion of the reduced nickel were derived from oxygen chemisorption. Prior to chemisorption as well as catalytic measurements the samples were reduced with H_2 for 2 h. Chemisorption of oxygen was carried out at 273 K by a pulse chromatographic method with helium as carrier gas. The specific nickel surface area was calculated using a surface area requirement of 0.48 ml O_2 per m^2 of nickel surface area (ref. 3).

The conversion of n-hexane as well as cyclohexane was carried out under normal pressure in H_2 using an integral flow reactor, the feed rates were 0.014 $mole \cdot g^{-1}$ catalyst$\cdot h^{-1}$ and 0.048 mole. g^{-1} catalyst$\cdot h^{-1}$, respectively. Ethane hydrogenolysis was performed in a pulse catalytic reactor. Reaction products were analyzed by gas chromatography.

RESULTS AND DISCUSSION
Degree of reduction and dispersion of Ni^0

Table 1 presents the reduction degree after reduction at 720 K for nickel exchanged zeolites Y with constant nickel content. As has been shown previously (refs. 4,5), the reduction degree decreases with increasing acidity of the zeolite support. This corresponds to the equilibrium:

$$Ni^{2+} + 2OZ^- + H_2 \rightleftharpoons Ni^0 + 2HOZ \qquad \text{(Equation 1)}$$

TABLE 1

Composition and reduction degree of Ni-exchanged modified zeolites Y

Sample	Reduction degree (720 K), %	
$Ni_{0,15}Na_{0,70}-Y$	56	
$(NH_4)_{0,09}Ni_{0,16}Na_{0,59}-Y$	30	
$Al_{0,06}Ni_{0,13}Na_{0,56}-Y$	26	increasing sodium exchange
$Al_{0,03}Ni_{0,17}Na_{0,57}-Y$	22	
$(NH_4)_{0,21}Al_{0,03}Ni_{0,16}Na_{0,38}-Y$	18	
$(NH_4)_{0,39}Ni_{0,16}Na_{0,29}-Y$	13	increasing acidity

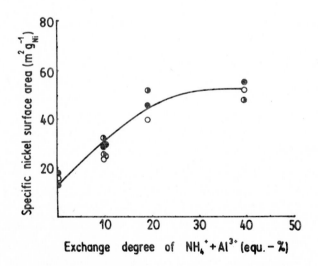

Fig. 1. Correlation between specific nickel surface area and NH_4^+/Al^{3+} exchange degree.
(o samples without pretreatment,
◐ samples pretreated for 2 h at 720 K in water vapor
● samples pretreated for 2 h at 720 K in air)

Furthermore, the samples differ not only in their reduction degree but also in the specific surface area of the metallic nickel obtained after reduction: With increasing exchange of the Na$^+$ ions (increased acidity) the specific surface area of the nickel also increases – as is evident from Fig. 1 – and remains nearly unchanged in the highly acidic samples. Since the content of exchanged Ni^{2+} ions in these samples was kept constant, the decrease of the reduction degree with increasing acidity indicates that higher dispersion in samples of higher acidity is connected with a smaller amount of Ni$^\circ$. This should also be reflected in the catalytic properties of these samples.

Conversion of cyclohexane on ion exchanged zeolites Ni-Y

Fig. 2 gives the conversion rates of cyclohexane to benzene (B) and to hydrogenolysis products (methane – M), resp., measured on modified Ni-Y zeolite samples at 650 K. The rates of both re-

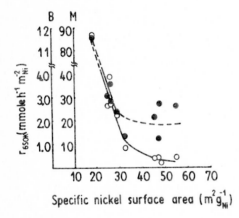

Fig. 2. Dependence of the dehydrogenation (●) and hydrogenolysis (○) activity on the specific nickel surface area.

actions decrease with increasing Ni$^\circ$ dispersion; in the range of large surface areas, however, this decrease is smaller for the dehydrogenation reaction, such that in these samples the conversion of cyclohexane to benzene proceeds with greater selectivity. The decrease in the reaction rate with increasing metal dispersion is understandable for the hydrogenolysis reaction which is known to be a "demanding reaction" but less so for the hydrogenation reaction ("facile reaction").

These results indicate that apart from the dependence on the particle size there must be another influence factor, which we consider, in agreement with other authors (refs. 6,7), to consist in the acidity of the support.

We have shown that in ion exchanged zeolites Ni-Y a correlation exists between the reduction degree, the dispersion of
the Ni$^{\circ}$ after reduction, and the acidity of the support. According to Equ. 1, a direct correlation between acidity and the
amount of Ni$^{\circ}$ in ion exchanged zeolites Ni-Y can be expected. In
order to separate these two factors, our investigations were
extended to zeolites Ni-Y obtained by impregnating zeolites of
different acidity with solutions of Ni$(NO_3)_2$.

<u>Reactions on ion exchanged and impregnated zeolites Ni-Y</u>

<u>Ethane hydrogenolysis.</u> As is demonstrated in Fig. 3, there
exist considerable differences between the catalytical activities

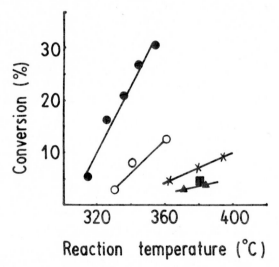

Fig. 3. Conversion of ethane vs. reaction temperature.
($\bullet$ 1,6NH$_4$5Ni/Na$_{1,0}$-Y, o 5Ni/(NH$_4$)$_{0,8}$Na$_{0,2}$-Y), (reduced for 2 h at
720 K, + 80 equ-% NaNO$_3$), $\times$ 5Ni/Ca$_{0,3}$Na$_{0,4}$-Y, $\blacktriangle$ 5Ni/(NH$_4$)$_{0,8}$
Na$_{0,2}$-Y, $\blacksquare$ 10Ni/(NH$_4$)$_{0,8}$Na$_{0,2}$-Y); the numbers mean wt% of Ni

of zeolites Y containing nickel (with the Ni-content of the
samples being nearly the same) in ethane hydrogenolysis: Whereas
the weakly acidic catalyst (sample 1,6NH$_4$ 5Ni/Na$_{1,0}$-Y) exhibits a
very high activity, the more acidic catalyst (sample 5Ni/
(NH$_4$)$_{0,8}$Na$_{0,2}$-Y) develops a slight catalytic activity only at
high temperatures, and the catalytic activity of a sample of
medium acidity (5Ni/Ca$_{0,3}$Na$_{0,4}$-Y) ranges between these two
samples. This effect of acidity on the catalytic properties of

metallic nickel (on which this reaction proceeds) can be demonstrated by varying the acidity of the samples after the reduction has been carried out. Such an additional modification of the acidity results in changes in the hydrogenolysis activity of the samples. Thus, the sample $5Ni/(NH_4)_{0,8}Na_{0,2}-Y$ becomes more active in hydrogenolysis reactions if the acidic centers are poisoned with $\geq$ 50 equ.% Na^+ ions (by impregnation with aqueous solutions of $NaNO_3$) (cf. Fig. 3). Obviously, the presence of acidic centers in the non-poisoned samples affects the electronic properties of the nickel particles and hence decreases the bonding strength of the carbon-nickel bond, which leads to the suppression of the hydrogenolysis reaction.

<u>Dehydrogenation of cyclohexane</u>. A completely analogous pattern as in the ethane conversion is found in the cyclohexane dehydrogenation to benzene. Whereas weakly and medium acidic Ni-containing samples are highly active in the dehydrogenation reaction $(r_B = 5,3$ and $1,9$ mmole$\cdot h^{-1}\cdot g^{-1}$ for samples $1,6NH_45Ni/Na_{1,0}-Y$ and $5Ni/Ca_{0,3}Na_{0,4}-Y$, resp., at a reaction temperature of 570 K), the strongly acidic samples are completely inactive under the reaction conditions chosen: The electronic properties and hence the catalytic properties of the metal are changed by electron shift from the metallic nickel to the acidic support. The rate-determining step in the dehydrogenation of cyclohexane is the desorption of benzene (ref. 8). Electron shift from nickel to a strongly acidic support increases the bonding strength of the benzene to nickel and hence decreases the reaction rate in the dehydrogenation of cyclohexane.

<u>Isomerization of n-hexane</u>. In this reaction, the metallic as well as the acidic components of the samples are active. Both functions determine not only the catalytic activity but also the selectivity parameters, i. e. the ratio of isomerization to cracking, the product distribution of the cracking products, and the relative composition of the obtained C_6 isomers. The differences in the product distribution on Ni-containing zeolite samples of high and low acidity are demonstrated in Fig. 4 by an example.

Samples of low acidity yield a high portion of 2,2 dimethylbutane (2,2-DMB) and hydrogenolysis products (C_{1-3}). The isomerization selectivity of these samples is relatively small. In contrast to the behaviour of weakly acidic samples, on strongly

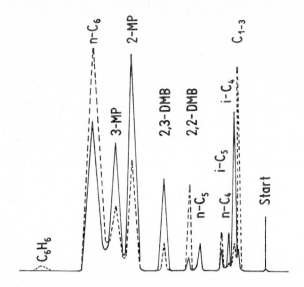

Fig. 4. Typical results of gas chromatography for samples of high (——) and low (----) acidity in the n-C$_6$ isomerization reaction.

acidic samples we obtained a greater amount of methylpentanes (MP) and 2,3-dimethylbutane (2,3-DMB).

Thus, hydrogenolysis activity of the nickel (in the same way as in the ethane conversion) is suppressed by a strongly acidic support. Also, these highly acidic samples fail to form benzene in n-hexane conversion, while benzene is formed on weakly acidic samples as catalysts. However, the hydrogenation/dehydrogenation activity of nickel is even in highly acidic samples sufficiently high to enable the bifunctional isomerization of n-hexane by a joint action with the Brønsted centers of the zeolite.

Generalization. The changes in activity and selectivity which can be achieved in the conversion of n-hexane by varying the acidity of the support are just as reversible as are the changes in the catalytic properties of the metal component Ni in the conversion of cyclohexane and ethane. They can be represented in summary by the scheme shown at the following page.

Obviously, in the catalyst samples presented here, this metal – support interaction is of considerably greater importance than the metal particle size. The electronic influence of the support also explains the equidirectional decrease of rates in the dehydrogenation and hydrogenolysis of cyclohexane (cf. Fig. 1).

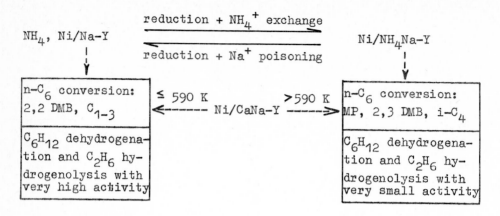

CONCLUSIONS

The catalytic behaviour of modified zeolites Ni-Y in n-hexane isomerization, cyclohexane dehydrogenation, and ethane hydrogenolysis indicates that the catalyst efficiency of the metal component can be controlled reversibly and stepless by changing the acidity of the support. These results can only be interpreted - in agreement with other authors (refs. 2,6,9) - by assuming an electronic interaction between support and metal component, which results in changes of the catalytic properties of the latter: Electron shift from the metal to the acidic centers of the support (Brønsted acidic centers) considerably affects the course of each individual reaction. Such an effect of the support on the metal component can be expected to be particularly evident with modified zeolites as a support, since the acidity of the zeolites can be varied in such a wide range (from neutral up to strongly acidic) as it is scarcely possible with other types of supports.

REFERENCES

1 J. C. Védrine, M. Dufaux, C. Naccache and B. Imelik, J. Chem. Soc., Farad. Trans. I, 74 (1978) 440.
2 C. Naccache and Y. Ben Taarit, Acta Phys. et Chem., Szeged 24 (1978) 22
3 N. E. Buryanova, A. P. Karnaukhov, L. M. Kefel, R. D. Ratner and D. N. Chernyavskaya, Kinet. Katal., 8 (1968) 868
4 E. D. Garbowski, C. Mirodatos and M. Primet, in P. A. Jacobs et al. (Eds.), Metal Microstructures in Zeolites, Studies in Surface Science and Catalysis 12 (1982) 235
5 H. Bremer, Nguyen Do Khue, W. Reschetilowski and K.-P. Wendlandt, Z. anorg. allg. Chem. in press
6 F. Figueras, R. Gomez and M. Primet, Adv. Chem. Ser. 121 (1973) 480
7 J. T. Richardson, J. Catal. 21 (1971)122
8 J. N. Sinfelt, J. L. Carter and D. J. C. Yates, J. Catal. 24 (1972) 283
9 R. A. Dalla Betta and M. Boudart, Proc. 5th Int. Congr. Catal., v. 1, p. 329, Amsterdam 1973

P.A. Jacobs et al. (Editors), *Structure and Reactivity of Modified Zeolites*
© 1984 Elsevier Science Publishers B.V., Amsterdam — Printed in The Netherlands

CORRELATION BETWEEN THE STRUCTURE OF REDUCED TRANSITION METAL ZEOLITE SYSTEMS AND THE CATALYTIC REACTIVITY IN BENZENE HYDROGENATION REACTIONS

V. KANAZIREV, V. PENCHEV, Chr. MINCHEV, U. OHLERICH[*], F. SCHMIDT[*]

Institute of Organic Chemistry of the Bulgarian Academy of Sciences

Sofia 1113, Bulgaria

[*]Institute of Physical Chemistry, University of Hamburg

Laufgraben 24, D-2000 Hamburg 13, West Germany

ABSTRACT

Nickel metal supported on zeolites was prepared by stoichiometric ion exchange or ion impregnation followed by dehydration in a stream of air or argon and reduction with molecular hydrogen. A, X, Y, and ZSM zeolites were used. The catalytic activity in benzene hydrogenation was found to depend mainly on the surface of the nickel particles larger than 2 nm and on the acidity of the support.

INTRODUCTION

Nickel zeolites are important catalysts in hydrocracking and shape selective cracking reactions (ref. 1,2). In order to optimize this metal zeolite system with respect to hydration/dehydration properties, the effect of the metal and the effect of the support must be understood and controlled separately. From this point of view, it is useful to study a model reaction such as the hydrogenation of benzene instead of such complex systems which occur in the industrial applications of metal zeolite systems.

Depending on the metal content, pretreatment, time and temperature of reduction, reducing agent, presence of other exchangeable cations, a different combination of the following metal species can be observed (ref. 3-7):

1.) Ni^{2+} in zeolite lattice positions

2.) Ni_x^{y+}-charged cluster and/or Ni^0 cluster in the sodalite cages

3.) Ni_x^{y+}-charged cluster and/or Ni^0 cluster in the supercage

4.) $(Ni^0)_i$ in the bulk of the zeolite crystal located in lattice distortions

5.) $(Ni^0)_e$ on the external surface of the zeolite

The early work on the activity of Ni-zeolites in benzene hydrogenation has been summarized by Minachev (ref. 3). In a previous paper it has been reported, that the activity of Ni-zeolites in benzene hydrogenation decreases in the sequence NiA > NiX > NiY (ref. 8).

The present investigation attempts to complete the previous results and to discuss the benzene hydrogenation on NiA, NiX, NiY and NiZSM in more detail.

METHODS

Some data of the investigated samples are listed in Table 1. Starting material was Linde A, X- and Y-zeolites prepared according to Kacirek et al. (ref. 9) and ZSM according to Maiwald et al. (ref. 10). In order to achieve comparable nickel content the NiZSM was obtained by ion impregnation whereas the other Ni zeolites were made by means of a stoichiometric ion exchange (ref.11). The thermal stability of the various samples was characterized by DTA (ref.12).

The degassing was performed in a flow reactor with an argon flow velocity of 60 cm^3/min.

Magnetic measurements have been performed using a Faraday balance in the temperature range 6-295 K and fields up to 1 T. Some of the results have been published earlier (ref. 13).

The hydrogenation of benzene was studied in a flow microreactor (LHSV 1 to 22 h^{-1}, hydrogen partial pressure 40 mbar, conversion lower than 5%) loaded with 0.05 to 0.45 mg zeolite. Product analysis was performed directly in the flow by GC methods.

RESULTS

Blanc experiments have shown that the sodium zeolites as well as the non-reduced nickel zeolites exhibit almost no activity in benzene hydrogenation. This demonstrates, that the activity of the reduced nickel zeolites is directly due to the formation of the metal phase.

Table 2 shows the reaction rates of the benzene hydrogenation related to the total nickel content. For similar conditions of pretreatment (air 773 K, 2 h) and reduction (723 K, 2 h) the numbers differ by a factor of less than 4 between A,X and ZSM zeolites. Consequently, the hydrogenation activity of Ni^0 zeolites seems to be due to the formation of the reduced metal phase (ref.13). However, this explanation fails in case of the tremendous difference in the activity between X and Y zeolites, which is more than one order of magnitude.

Keeping in mind, that the pretreatment conditions influence the cation location and, hence, the degree of reduction and the dispersion of the metal, the conditions of the pretreatment were varied. If the samples were dehydrated in argon at 823 K, the activity of X and Y zeolites is higher than in other

TABLE 1

Sample characterization

Sample	$\frac{SiO_2}{Al_2O_3}$	Exchange capacity for Ni,wt%	Ni content wt %	$T_{stab.}$ K
NiNaA	2.0	19.8	6.5	993
NiNaX	2.4	18.3	7.3	993
NiNaY	5.2	12.0	7.7	1073
NiZSM-5	75.6	0.94	4.9	1273

T_{stab}: Thermal stability from DTA measurements

TABLE 2

Activity in benzene hydrogenation as a function of pretreatment
Reduction: T = 723 K, 2 h, Reaction: T = 463 K

Sample	Pretreatment	Activity $\dfrac{mmol\ C_6H_6 \cdot 10^{-5}}{g_{Ni} \quad sec}$
NiA	air 773 K, 2 h	241.1
NiZSM-5	"	85.6
NiX	"	60.1
NiY	"	0.46
NiX	Ar, 823 K, 4 h	87.8
NiY	"	0.99
NiY	only H_2, 723 K, 2 h	0.21

TABLE 3

Activity in benzene hydrogenation as a function of the temperature of reduction (pretreatment: Ar, 4h, 823 K)

Line	Sample	Activity at 463 K	Temperature of Reduction				
1	NiX						
2			573	623	673	723	773
3		$\dfrac{\text{mmol } C_6H_6 \cdot 10^{-5}}{g_{Ni} \quad sec}$	3.6	11	55	88	
4		$\dfrac{\text{mmol } C_6H_6 \cdot 10^{-7}}{m^2_{Ni(>.5 \text{ nm})} sec}$		5.9	27	49	
5		$\dfrac{\text{mmol } C_6H_6 \cdot 10^{-7}}{m^2_{Ni(1 \text{ nm})} sec}$		6.4	35	73	
6		$\dfrac{\text{mmol } C_6H_6 \cdot 10^{-6}}{m^2_{Ni(>2nm)} sec}$		8	11	15	
1	NiY						
2			573	623	673	723	773
3		$\dfrac{\text{mmol } C_6H_5 \cdot 10^{-5}}{g_{Ni} \quad sec}$		0.6	0.8	1.0	1.3
4		$\dfrac{\text{mmol } C_6H_5 \cdot 10^{-8}}{m^2_{Ni(>.5nm)} sec}$		4.2	4.9	7.1	10
5		$\dfrac{\text{mmol } C_6H_5 \cdot 10^{-8}}{m^2_{Ni(1 \text{ nm})} sec}$		5.4	6.2	10	15
6		$\dfrac{\text{mmol } C_6H_5 \cdot 10^{-7}}{m^2_{Ni(>2nm)} sec}$		1.9	2.3	2.3	3.4

cases (see Table 2). In addition, the activity of NiY is lower, if the reduction is performed without any thermal pretreatment. These results demonstrate that the activity depends on the pretreatment. However, this influence
is not sufficient to explain the differences between NiX and NiY zeolites.

The temperature of reduction is one of those parameters which strongly
affect the metal dispersion and, in turn, the hydrogenation activity. The influence of temperature of reduction on the activity and on the metal dispersion
was investigated in a wide range of temperatures. Table 3 shows the results
for NiX and NiY zeolites. The NiX and NiY samples contain comparable amounts of
Ni and are of the same structure. They differ in the Si/Al ratio and, hence, in
the number of acidic centers (ref. 14). In case of NiY, the increase of the reduction temperature from 623 K to 773 K results only in a twofold increase in
activity, whereas in case of NiX this increase between 623 K and 723 K amounts
to a factor of 8.

Summarizing, the increase in the activity in benzene hydrogenation is in the
sequence NiA > NiZSM > NiX > NiY.

Guilleux et al. (ref. 15) as well as Ione et al. (ref. 16) reported a low
level of activity in NiX and NiY as compared to Ni/SiO_2. The present study is
in agreement with these observations and shows, that NiA and NiZSM zeolites
behave more like the Ni/SiO_2 system. This seems to indicate, that the degree
of reduction and the dispersion of the metal has an important influence on the
activity.

DISCUSSION

It is well known, that various forms of the metal can occur (see 'introduction' 1.) to 5.)) (ref. 4,5,13,17,18) and that the degree of reduction in
NiX and especially in NiY zeolites is rather low. The particle size distributions of the same samples of NiA, NiX and NiY, which were used for the present
study have been published earlier (ref. 13). As has been shown in the previous
paper, the weight content of particles with dimensions between 2 nm and 8 nm
decreases in the sequence NiA > NiX > NiY, which qualitatively agrees with the
decrease in activity.

In case of NiX, on the other hand, the increase of the temperature of reduction from 623 K to 723 K causes a fourfold increase of the amount of particles with 2 nm to 8 nm (ref. 13). This is in line with the catalytic data
(see Table 3, line 3). In contrast to NiX, the dependence of the particle sizes
(in the range 2 nm to 8 nm) on temperature of reduction is much less in NiY.
This is also in agreement with the catalytic data.

Therefore, the activity of the NiX and NiY zeolites on benzene hydrogenation
was not only related to the total nickel content in the sample (see Table 3,

line 3) but also to the surface of the metal particles, larger than 0.5 nm (see Table 3, line 4) as well as to the surface of particles of 1 nm (see Table 3, line 5) and to the surface of the nickel particles between 2 nm and 8 nm, which are present in the sample (see Table 3, line 6).

The surprising results show, that the activity of the Ni zeolites neither significantly depends on the value of the total surface area of the metal nor remarkably depends on the surface area of the 1 nm particles, but dominantly on the surface of the particles between 2 nm and 8 nm. Only in the latter case the enormous differences in increase of activity with increasing temperature of reduction disappear (see Table 3, line 6). This is an interesting observation.

The differences in the absolute values of reaction rates between NiX and NiY (which are still left in both types of zeolites after the relation of the reaction rate to the surface area of the large particles), together with the small increase of activity with increasing temperature of reduction indicate, that influences other than the metal surface area are also important in this system.

In fact, Ni zeolites are typical representatives of bifunctional catalysts. It is not surprising to expect support effects. The support effects discussed so far were related to the influence of the zeolite matrix on the metal dispersion (this paper) or to the state of the metal (charged clusters or electron transfer between matrix and metal (ref. 19)).

As reported by Figueras et al. (ref. 19) the benzene hydrogenation activities of Pd-zeolites increase with their acidic properties. The results have been explained by interaction of palladium with the zeolite matrix, which changes the electron configuration of the metal.

However, the acidity of the zeolite may also directly affect the catalytic reaction. Therefore, the strength as well as the number of the various kinds of acid centers in the reduced Ni-zeolites have to be taken into account. The number of acid centers is larger in NiX compared to NiY (ref. 14).

If we relate the catalytic activity in benzene hydrogenation (see Table 3, line 6) to the number of acid centers, which are already present in the NiX and NiY zeolites (this is a constant value for every temperature of reduction) and in addition to those acid centers, which are produced during reduction, a clear correlation between activity and acidity is found. The change in acidity with increasing temperature of reduction is caused by the additional acidity according to:

$$Ni^{2+}\text{-zeolite} + H_2 \rightarrow Ni^0 + (2H^+)\text{-zeolite}.$$

Therefore, the amount of protons is equal to the amount of nickel metal formed. With increasing temperature of reduction, the degree of reduction and hence the acidity increases.

CONCLUSION

To explain the results discussed above, the following model is proposed:
During reduction local distortion of the zeolite framework occurs. The metal particle growth is controlled by the formation of these defects (ref. 13). Particles larger than the cage dimensions are located in these places and are surrounded by an undestroyed matrix. Therefore, the sieving effect of the matrix may still be valid even for catalysis on these large particles. The walls of the enlarged cages contain acid centers. The number of these acid centers is larger in NiX than in NiY. This explains the differences in activity between NiX and NiY. The increase of the specific activity related to the surface of the large particles with increasing temperature of reduction is due to the increase of the number of acid centers formed upon reduction. The acidic centers are active sites for hydrocarbon activation. Hydrogen activation (and possibly also benzene activation) occurs on the metal surface nearby.

ACKNOWLEDGEMENTS

Financial support from Deutsche Forschungsgemeinschaft is gratefully acknowledged. One of us (F.S.) is indebted to the Fonds der Chemischen Industrie for support of this work.

REFERENCES

1 H. Heinemann, Catal. Rev., Sc. & Eng., 23 (1981) 315
2 E. Gallei, Chem. Ing. Tech., 52, (1980) 99
3 Kh.M. Minachev and Ya.I. Isakov, in J.A. Rabo (Ed.) Zeolite Chemistry and Catalysis, ACS Monograph 171 (1976) 552
4 J.B. Uytterhoeven, Acta Phys. Chem., 24 (1978) 53
5 D. Delafosse, in B. Imelik et al. (Eds.) Catalysis by Zeolites, Elsevier, 1980, 235pp.
6 P.A. Jacobs, in P.A. Jacobs et al. (Eds.) Metal Microstructures in Zeolites, Elsevier, 1982, 71pp.
7 F. Schmidt, in P.A. Jacobs et al. (Eds.) Metal Microstructures in Zeolites, Elsevier, 1982, 191pp.
8 V. Penchev, N. Davidova, V. Kanazirev, Chr. Minchev, Y. Neinska, Adv. Chem. Ser., N 121 (1973) 461
9 H. Kacirek and H. Lechert, J. Phys. Chem., 79 (1975) 1589.
10 W. Maiwald, W.D. Basler and H. Lechert, Proc. 5th Int.Conf.Mol.Sieves, Heyden, 1980, 562pp.
11 F. Schmidt, H. Kacirek, W. Gunßer, Chr. Minchev, V. Kanazirev, L. Kosova, V. Penchev, Heterogeneous Catalysts, 2, Varna, 1979, 331pp.
12 V. Penchev, Chr. Minchev, V. Kanazirev, I. Tsolovski, Adv. Chem. Ser., N 102, 1971, 434pp.
13 F. Schmidt, Th. Bein, U. Ohlerich and P.A. Jacobs, in D.E.W. Vaughan (Ed.) Proceedings of the Sixth Int. Conf. on Zeolites, Butterworth, 1984, in press

14 J.W. Ward in J. Rabo (Ed.) Zeolite Chemistry and Catalysis, ACS Monograph 171 (1976) 118
15 M.F. Guilleux, J. Jeanjean, M. Bureau-Tardy and G. Djega-Mariadassou, VIe Symp. Ibero Americano de Catalise, Rio de Janeiro, 1978
16 K.G. Ione, V.N. Rommanikov, A.A. Davidov and L.B. Orlova, J. Catal. 57 (1979) 126
17 Chr. Minchev, V. Kanazirev, L. Kosova, V. Penchev, W. Gunßer and F. Schmidt, Proc. 5th Int. Conf. Mol. Sieves, Napoli, Heyden, 1980, 355pp.
18 F. Steinbach, J. Schütte, R. Krall, Chr. Minchev, V. Kanazirev and V. Penchev, in D.E.W. Vaughan (Ed.), Proc. of the Sixth Int. Conf. on Zeolites, Butterworth, 1984, in press
19 F. Figueras, R. Gomez and M. Primet, Adv. Chem. Ser., 121 (1973) 480

P.A. Jacobs et al. (Editors), *Structure and Reactivity of Modified Zeolites*
© 1984 Elsevier Science Publishers B.V., Amsterdam — Printed in The Netherlands

CHARACTERIZATION OF X-TYPE ZEOLITES CONTAINING METALLIC NICKEL PARTICLES BY DIELECTRIC RELAXATION

J.C. CARRU[1], D. DELAFOSSE[2] AND M. BRIEND[2]

[1]C.H.S. - LA CNRS 287 - Equipe "Nouveaux Matériaux" - Bâtiment P3
Université de Lille 1 - 59655 Villeneuve d'Ascq cédex (France)

[2]E.R. 133 Réactivité de surface et structure, Laboratoire de Chimie des Solides,
Université P. et M. Curie, 4, Place Jussieu - 75230 Paris (France)

ABSTRACT

We have formed metallic nickel particles in sodium X zeolite by two different procedures : molecular hydrogen and atomic hydrogen. The size characterization and the repartition of the particles were carried out using different techniques and from the analysis of the dielectric spectra it was possible to obtain some complementary information concerning the non reduced cations and the behaviour of the protons.

INTRODUCTION

The characterization of zeolites containing cations reduced to the metallic state necessitates the use of numerous physical chemistry techniques. In a previous study (ref.1), we have shown that dielectric relaxation used in a wide frequency range (8 decades) can contribute to zeolite characterization. To explain our results we considered 3 hypothesis :

i) the existence of some heterogeneity in the repartition of the non reduced cations,

ii) the existence in the framework of small metallic clusters having a residual electric charge,

iii) the existence of protons in interaction with the zeolite structure and the metallic particles.

The aim of this study is to complete that work by evaluating the importance of the different possible hypothesis. For this purpose, we made metallic nickel particles by two different techniques. The principal interest of these procedures is that they give rise to particles having different characteristics as regards size, repartition in the zeolite matrix and consequently catalytic behaviour (ref.2,3).

EXPERIMENTAL

The starting material was prepared by conventional exchange of NaX (from Union Carbide Corp.) with a 0.1 normal solution of Ni^{2+} nitrate in order to exchange about 20 % of the sodium ions. The composition of the sample referred to as Ni_8X and determined by chemical analysis, was as follows for a unit cell :

$$Ni_8X : Ni_{8.5} Na_{69} Al_{86} Si_{106} O_{384} nH_2O$$

Reduction treatments

All the samples were pretreated overnight in vacuum either at 673 K or 773 K. After activation, two different methods (ref.4,5) were used in order to reduce the nickel ions :

    i) by molecular hydrogen (H_2) with different conditions of temperature and duration to vary the degree of reduction,

    ii) or by atomic hydrogen ($H^{\cdot}$) for 6 hours at 273 K for complete reduction.

The notation of these samples will be Ni_8X followed by the degree of reduction and H_2 or H . For the sake of clarity, they will be referred to as sample S1, S2, S3 ... (see table 1).

Experimental methods

Various methods were used to characterize the samples such as hydrogen uptake, ferromagnetic resonance, temperature programmed desorption etc. Some results obtained with these methods are listed in table 1. Dielectric relaxation, which is the principal method utilized in this paper, has been described previously (ref.1).

RESULTS

We give in table 1 some characteristics of the reduced samples studied in this paper.

TABLE 1

Characterization of the Ni_8X reduced samples

Samples	Ref.	Pretreatment temperature	Treatment method.	Treatment Temp.	Treatment duration	Reduction %	Particles Size	Particles Dispersion
Ni_8X25H_2	S1	673 K	H_2	523 K	70 h	25	25 Å	
Ni_8X42H_2	S2	673 K	H_2	595 K	17 h	42	40 Å	
Ni_8X70H_2	S3	773 K	H_2	773 K	2h 30	70	70 Å	bi-dispersion
	S4	673 K	H_2	723 K	27 h	100		
Ni_8X100H_2	S5	773 K	H_2	723 K	20 h	100	100-200Å	
	S6	773 K	H_2	773 K	15 h	100		
Ni_8X100H	S7	673 K	H	273 K	6 h	100	10 Å	mono-dispersion

From the measurement of the complex permittivity ε^*, it is possible to obtain different curves (ref. 1). For example, we give in figure 1, the imaginary part ε'' of ε^* as a function of the frequency, in logarithmic scales. We can see, in particular, for the totally reduced samples, that the two reduction methods lead to two different spectra. As it is possible to describe in a first approximation the dielectric spectra with 4 parameters, we have given them in table 2 for all the samples used in this study. For the non reduced ones the values correspond to a dehydration at 673 K for 24 hours.

From these values, it can be seen that the position of the dielectric domain (given by F_c) is the same in a first approximation for the reduced samples and the non reduced Ni_8X. The same remark can be made for the amplitude $\Delta\varepsilon'$ except for the sample reduced by H·. With regard to the values of the activation energy ΔU and the spreading in frequency of the domain α , they are always higher for all the reduced samples than for Ni_8X. Finally, for the samples completely reduced by H_2 (S4 to S6) the different pretreatment and reduction conditions give rise to slight differences in the values of the dielectric parameters.

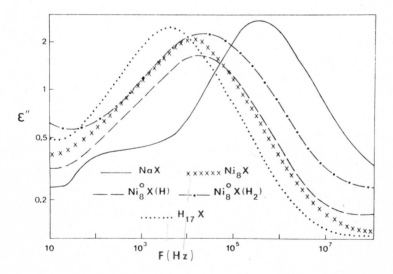

Fig. 1 : Dielectric spectra at T = 298 K of NaX, Ni_8X, $H_{17}X$, S4 and S7 samples.

TABLE 2

Values of the dielectric parameters for all the samples studied

Non reduced zeolites	Reduced zeolites	Fc(Hz)	ΔU kcal/mole	$\Delta \varepsilon'$	α (degrees)
NaX		360×10^3	7.4	8.2	28
Ni_8X		13×10^3	8.7	8	35
$H_{17}X$		4×10^3	12	8.8	32
	S1	11×10^3	10.8	8.3	40
	S2	10×10^3	11	8.4	41
	S3	20×10^3	9.4	9.8	37
	S4	24×10^3	10.8	8.4	41
	S5	22×10^3	10.1	8.6	39
	S6	24×10^3	9.9	9.2	37
	S7	17×10^3	9.4	6.6	40

DISCUSSION

Origin of the dielectric relaxation

In the frequency range studied, the electrical polarization observed is due to the jumps of the cations in the cages of the zeolite matrix. The mean frequency of the jumps is linked to the critical frequency F_C of the dielectric domain (ref.6). Moreover, F_C is all the higher as the number of cations in SIII sites is greater.

Reproducibility of the samples

After having recorded numerous dielectric spectra on reduced samples, it is possible to obtain reproducible results if the following parameters are controlled carefully :

- thickness of the bed (shallow bed conditions)
- speed of the temperature rise ($\leqslant$ 60 K/hour)
- temperature of pretreatment and reduction.

Nevertheless, the reproducibility of the reduction is not as good with the H˙ method as with H_2 . This is probably because the former technique is more sophisticated, so there are more parameters to monitor.

Homogeneity

The information on the homogeneity of the samples is given by the shape of the spectra. It is quantified in first approximation with the parameter α. As it can be seen from table 2, it varies from 35 for the non reduced samples

to 41 for a sample completely reduced by H_2. This important variation is linked
to a modification of the repartition of the cations (Na^+ and Ni^{2+}) in the framework during the reduction process. In fact, to create metallic aggregates, the
nickel cations must migrate from the inner sites (S_I , $S_{I'}$) to sites in the
supercages (S_{II} , S_{III}) and then from one supercage to another. For electrostatic reasons this process causes a rearrangement of the sodium ions which are
in majority in our samples. Moreover, as it has already been established (ref.7),
there are protons in the reduced samples.
All these phenomena give rise to the observed heterogeneity.

Presence of protons

To ensure the stoichiometry, the presence of protons are needed in the reduced samples. Therefore, the totally reduced samples contained only sodium (80 %)
and protons (20 %) as ions. Under these conditions, it is possible to compare
the dielectric spectrum of this sample with that of a 20 % proton exchanged NaX
zeolite (denoted $H_{17}X$ in table 2). The two spectra are quite different as are
the parameters F_c, U, α(see table 2). This can be explained by the fact that
the protons do not have the same behaviour in the reduced samples as in $H_{17}X$.
In this view two cases can be distinguished depending on the size of the metallic
particles :

i) samples containing small particles : this is the case when reducing
with H˙ (see table 1). There is some evidence (ref. 2,8) that, in these samples,
hydrogen species are strongly bonded to the small particles.

ii) Samples containing big particles : this is the case with a reduction
by H_2 (see table 1). Here the protons are not bonded to the particles (ref. 2).
On the other hand a drop of SiOH band intensity can be observed (ref. 9).
A dehydroxylation process probably explains these results.
So, in the two cases considered the hydrogen present in the reduced samples did
not behave,from a dielectric point of view, as mobile ion in the framework
like in $H_{17}X$.

Crystallinity

The crystallinity of the samples can be checked by the value of $\Delta\varepsilon'$. In fact,
when the structure collapses, there is no more dielectric relaxation and $\Delta\varepsilon'$
tends towards zero. For all the samples studied,$\Delta\varepsilon'$ is higher than 8 which is
normal, except for the one reduced by H˙. In that case only, there is a loss of
crystallinity. Nevertheless, it is probably a local breakdown of the framework
as the X-ray diffraction patterns did not show a noticeable loss of crystallinity. This phenomenon is certainly due to the reduction mechanism by H˙ because
we have tested the plasma with a NaX zeolite containing no nickel and obtained
a normal value of 8.8 for $\Delta\varepsilon'$. This is clearly a disadvantage of the H˙ method

which has however the overwhelming advantage of giving small, monodispersed
metallic particles (ref.5).

Electrically charged clusters

The existence of small aggregates (essentially in the sodalite cages) having
a residual electric charge has been proposed by different authors (ref. 10-12).
In our samples, such species would exist after the reduction by $H^{\cdot}$ rather than
by H_2. As the dielectric spectra of all the reduced samples have the same parameter F_C, in a first approximation, it is probable that the number of charged
clusters is negligible with regard to the number of the non reduced cations.

Difference in pretreatment and reduction conditions

We have prepared different totally reduced samples by varying the H_2 reduction conditions (see table 1). The corresponding spectra have not very different
characteristics. Nevertheless, it seems that a pretreatment at 773 K is preferable to one at 673 K, because the homogeneity of the former sample is better
as indicated by the value of α (see table 2). This could be related to the difference of repartition of the Ni^{2+} cations prior to the reduction process
(ref.13).

CONCLUSION

We have presented the characterization by dielectric relaxation of X zeolites
containing nickel particles. Firstly, from this study, it is possible to determine the conditions giving rise to reproducible samples. Secondly, we have
given evidence for the existence of two effects linked to the reduction process :

    - the heterogeneity of the repartition of the non reduced cations

    - the presence of protons in the reduced samples having different beha-
viour from those in non reduced ones.

REFERENCES

1 J.C. Carru and D. Delafosse, Studies in surface science and catalysis, Vol.12,
   Elsevier, Amsterdam, 1982, pp. 221-227
2 G.N. Sauvion, Ph. D. Thesis, University of Paris, 1983
3 G.N. Sauvion, S. Djemel, J.F. Tempère, M.F. Guilleux and D. Delafosse, Studies
   in surface science and catalysis, Vol. 5, Elsevier, Amsterdam, 1980, pp. 245-
   249
4 M.F. Guilleux, D. Delafosse, G.A. Martin and S.A. Dalmon, J.C.S. Faraday I,
   75, 1979, pp. 165-171
5 M. Che, M. Richard and D. Olivier, J.C.S. Faraday I, 76, 1980, pp. 1526-1534
6 P. Tabourier, J.C. Carru and J.M. Wacrenier, Zeolites, 3, 1983, pp. 50-56
   Earlier papers are cited in this reference.
7 P.A. Jacobs, Carboniogenic activity of zeolites, Elsevier, Amsterdam, 1977,
   p. 194
8 G.N. Sauvion, J.F. Tempere, M.F. Guilleux and D. Delafosse, submitted for
   publication in J. Catal.

9 G. Schulz-Ekloff, Communication
   to the 6th Int. Conf. on zeolites, Reno (U.S.A.), July 10-15, 1983
10 H. Beyer, P.A. Jacobs and J.B. Uytterhoeven, J.C.S. Faraday I., 72, 1976
   pp. 674-685
11 J.C. Vedrine, M. Dufaux, C. Naccache and B. Imelik, J.C.S. Faraday I., 74,
   1978, pp. 440-449
12 Ch. Minchev, V. Kanazirev, L. Kosova, V. Penchev, W. Gunsser and F. Schmidt
   in L.V.C. Rees (Ed.), Proc. 5th Int. Conf. on zeolites, Naples, June 2-6,
   1980, Heyden, London, 1980, pp. 355-363
13 D. Delafosse, Studies in surface science and catalysis, Vol. 5, Elsevier,
   Amsterdam, 1980, pp. 235-243

P.A. Jacobs et al. (Editors), *Structure and Reactivity of Modified Zeolites*
© 1984 Elsevier Science Publishers B.V., Amsterdam — Printed in The Netherlands

FEATURES OF AN ELECTRON STATE AND CATALYTIC PROPERTIES OF THE L-TYPE PLATINUM-CONTAINING ZEOLITE

S.V. GAGARIN[1], V.S. KOMAROV[2], I.I. URBANOVICH[3], T.I. GINTOVT[3] and Yu.A. TETERIN[4]

[1]Institute of Combustible Minerals, Moscow (USSR)
[2]Institute of General and Inorganic Chemistry, Minsk (USSR)
[3]Institute of Physico-Organic Chemistry, Minsk (USSR)
[4]Kurchatov Institute of Atomic Energy, Moscow (USSR)

ABSTRACT

The influence of Re and W additions on the platinum valence state and the catalytic properties of the L-type Pt-containing zeolite has been studied. Due to the cluster nature of a deposited metal, quantum effects have been found to occur in the X-ray electron spectra of the Pt/KL catalysts. The analysis has shown that the RePt/KL and WPt/KL catalysts contain rhenium and tungsten, in metallic as well as in partially or completely oxidized states. High activity of platinum-zeolite catalysts is accounted for by an uncommon electron state of platinum, which varies on introduction of modifying elements.

INTRODUCTION

The dehydrocyclization of paraffins attracts the attention of investigators as one of the most important reactions of the catalytic reforming. In the process of reforming only an insignificant part of paraffins is subjected to conversion. In this connection the maximum participation of n-alkanes in dehydrocyclization is one of the main problems in the improvement of the process under consideration. The L-type zeolite because of its structural features is a promising support for catalysts used in aromatization of the C_6-hydrocarbons of normal structure. By the introduction of platinum into the zeolite a polydisperse contact displaying high activity in dehydrocyclization of n-hexane and benzene hydrogenation may be obtained (ref.1,2). The state of Pt in Pt/KL is characterized by the absence of electron deficiency, which is confirmed by IR-spectroscopy of the adsorbed CO and is caused, according to the authors suggestion (ref.1), by an effect of the zeolite field and structure on platinum particles. The introduction of modifying elements may be expected to result

in a change in both the platinum electron state and the properties of the catalytic system as a whole. The aim of the present
paper is to find out the causes affecting an increase in catalytic activity of the platinum-containing L-zeolite in the potassium form (KL) in dehydroyclization of n-hexane on introduction
of W and Re. Besides n-hexane, the conversion of possible intermediate products of the reaction, hexene-1 and hexadiene-2,5, has
been studied.

METHODS

Platinum, rhenium and tungsten were incorporated into the KL
-zeolite by impregnation with aqueous solutions of chloroplatinic
acid, ammonium perrhenate and ammonium paratungstate, respectively.

The activation procedure of the catalysts is described in
(ref.2).

The aromatization of the C_6-hydrocarbons n-hexane, hexene-1
and hexadiene-2,5 has been tested in a microcatalytic apparatus
in the flow of helium at 700 K. The gas chromatographic analysis
of the products of the reaction was carried out with a column,
filled with 20% of the ester of tri-ethylene glycol and n-buta-
noic acid on Celite-545.

IR-spectra of pyridine adsorbed on catalysts were measured
with a UR-20 spectrometer at 470 K. Electron microscopic investigations were carried out with an EMV-100LM microscope using the
carbon replica method. X-ray electron spectra (XPS) of samples
were taken by means of a Hewlett Packard HP595A spectrometer applying photoionization of electrons by monochromatized X-ray radiation of AlK and using a HP2100A computer.

RESULTS AND DISCUSSION

Table 1 presents the results of the conversion of n-hexane,
hexene-1 and hexadiene-2,5. From the Table it follows that at a
short contacting time under the indicated conditions the aromatization of n-hexane practically does not proceed. Small quantities of benzene are formed only from unsaturated hydrocarbons. As
the contact time increases, the dehydrogenizing and dehydrocycli-
zing activity of the platinum-zeolite catalyst as well as the
yield of products of the C_5-cyclization and the general conversion of the individual hydrocarbon under investigation increases
too, the yield of the C_5-C_6 isomers remaining approximately

constant. Negligible rates of isomerization, crackingand C_6-cy-
clization over Pt/KL are explained by the low-acidity nature of
the given catalyst. A modification of the platinum-zeolite contact with tungsten (Table 1) leads to an increase in cracking
and isomerization yields, intensifying the dehydrogenation and
dehydrocyclization function. An incorporation of rhenium into the
platinum-zeolite catalyst (Table 1) causes a more noticeable increase in the cracking dehydrogenating and dehydrocyclizating
activity. Simultaneously, the isomerizing and C_5-cyclizing functions of the contact are changed, however, to a minor extent,
compared to the platinum-tungsten zeolite.

Thus, under the investigated conditions the conversion of
C_6-hydrocarbons over Pt/KL results largely in formation of benzene, the quantity of which increases with the degree of unsaturation of a molecule and the time of contact. The incorporation of
tungsten or rhenium appreciably intensifies the dehydrogenating
and dehydrocyclizating functions of the catalyst causing a certain increase in the product yields from cracking, isomerization
and C_5-cyclization.

It is to be noted that such changes in the distribution of
products obtained from the C_6-hydrocarbon conversion are indicating the predominant intensification of the metallic function of
the bifunctional platinum-zeolite contact on introduction of modifying elements. This fact seems to be associated with additives
affecting the state of platinum in the catalyst.

The XPS study showed that platinum contained in the L-zeolite is in a highly dispersed state. The spectra of the platinum-
containing contact show quantum dimensionality effects specified
by the cluster nature of the deposited metal. They are reflected
in the narrowing of a valence band, a density decrease in the
states at the Fermi level, a decrease in the spin-orbital splitting of external 5d electrons and in the availability of a narrow (about 1 eV) forbidden band between the occupied and vacant
levels of platinum. An insignificant chemical shift of the Pt
$4f_{7/2}$ line in comparison with platinum-containing faujasites is
an important feature, which indicates a lower degree of electron
deficiency of a metal in the given catalyst than in the case of
Pt deposited on the Y-type zeolite. The data obtained are in
agreement with the conclusions of (ref.1). However, on introduction of Mo in the Pt/Y catalyst, the Pt-electron structure becomes similar to that of a bulk metal (ref.3).

TABLE 1

Conversion of C_6-hydrocarbons over 0.5 wt.% Pt/KL

(The flow rate of the carrier gas is 20 ml/min; educt concentration: 2g/l)

Additive, wt.%	Weight, mg	Hydrocarbon	Composition of reaction products, wt.%							Conversion, wt.%
			nC_1-C_5	iso- C_5 - C_6	hexane	hexenes	HD+[1] MCPD	MCP+[2] MCPen	benzene	
-	20	hexane	0.05	0.05	99.7	0.1	-	0.03	-	0.3
		hexene-1	0.1	0.1	0.6	93.0	3.2	3.1	0.05	7.0
		hexadiene-2,5	0.3	0.1	-	4.1	91.2	4.1	0.1	8.8
	100	hexane	0.1	0.2	93.0	4.2	1.7	7.7	3.1	7.0
		hexene-1	0.1	0.3	12.1	50.1	22.8	8.9	6.4	49.9
		hexadiene-2,5	0.3	0.3	0.5	21.1	47.5	17.6	12.7	52.5
	280	hexane	0.2	0.1	68.9	5.5	0.6	10.0	14.7	31.1
		hexene-1	0.3	0.1	8.1	38.8	24.1	12.3	16.3	61.3
		hexadiene-2,5	0.6	0.3	0.4	23.3	31.6	16.4	27.2	68.4
0.08 W	280	hexane	3.1	2.6	46.2	23.6	2.7	5.9	15.9	53.8
		hexene-1	3.8	2.9	11.1	12.8	18.1	6.3	25.1	67.2
		hexadiene-2,5	4.1	3.4	-	18.5	30.1	8.8	36.1	69.9
0.1 Re	280	hexane	4.1	1.7	33.9	35.1	3.9	2.5	18.8	66.1
		hexene-1	4.7	1.9	12.1	26.2	22.1	3.7	29.3	73.8
		hexadiene-2,5	5.4	1.3	2.9	20.0	20.0	4.6	45.6	80.0

[1] hexadienes + methylcyclopentadienes;   [2] methylcyclopentane + methylcyclopentenes

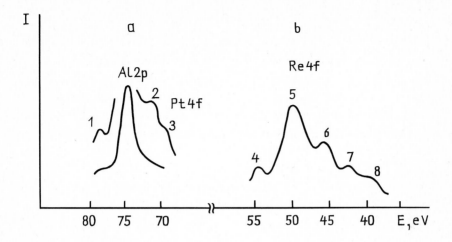

Fig. 1 X-ray photo electron spectrum of RePt/KL
    a - Pt 4f core levels; 1 - 3 - structure specified by
        the presence of platinum in the catalyst;
    E - binding energy;
    b - Re 4f core levels; 4 - 8 structure, specified by
        the presence of rhenium in the catalyst.

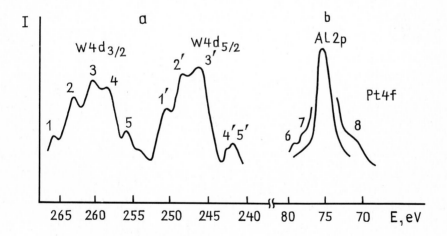

Fig. 2 X-ray photo electron spectrum of WPt/KL
    a - W 4d core levels; 1 - 5 and 1'- 5'- structure, spe-
        cified by the presence of tungsten in the catalyst;
    E - binding energy
    b - Pt 4f core levels; 6 -8 - structure, specified by
        the presence of platinum in the catalyst.

XPS spectra of the modified platinum-containing catalysts on the KL-basis are characterized by the formation of additional species of platinum, as well as by a variety of the valence states of the additives. In the case of a platinum-rhenium catalyst, the spectrum analysis of the Re 4f core levels indicates the presence of rhenium species in metallic, partially oxidized (Re^{4+}) and totally oxidized (Re^{7+}) state, as well as the location of the Re^{n+} ions in cation positions, their electrons exhibiting a considerable chemical shift due to relaxation effects (Fig.1). Platinum in the RePt/KL catalyst is present in three states: Pt^{0}, Pt^{+} and Pt^{n+}, as evidenced by three peaks of Pt 4f-electrons possessing different binding energies It should be emphasized that with rhenium a substantial structural change in the valence state region of platinum occurs. The X-ray electron spectra of the platinum-rhenium L-zeolite indicate the possibility of both Re and Pt mixed clusters and an isolated deposition of metals.

On the basis of the analyzed spectra of core 4f and 4d electrons as well as the valence region of the catalyst, species of

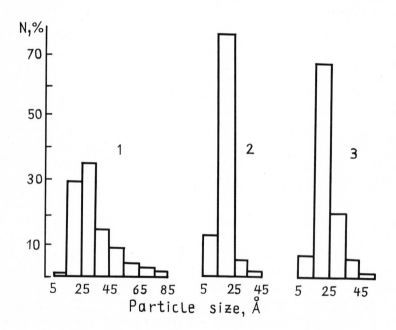

Fig. 3 Histograms of samples of the KL-base catalysts.
1 - Pt/KL; 2 - WPt/KL; 3 - RePt/KL.

two types of W^0, W^{4+}, W^{5+} and W^{6+} have been shown to be present
in the platinum-tungsten zeolite, Fig.2. In this case accumula-
tion of metallic tungsten species on the surface or in the zeolite channels is unprobable because of the facility of its oxidation by hydroxyl groups. Therefore it is possible that PtW alloy
particles of cluster character are formed. The latter is confirmed by the structure of the valence state region of the deposited
metals.

The data of electron-microscopic study of Pt/KL, RePt/KL and
WPt/KL agree with the results of XPS. Fig.3 presents histograms
of the catalysts samples. The sample Pt/KL contains platinum in
a highly dispersed state. The average size of Pt particles is
44 $\overset{o}{A}$. Histograms of modified contacts indicate a significant decrease in the particle size of platinum in the presence of rhe-
nium or tungsten. The maximum dispersity of platinum (d_{Pt}= 23 $\overset{o}{A}$)
is observed in the case of a platinum-tungsten catalyst, and a
platinum-rhenium sample has a similar dispersity (d_{Pt}= 29 $\overset{o}{A}$). The
electron-microscopic studies confirm the assumption that Pt interacts with modifying additives which results in changing its e-
lectron state.

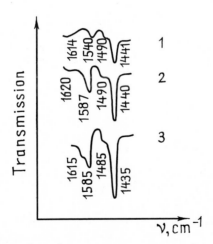

Fig.4. IR-spectra of pyridine adsorbed on the KL-based catalysts.
       1 - Pt/KL; 2 - WPt/KL; 3 - RePt/KL.

We may assume that rhenium and tungsten contained in Pt/KL
may change the acidic properties of the zeolite catalyst. Fig.4
shows IR-spectra of pyridine adsorbed on the samples of bimetal-
lic contacts. From Fig.4 it follows that the introduction of modifying elements increases the absorption intensity of the

bands, related to coordinately-bonded pyridine (ref.4). Therefore it is to be noted that the modification of the platinum-containing KL-zeolite with rhenium or tungsten intensifies the electron--acceptor properties of the catalysts. A similar phenomenon may promote a change in the adsorption of hydrocarbons on metal-zeolite contacts resulting in a specific change in catalytic properties (ref.5).

Thus from the data obtained on the catalytic study and physicochemical properties of the L-zeolite-based catalysts the conclusion may be drawn that an unusual state of platinum in KL determines the activity of the Pt/KL catalyst in the aromatization of n-hexane, hexene-1 and hexadiene-2,5. The introduction of rhenium and tungsten additives favours a substantial increase in the catalytic activity due to the change of electron properties and platinum dispersity and promotes the intensification of electron-acceptor properties of the zeolite catalyst.

REFERENCES

1 C. Besoukhanova, J. Guidot, D. Barthomeuf, M. Breysse and J.R. Bernard, J. Chem. Soc., Faraday Trans. 1, 77 (1981) 1595--1604.
2 I.I. Urbanovich, T.I. Gintovt, N.S. Kozlov and N.A. Akulich, Heterogeneous catalysis, Proceedings of the All-Union Conference, Novosibirsk, Vol. 1, 1982, pp. 50-53 (Russ.).
3 T.M. Tri, J.P. Candy, P. Gallezot, J. Massardier, M. Primet, J.C. Vedrine and B. Imelik, J. Catalysis, 79 (1983) 396-409.
4 J.W. Ward, in Zeolite Chemistry and Catalysis (J.A. Rabo, Ed.) ACS Monograph 171, 1976, p. 226.
5 Yu.I. Ermakov, V.A. Zakharov and B.N. Kuznetsov, Fixed complexes on oxidic supports in catalysis, Nauka, Novosibirsk, 1980, 248 pp.

P.A. Jacobs et al. (Editors), *Structure and Reactivity of Modified Zeolites*
© 1984 Elsevier Science Publishers B.V., Amsterdam — Printed in The Netherlands

STUDY OF THE DISPERSION OF ZEOLITE SUPPORTED NICKEL IN DEPENDENCE ON THE ZEOLITE TYPE AND THE REACTION MEDIUM

N.P. DAVIDOVA, M.L. VALCHEVA and D.M. SHOPOV

Institute of Kinetics and Catalysis, Bulgarian Academy of Sciences, Sofia 1040 (Bulgaria)

ABSTRACT

The particle size distributions of zeolite supported nickel prepared by (i) mechanical mixing of zeolite and nickel oxide, (ii) impregnation via precipitation and (iii) stoichiometric ion-exchange using zeolites of the types A, X, Y, Mordenite and ZSM-5 were evaluated from electron micrographs. Generally, three maxima around 4,8 and 14 nm are obtained after reduction. A redispersion of the nickel is observed under the influence of water or carbon monoxide on the samples reduced in hydrogen. The most probable size of the smaller particles depends on the Si:Al ratio of the zeolite and decreases in the order A → X → Y → Mordenite → ZSM-5.

INTRODUCTION

X-ray studies of the formation and state of the metallic phase in nickel zeolite catalysts showed reversible dispersion and sintering processes in dependence on the type of the interacting gas phase (refs.1-3). In the present study the particle size distributions of nickel on a series of differently prepared nickel-zeolite catalysts resulting after reduction and subsequent treatment with water, carbon monoxide and synthesis gas are determined from electron micrographs.

METHODS

‹The test samples were prepared by mechanical mixing of the components (NiO+Z), deposition by precipitation (Ni(OH)$_2$→NiO/Z) and ion exchange (NiCaZ or NiNaZ). The corresponding procedures are described in (refs.1,2). As carriers zeolites with molar ratios SiO_2/Al_2O_3 of 1.9, 2.5, 5.0, 10 and 40 were used. The calcium forms of the zeolites A, X and Y with maximum extent of ion exchange with Ca^{2+} are used. The composition and preparation mode of the catalyst test samples are presented in Table 1.

TABLE 1

Composition and preparation mode of the catalyst test samples

Sample	SiO_2/Al_2O_3	Ni content wt %	Preparation mode
NiO+CaA	1.9	3.0	mechanical mixing
NiO+CaX	2.5	3.0	- ,, - ,, -
NiO+CaY	5.0	3.0	- ,, - ,, -
NiO+NaM	10.0	3.0	- ,, - ,, -
NiO+NaZSM-5	40.0	3.0	- ,, - ,, -
NiO/CaA	1.9	3.0	deposition
NiO/CaX	2.5	3.0	- ,, -
NiO/CaY	5.0	3.0	- ,, -
NiO/NaM	10.0	3.0	- ,, -
NiO/NaZSM-5	40.0	3.0	- ,, -
NiCaA	1.9	3.0	ion exchange
NiCaX	2.5	3.1	- ,, -
NiCaY	5.0	2.9	- ,, -
NiNaM	10.0	0.8	- ,, -

The preliminary reduction treatment of the catalysts, the conditions of the methanation reaction and the study of the effect
of the separate components of the reaction medium are described
in detail in (refs.1-3).

Specimens for electron microscopy were prepared from aqueous
suspensions of the catalyst. The micrographs were taken at U=80 kV
and magnifications of 12000, 30000, 60000, 80000 and 120000. From
the photographic plates 4-times enlarged copies were made. 8 - 10
photographs at various magnifications each having an average of
300 to 500 particles were evaluated for each sample. The particle
sizes and the size distribution of the particles were determined
using standard procedures of the mathematical statistics (refs.
4,5).

The X-ray analysis of powder samples was performed on a
"Philips" analyzer at U=36 kV using a copper anode. The integral
intensity of (111)Ni is determined by the procedure described in
(refs.2,3) and is related to 1 g-atom of nickel. The peak profile
of (111)Ni is recorded using a program for the calculation of the
Simpson's integral.

RESULTS AND DISCUSSION

A non-homogeneous size distribution of the particles was established for all samples. Hence, the distribution function was
examined in separate particle size intervals in order to determine
their limits and to calculate the most probable particle size for
each interval.

In Fig.1 the size distribution curves for zeolite A samples
obtained by different preparation procedures are shown. Fig.2
gives size distribution curves for the three limiting cases of the
examined zeolites with SiO_2/Al_2O_3 ratios of 1.9, 10 and 40 for
samples obtained by deposition (Fig.2a) and by mechanical mixing
(Fig.2b). Three sub-intervals are clearly distinguished in the
whole size range: 0-5 nm, 6-10 nm and 11-24 nm. The statistical
weight of the particles in the three sub-intervals related to the
total particle number is different and depends on the zeolite type
and on the sample preparation mode. In all cases the statistical
weight of the particles sized within the 0-5 nm interval for the
same SiO_2/Al_2O_3 ratio is higher for the samples obtained by deposition than by mechanical mixing, whereas for the 6-10 nm and
11-24 nm intervals the reverse case is observed.

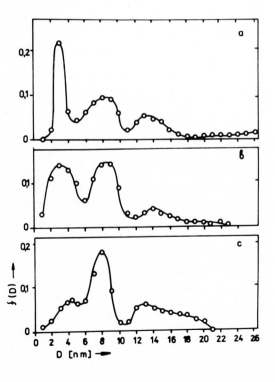

Fig.1. The size distri-
bution curves for samples
obtained on the basis of
zeolite type A by diffe-
rent procedures of pre-
paration.
(a) Ion exchange
(b) Deposition
(c) Mechanical mixing

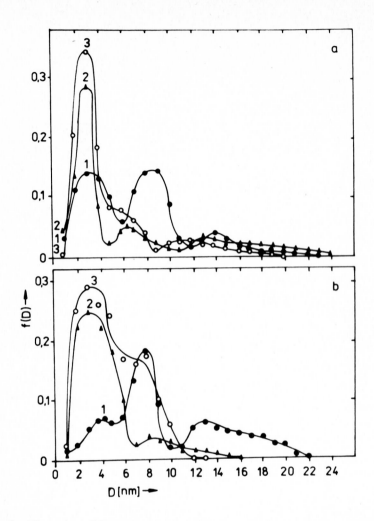

Fig.2. The size distribution curves for samples obtained by deposition (a) and by mechanical mixing (b) with a SiO_2/Al_2O_3 ratio = 1.9 (curve 1), 10 (curve 2) and 40 (curve 3)

For samples obtained by the same preparation mode, the distribution curve depends on the SiO_2/Al_2O_3 ratio, the statistical weight of the particles sized within 0-5 nm being proportional and that of the particles sized within 6-10 nm being inversely proportional to the SiO_2/Al_2O_3 ratio.

The effect of the SiO_2/Al_2O_3 ratio on the most probable particle size in the intervals 0-5 nm and 6-10 nm for samples obtained by deposition and mechanical mixing is shown in Fig.3. It follows from Figs.2 and 3 that on raising the SiO_2/Al_2O_3 ratio the relative

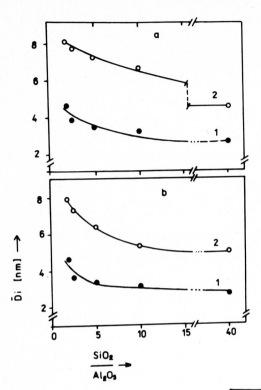

Fig.3. Effect of the molar ratio SiO_2/Al_2O_3 on the most probable particle size in the intervals 0-5 nm (curve 1) and 6-10 nm (curve 2) for the samples obtained by deposition (a) and mechanical mixing (b).

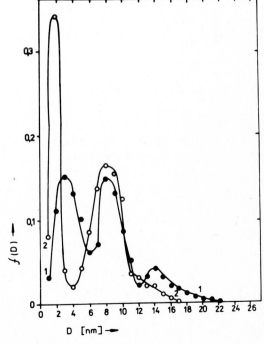

Fig.4. Effect of the interaction between CO and H_2 on the size distribution curve of the particles on the sample NiO/CaA in reduced state (curve 1) and after the interaction (curve 2).

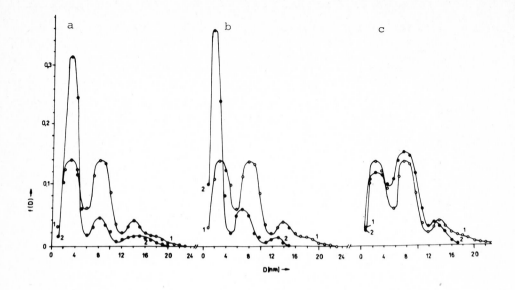

Fig.5. Effect of the separate components of the reaction medium on the size distribution of the particles in comparison to the reduced state (curves 1-a,b,c). (a) Effect on H_2O (curve 2-a). (b) Effect of CO (curve 2-b). (c) Effect of CH_4 (curve 2-c).

weight of the fraction 0-5 nm increases, whereas the most probable average size in the same interval decreases, irrespective of the preparation mode. For the fraction 6-10 nm there is a decrease in both statistical weight (steeper for deposited samples) and most probable average particle size.

The size distribution curve of the particles, shown in Fig.4, is recorded after the interaction between CO + H_2 on the catalyst sample NiO/CaA. There is a sharp decrease in the statistical weight of the fraction 0-5 nm, as well as a decrease in the most probable average size in this range resulting from the simultaneous effect of all components of the reaction medium (reacting substances and reaction products). The effect of the separate components of the reaction medium on the size distribution curve of the particles is shown in Fig.5(a,b,c). Water and carbon monoxide produce a sharp increase in the statistical weight of the fraction 0-5 nm at the expense of the fractions 6-10 nm and 11-24 nm. CH_4 causes the following redistribution: the statistical weight of the fractions 0-5 nm and 11-24 nm decreases and that of the fraction 6-10 nm increases.

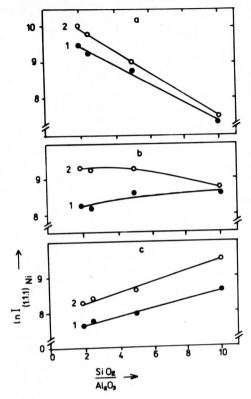

Fig.6. Effect of the reaction medium on the integral intensity of the signal (111)Ni, related to 1 g-equivalent nickel, in reduced state (curves 1-a,b,c) and after interaction of CO and H_2 (curves 2-a,b,c). (a) For the samples obtained by mechanical mixing. (b) By deposition. (c) By ion exchange.

The X-ray study of the effect of the reaction medium on the microstructure of the metal (ref.3) revealed that CH_4 causes an increase in the integral intensity of Ni(111), which we related to the particle agglomeration. Fig.6 shows that in all cases after the reaction between CO and H_2 the integral intensity of Ni(111) is higher than that of the reduced forms, which can be attributed to the effect of CH_4. The change in the integral intensity of Ni(111) caused by the reaction medium H_2 and CO/H_2 respectively as a function of the SiO_2/Al_2O_3 ratio is similar for samples obtained by the same praparation mode.

It can be assumed that the observed redispersion effects (Fig.5a,b; Fig.4), resulting in larger fractions of the small particles, are due to mobile, migrating species formed by the interaction of nickel and the gas media water and carbon monoxide, respectively. The reason for the preferred maxima around 4, 8 and 14 nm in the size distribution curves is not clear, but should be related to the structure of the zeolite support. A progress in the knowledge on the mesopore structure of zeolites would be helpful for the elucidation of this effect.

REFERENCES

1 N. Davidova, M. Valcheva and D. Shopov, Metal Microstructures in Zeolites: Studies in Surface Science and Catalysis, vol. 12, 1982, Els.Sci.Publ.Co., Amsterdam, p. 253.
2 N. Davidova, M. Valcheva and D. Shopov, Proceedings of the 1982 Conference "Catalysis by Zeolites", Liblice, CSSR, p. 44.
3 N. Davidova, M. Valcheva and D. Shopov, Zeolites, 1 (1981) 72.
4 A.A. Borovkov, Course of theory of probability (in Russian), Moscow, "Nauka", 1972.
5 G. Korn and T. Korn, "Mathematical Handbook for Scientists and Engineers", Moscow, "Nauka", 1977 (in Russian).

P.A. Jacobs et al. (Editors), *Structure and Reactivity of Modified Zeolites*
© 1984 Elsevier Science Publishers B.V., Amsterdam — Printed in The Netherlands

MAGNETIC INVESTIGATION OF METALLIC IRON CLUSTERS IN ZEOLITES

J.M. JABLOŃSKI

Institute of Low Temperature and Structure Research, Polish Academy of
Sciences, P.O.Box 937, 50-950 Wroclaw (Poland)

ABSTRACT

Complex iron compound obtained by the reaction of $K_4[Fe(CN)_6]$ with exchanged
Fe^{+2} ions in Y zeolite was reduced with dihydrogen at temperatures 573 - 753 K,
giving highly dispersed metallic iron with particle sizes ranging from atomical-
ly dispersed iron at 573 K to superparamagnetic (few nm in diameter), and normally ferromagnetic particles at higher temperatures.

INTRODUCTION

The reduction of Fe^{+2} and Fe^{+3} - ions introduced into zeolites by exchange
was investigated by various methods e.g. XPS, ESR, Mössbauer spectra, and by
magnetic methods (refs 1-4). It has been shown that Fe^{+3} - ions located in cationic sites undergo partial autoreduction to Fe^{+2} during the dehydration in vacuo.
The reduction by dihydrogen of zeolites containing Fe^{+2} and Fe^{+3} - ions in cationic sites does not lead to metallic iron, even at temperatures as high as 873 K
(ref.4). The earlier report (ref.1) claiming a partial reduction of iron ions
to metallic iron at these conditions seem to refer to only trace amounts of
zero-valent iron. It has been established however, that complex iron compounds
(carbonyls, ferrocyanides) contained in zeolites can be reduced by dihydrogen at
mild conditions (ref.5), or decomposed (ref.6), giving metallic iron with high
degree of dispersion. Since zeolites containing metallic iron proved active in
the synthesis of hydrocarbons from carbon monoxide and hydrogen (ref.7), the investigation of the reduction of ferrocyanide complexes of iron deposited in
zeolites Y was undertaken in this study.

EXPERIMENTAL

Ammonium form of Y zeolites, manufactured at the Institute of Industrial
Chemistry, Warsaw with the Si/Al ratio 2.5, was used in preparation. 10 g of
dry NH_4Y zeolite was added to 500 ml of distilled water brought to pH = 4 by the
addition of hydrochloric acid. The obtained water-zeolite suspension was deoxygenated by blowing argon through it during several hours, and then after the removal of water the zeolite was immersed in 1000 ml deoxygenated 0.025 mol/l
$FeSO_4$ solution at pH = 4. All operations were made in argon atmosphere. The ion
exchange was carried out during 6 hours with occasional stirring, and then after
removal of the solution the zeolite was washed with deoxygenated, acidified to

pH = 4 , distilled water until the disappearance of Fe^{+2} - ions. Then the deoxygenated solution of $K_4[Fe(CN)_6]$ (7 g in 500 ml of water) was added to the zeolite and the suspension remained in argon atmosphere under occasional stirring during 6 hours. After several washings with distilled water the product was filtered and dried under vacuum during 24 hours.

The reduction was carried out in hydrogen stream at temperatures in the range of 573-753 K, and the samples for the magnetization measurements were taken without any contact with air, degassed under 10^{-5} Tr vacuum at reduction temperature (1 - 2 hours), and sealed off under vacuum. Magnetic measurements were made by Faraday method in the field strengths up to 0.7 Tesla (ref.8). In order to determine the reduction degrees (content of metallic iron), the samples after first magnetization measurements were heated at temperatures 50 K higher than those of reduction during a long time with intermittent magnetization measurements, until no changes in sample magnetization occurred. The total iron content was determined by gravimetric methods as iron oxychinolate.

RESULTS AND DISCUSSION

Iron ions introduced into Y zeolite by exchange and treated then with ammonium ferrocyanide solution, are, according to (ref.5) at least partially reducible with dihydrogen to metallic iron, even at the temperature as low as 670 K. Our sample, prepared as described above and treated with hydrogen 5 h at 570 K and subjected to magnetic measurements after degassing, showed no susceptibility increase, and no field dependence of susceptibility, (as it should in the presence of ferromagnetic iron) (Fig. 1). In fact, there occurred even some small decrease of the susceptibility at 78 K. Only after 30 h of heating at 623 K in vacuo, the magnetization of this sample increased by almost two orders of magnitude, and its values approached saturation at higher fields, which is a proof of the presence of ferromagnetically ordered iron crystallites. Since the magnetization per unit mass amounts to slightly over 2.2 at 78 K, one can estimate the content of metallic iron as at least about 1.0 % (weight), while the total iron content in the sample was 5.2 % (weight).

The absolute value of the susceptibility of the sample after reduction and its decrease to the value slightly less than that before reduction (Fig. 2), seem to indicate that some part of previously exchanged Fe^{+2} ions in zeolite was reduced, and that the reduced Fe^0 species was in the state of atomic dispersion. Since the magnetic moment of paramagnetic Fe^0 atoms is much less (μ = 3.15 BM) than that of the Fe^{+2} and Fe^{+3} - ions (5.91 and 5.15 - 5.66 BM, respectively), only the reduction of ions to atomic Fe^0 can result in the decrease of the susceptibility. If only diamagnetic $Fe(CN)_6^{-4}$ were reduced to atomic Fe^0 , an increase of susceptibility would result. The difference between magnetization curves at 29 K and 78 K of samples after heating at 623 K is larger than that

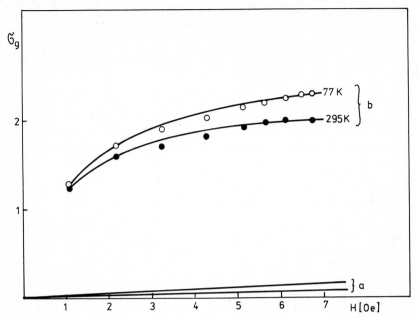

Fig. 1. Specific (per gram) magnetization of the sample FeY zeolite reduced 5 h at 573 K with hydrogen: a) after reduction and degassing, b) after subsequent heating 30 h at 623 K in vacuum.

for massive ferromagnetic iron. Thus one can infer that there is still some small amount of small particles of iron after that heating whose aggregate magnetization is anomalous (superparamagnetic particles) and thus the saturation magnetization obtained from the graph is too low.

Samples reduced at higher (623 K and 673 K) temperatures showed magnetization curves similar to those commonly obtained for systems containing both larger and very small ferromagnetic particles (ref.9), and are represented in the reduced form: $\frac{\sigma}{\sigma_\infty} - \frac{H}{T}$ in Fig. 3. (The σ_∞ (saturation magnetization) values are obtained from magnetization curves after long heating at temperature 50 K higher than that of reduction, assuming that after this heating all iron is contained in normal ferromagnetic form). Using the procedure described in (ref.9), it was shown that the experimental (reduced) magnetization curves can be expressed as those of the systems composed of particles with size distributions having two or three narrow modes: one at the larger sizes and other at particle sizes in the superparamagnetic range. The general formula for such curves has the form

$$\frac{\sigma}{\sigma_\infty} = n_0 + n_1 L \left(\frac{\mu_1}{k} \cdot \frac{H}{T} \right) + n_2 L \left(\frac{\mu_2}{k} \cdot \frac{H}{T} \right) \quad , \tag{1}$$

where L is the Langevin function, μ_1 - magnetic moment of ferromagnetically

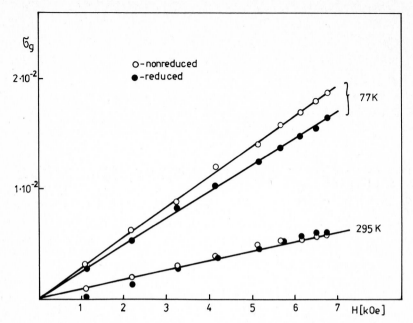

Fig. 2. Magnetization at 78 K and 295 K of sample FeY zeolite unreduced and reduced at 573 K and degassed.

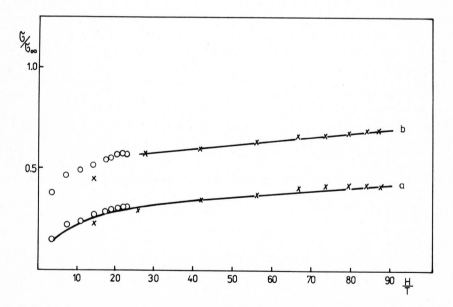

Fig. 3. Reduced magnetization curves of samples reduced with hydrogen at 623 K and 673 K. Continuous lines - magnetization curves calculated by formula (1) with parameters shown in Table 1, circles and crosses - experimental data at 295 K and 78 K, respectively.

ordered larger particles, μ_2 - the same for smaller particles, H - the magnetic field strength, k - Boltzmann constant, n_1, n_2 - corresponding fractions of particles. The first term refers to "massive ferromagnetic" particles reaching saturation at lower fields; in the case of iron they are all particles larger than about 5000 atoms, i.e. with the radius above 25 nm. Fractions of particles with average magnetic moments μ_1 and μ_2, giving best fit to the experimentally obtained reduced magnetization curves are given in Table 1.

TABLE 1

No	Reduction conditions	n_o	n_1	n_2	μ_1 μ_2 $[A \cdot m] \cdot 10^{21}$	Atoms per particle for μ_1	for μ_2	Corresponding radii of particles
1	5 h , 623 K	-	0.30	0.70	50 1.0	2440	49	1.9 nm, 0.8 nm
2	5 h , 673 K	0.50	0.50	-	2.05 -	100	-	} 1.06 nm
3	10 h , 673 K	0.56	0.44	-	2.05 -	100	-	

As seen from this table, magnetization curves of the samples reduced at 623 K can be obtained assuming no iron particles above 5nm in diameter, and the fraction of small particles with the diameter about 1.6 nm is large. Samples reduced at 673 K however, show magnetization curves indicating the large fraction of particles with sizes exceeding those superparamagnetic ones, and these fractions differ only slightly for different reduction times. There are no great differences also in the reduction degree achieved at the temperatures 573, 623 and 673 K, as can be judged from the values of saturation magnetization of samples. It seems probable that the extent of reduction depends mainly on the fraction of exchanged iron ions rendered reducible by the reaction with ferrocyanide ion, that in turn is strongly dependent of preparation conditions. The problem of reduction degree and attempts at confirmation of the magnetic particle granulometry by high resolution electron microscopy will be the subject of the future study.

ACKNOWLEDGMENT

The author is greatly indebted to Professor Władysław ROMANOWSKI for enlightening discussions and the help in preparation of this paper.

REFERENCES

1 Kh.M. Minachev, G.V. Antoshin and E.S. Shpiro, Izv. A.N.SSSR, Ser. Khim., 5 (1974) 1013.
2 B. Wichterlova, L. Kubelkova, J. Novakova and P. Jiru, in "Metal Microstructures in Zeolites", P.A. Jacobs et al. eds., Elsevier, Amsterdam 1982, 143 pp.
3 R.L. Garten, W.M. Delgass and M. Boudart, J. Catal., 18 (1970) 90.
4 J.M. Jabłoński, in preparation.
5 J. Scherzer and D. Fort, J. Catal., 71 (1981) 111.
6 Th. Bein, P.A. Jacobs and F. Schmidt, in "Metal Microstructures in Zeolites", P.A. Jacobs et al. eds., Elsevier, Amsterdam 1982, 111 pp.
7 D. Ballivet-Tkatchenko and I. Tkatchenko, J. Mol. Catal., 13 (1981) 1.
8 J.M. Jabłoński, J. Mulak and W. Romanowski, J. Catal., 47 (1977) 147.
9 W. Romanowski, Polish J. Chem., 54 (1980) 1515.

P.A. Jacobs et al. (Editors), *Structure and Reactivity of Modified Zeolites*
© 1984 Elsevier Science Publishers B.V., Amsterdam — Printed in The Netherlands

HYDROGENATION OF CO AND CO_2 OVER STABILIZED NiY CATALYSTS

V. PATZELOVÁ, A. ZUKAL, Z. TVARŮŽKOVÁ and O. MALÍČEK

J. Heyrovsky Institute of Physical Chemistry and Electrochemistry, Czechoslovak Academy of Sciences, CS-121 38 Prague 2 (Czechoslovakia)

ABSTRACT

The catalytic activities in the methanation of CO and CO_2 were studied on two catalysts prepared by reduction of stable forms of NiY zeolites. The samples differed only in the temperature of stabilization of the zeolite support, i.e. in the degree of dealumination of the zeolite and thus in the volume of the secondary pores formed during the stabilization. The content of elementary nickel did not differ significantly. It was found that the studied samples exhibit almost identical selectivity in the production of CH_4 in the methanation reaction of CO (higher than 90%); the conversion of CO for the sample with higher secondary porosity is systematically higher. The higher conversion on this catalyst is explained by increased kinetics of the reaction because of the greater number of secondary transport pores.

In the methanation reaction of CO_2 both samples exhibit increased production of CO compared with catalysts based on nickel on an amorphous support. This production is especially marked for sample with lower secondary porosity. It has been suggested that this effect can be attributed to increased adsorption of CO_2 on these samples as a result of the physisorption activity of the support.

INTRODUCTION

In the last few years, interest has increased considerably in the catalytic hydrogenation of carbon oxides, reflected in a marked increase in the number of publications dealing with this subject (refs. 1-8). The properties of group VIII metals applied to various supports with well-developed porous structure have been studied systematically. The first work (ref. 9) dealing with the use of synthetic zeolites as supports for dispersed metals was published in 1960. Catalysts based on zeolites exhibit special properties given by the specificity of the zeolite support: high dispersion of the applied metal, selectivity due to molecular dimensions and shape of the pores and high resistance to sulphur and nitrogen compounds present in the reaction mixture. Bhatia et al. (ref. 10) first described the use of Y zeolite as a support for catalytically active nickel in the methanation reaction of carbon monoxide. The insufficient stability of the zeolite structure under the reaction conditions is a certain drawback of catalysts prepared by reduction of NiY zeolite. This property is understandable in the light of the fact that reduction of the Ni cation leads to formation of the decationized zeolite form whose structural amorphization under hydrothermal conditions, has been demonstrated many times e.g. ref.11 . Stabilization of the zeolite under "self-steaming" conditions-before introduction of the Ni cation - not only prevents lattice amorphization but also, as has been illustrated (refs.12,13), increases the rate of reduction of the Ni cation; the Ni^{o} clusters formed are homodisperse spherical species located in the zeolite cavities.

This work was carried out in order to study the effect of the degree of dealumination of the zeolite , given by the stabilization temperature, on the catalytic activity in the methanation reaction of carbon oxides.

EXPERIMENTAL

The parent NaY zeolite was obtained from the Institute of Petroleum and Hydrocarbon Gases, Czechoslovakia. Standard ion exchange was employed for preparation of the $NaNH_4Y$ form. The sodium - - ammonium form was stabilized in the manner described in an earlier work (ref.14). Ion exchange of the residual Na ions in the stabilized zeolite was carried out using a 0.05 M $NiCl_2$ solution. Ni^{2+} was reduced at T = 693 K over 18 h. The degree of reduction was calculated from the amount of Ni^{2+} cations washed out during ion exchange with O.1N $CaCl_2$ after reduction. The content of extra-lattice aluminium was found by titration after extraction with O.1N HCL. Table 1 gives the characteristics of the studied catalysts.

TABLE 1

Sample	Chemical composition	Ni^{o} (wt.%)	Stabil. Temp.(K)
1	$Na_5 \cdot Ni_{11} \cdot / (AlO_2)_{50} \cdot (SiO_2)_{135} / \cdot (Al_2O_3)_{3.6}$	5.0	823
2	$Na_{12} \cdot Ni_9 / (AlO_2)_{26.5} \cdot (SiO_2)_{138} / \cdot (Al_2O_3)_{13.6}$	3.9	1043

CO and CO_2 (99.3% pure) were provided by Scot Environmental Technology Inc.; the hydrogen used was electrolytically pure. Catalytic tests were carried out in a fixed-bed flow reactor. The composition of the reaction product was determined gas chromatographically using Carbosieve B (Supelco Inc.) as packing in the separation column. The catalytic activity was tested at flow-rates of 20 to 120 $cm^3.min^{-1}$ of a mixture of H_2 + CO in a ratio of H_2: CO = 4 : 1, for reduction of CO_2 the ratio was $H_2:CO_2$= 6:1, at various temperatures from 573 to 790 K.

The porous structure of the studied samples prior to reduction of the Ni cations was characterized by the shapes of the benzene adsorption isotherms, at a temperature of 293 K. The adsorption capacity, (a_o) determined using argon at T = 185 K on a quartz microbalance confirmed the stability of the zeolite lattice. The a_o values were measured before Ni reduction,after reduction and after long-term tests:
Sample 1 : (10.1; 10.0; 9.8) mmole $Ar.g^{-1}$
Sample 2 : (8.5; 8.1; 8.1) mmole $Ar.g^{-1}$

RESULTS

It followed from the long-term experiments that a decrease in conversion or a change in the selectivity did not occur during 60 h on the tested catalysts. This fact, together with data on the adsorption capacities after long-term tests confirmed that carbonaceous deposits are not formed on these catalysts and that amorphization of the zeolite lattice also does not occur. The values given below correspond to results measured after 6 h tests.

CO Reduction

In addition to methane, a small amount of CO_2 is formed on both catalysts, appearing as a decrease in selectivity for methane at temperatures higher than 650 K. The CO conversion for sample 1 is systematically lower than that one for sample 2, see Fig.1.

CO_2 Reduction

In contrast to the previous experiment, the temperature dependence of conversion for both catalysts is similar, as depicted in Fig. 2. There are, however, considerable differences in the composition of the reaction products, see Fig.3. On catalyst 1, CO is formed primarily in the temperature region up to 720 K. On catalyst 2, the dominant product in the whole range of tested temperatures is methane and the CO concentration does not exceed 15%.

Adsorption isotherms

It follows from Fig. 4 that the structures of the two catalysts differ considerably in the region of secondary porosity. In agreement with results published in another work (ref. 15), the volume of the secondary pores, formed during stabilization of the

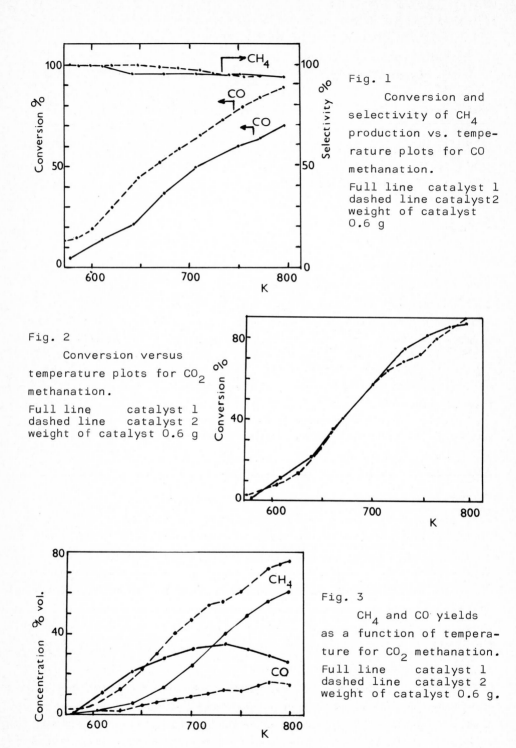

Fig. 1
Conversion and selectivity of CH_4 production vs. temperature plots for CO methanation.
Full line catalyst 1
dashed line catalyst 2
weight of catalyst 0.6 g

Fig. 2
Conversion versus temperature plots for CO_2 methanation.
Full line   catalyst 1
dashed line   catalyst 2
weight of catalyst 0.6 g

Fig. 3
CH_4 and CO yields as a function of temperature for CO_2 methanation.
Full line   catalyst 1
dashed line   catalyst 2
weight of catalyst 0.6 g.

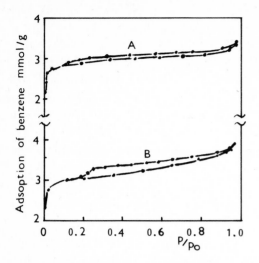

Fig. 4

    Adsorption isotherms
of benzene at T = 293 K.
Before the measurements
the samples were dehydra-
ted at T = 623 K for 4 h,
$p = 1.3 \ 10^{-2}$ Pa.

A... sample 1

B... sample 2

zeolite lattice, is proportional to the degree of dealumination of
the skeleton. This secondary porosity is formed as a result of
decomposition of a certain number of cubo-octahedra in the monocrystal; the freed Si atoms are localized in positions freed by the
aluminium atoms that migrate into the zeolite cavities. The degree
of crystallinity and size of the monocrystal are not influenced by
this process. Fig. 4 also indicates that the secondary pores formed have the same diameter and differ only in number. It follows
from the rate of equilibrium establishment that the secondary
pores increase the rate of adsorbate diffusion proportionally to
their number and that they can thus be characterized as transport
pores.

DISCUSSION

    As demonstrated in the work of Dalmon et al. (ref. 16), and
by other authors (refs 17,18), CO hydrogenation proceeds via CO
dissociative adsorption with formation of superficial nickel carbide; the carbon bonded in this way is very reactive and is rapidly reduced by the hydrogen present, according to the mechanism
(ref.16):

$$CO_{gas} \rightarrow Ni_4CO_{ads} \rightleftharpoons Ni_3C_{surf.} + NiO_{ads} \ ; \quad 3H_2 + C_{surf.} + O_{ads} \rightarrow CH_4 + H_2O$$

The rate determining step in the reaction is the breaking of the
C-O bond in CO (ref.19). The subsequent reaction with hydrogen is
very fast (refs.20,21). It can be assumed on the basis of these
results that the carbon dioxide whose formation has been observed
on both our catalysts in the higher temperature region is a product

of the water gas-shift reaction $CO + H_2O \rightleftharpoons CO_2 + H_2$, as its concentration increases with the water content in the reaction product and is indirectly proportional to the CO concentration in the reaction mixture. For these reasons we consider this reaction as more probable than oxidation of CO by chemisorbed oxygen (ref. 8). It follows from the results of TPD measurements (ref. 20) that the reaction $H_a + O_a \rightarrow OH_a$ occurs in excess H_2 much faster than the reaction $CO_a + O_a \rightarrow CO_2$.

Most works dealing with the reduction of CO_2 on nickel (refs. 16,8,21) consider hydrogenation via partial dissociation of CO_2 to be the first step and subsequent dissociation of CO. Hydrogenation occurs through superficial Ni carbide. CO is formed as a reaction side-product by recombination of carbon and oxygen; its presence decreases the yield of methane because of preferential sorption of CO on Ni (ref.5). This negative influence of CO on the production of methane during reduction of CO_2 has been mentioned also by other authors (refs.16,21).

In evaluation of the results of carbon oxides reduction on the catalysts prepared here, it should be considered that the studied samples do not differ significantly in nickel content or in the size of the Ni clusters, as follows from the results of FMR measurements (ref.22). The only marked difference is in the structure of the zeolite support and depends on the different numbers of secondary pores as a result of different stabilization conditions; in sample 2 the number of secondary pores is roughly two-fold.

The systematically higher degree of CO conversion on sample 2 in the temperature dependence indicates that the activation energy for CH_4 formation on this catalyst is lower as follows: catalyst 1:E_{act}= 69.5 kJ.mole^{-1}; catalyst 2:E_{act}= 47.5 kJ.mole^{-1}. The observed decrease in the output concentration of methane with increasing space velocity on both catalysts indicates that the production of methane apparently depends on kinetics factors. The higher number of transport pores in catalyst 2 increases the rate of intracrystallic diffusion of the reactants and reaction products, which has a positive effect on the course of the reaction. The selectivity of methane formation on the catalysts tested here in the reduction of carbon dioxide is much lower than in the reduction of CO. This observation is rather unique, as, with the exception of one study (ref.6), most previous studies, see e.g. refs.21,23, reported little or no CO as a by-product of CO_2 methanation on various supports loaded with Ni.

TABLE 2

The Selectivity of CO Formation for Various Reaction Temperatures.

Sample	Reaction Temperature (K)								
No	573	606	627	640	670	700	750	779	790
1	80.0	80.0	78.4	77.0	68.1	57.4	40.2	34.5	31.1
2	0	0	12.8	13.7	14.3	16.8	16.8	17.8	18.2

The values given in Table 2 differ for catalysts 1 and 2 significantly. As the conversion of CO_2 on the two types is very similar (see Fig.2), it appears that - assuming partial dissociative adsorption of CO_2, desorption of the CO formed on catalyst 1 predominates over dissociation especially at lower temperatures. This explanation is not in agreement with that given by other authors (ref. 5) who demonstrated that the adsorbed CO_2 reacts almost completely to form CH_4. These experiments were, however, carried out on catalyst with supports exhibiting negligible adsorption of CO_2. In our experiments, the zeolite support increases the sorption of CO_2 and thus affects the course of the reaction. The greater content of CH_4 in the reaction product for catalyst 2 is probably conected with the increased rate of intracrystalline diffusion, permitting faster transport of the reactant and the reaction products.

CONCLUSIONS

1. The stabilized form of Y zeolite used as a support of catalytically active nickel increases the resistance of the catalyst to amorphization and decreases the formation of carbonaceous deposits.
2. The catalytic activity of the thus-prepared catalysts in the methanation reaction of carbon monoxide is proportional to the number of secondary pores in the zeolite support, which is given by the degree of dealumination of the lattice.
3. The production of CO in the reduction of CO_2, differentiating this type of catalyst from other types containing nickel on amorphous supports, is probably connected with increased sorption of CO_2 on the zeolitic support. The higher production of CH_4 on the catalyst with more highly developed secondary porosity is discussed as a consequence of improved kinetics.

REFERENCES

1 M.A.Vannice, J.Catal.,37 (1975) 449.

2   M.A. Vannice, J.Catal., 40 (1975) 129.
3   M.A. Vannice, J.Catal., 44 (1976) 152.
4   M.A. Vannice, J.Catal., 50 (1977) 226.
5   J.L.Falconer and A.E. Zagli, J.Catal., 62 (1980)280.
6   G.D. Wheatherbee and C.H. Bartholomew, J.Catal., 68 (1981) 67.
7   J.Falbe and C.D. Frohning, J.Mol.Catalysis, 17 (1982) 117.
8   R.P.A. Sneeden, J.Mol.Catalysis, 17 (1982) 349.
9   P.B. Weisz and V.J. Frilette, J.Phys.Chem, 64 (1960)382.
10   S. Bhatia, J.F. Mathews and N.N. Bakhshi, Acta Phys. et Chem.
     Szeged, 24 (1978)83.
11   V. Patzelová, V. Bosáček and Z. Tvarůžková, Acta Phys. et Chem.
     Szeged, 24 (1978) 257.
12   V. Patzelová, Z. Tvarůžková, V. Bosáček and K. Mach, Proc. 3rd
     Czech-italian Symposium on Catalysis, Prague 1981.
13   V. Patzelová, Z. Tvarůžková, K. Mach and A. Zukal, in P.A.
     Jacobs, N.I. Jaeger, P. Jírů and G. Schulz-Ekloff (Eds.),
     Studies in Surface Science and Catalysis 12, Elsevier, Amster-
     dam 1982, p.151.
14   Z. Tvarůžková, V.Patzelová and V. Bosáček, React.Kinet.Catal.
     Lett., 6 (1977)433.
15   U. Lohse and V. Patzelová, "in preparation".
16   J.A. Dalmon and G.A. Martin, J.Chem.Soc.Faraday Trans., I.,75
     (1979)1011.
17   P.R. Wentrcek, B.J. Wood and H. Wise, J.Catal., 43 (1976) 363.
18   M. Araki and V. Ponec, J.Catal., 44 (1976)439.
19   R.A. Dalla Betta and M. Shelef, J. Catal., 49 (1977) 383.
20   A.E. Zagli, J.L. Falconer and C.A. Keenan, J.Catal., 56 (1979)
     453.
21   G.A. Mills and F.W. Steffgen, Catal.Rew., 8 (1973) 159.
22   V.Patzelová and K. Mach, "in preparation".
23   T.V. Herwijnen, H. Van Doesburg and W.A. de Jong, J.Catal.,
     28 (1973) 391.